ENVIRONMENTAL MEDICINE

ENVIRONMENTAL MEDICINE

Edited by

G. Melvyn Howe, MSc, PhD

and

John A. Loraine, DSc, MB, PhD, FRCPE

WILLIAM HEINEMANN MEDICAL BOOKS LIMITED
LONDON

First Published 1973

[i.e. 1974]

© G. Melvyn Howe and John A. Loraine

ISBN 0 433 32655 7

Text set in 10/11 pt. Monotype Times New Roman, printed by letterpress,
and bound in Great Britain at The Pitman Press. Bath

Contents

List of Illustrations

ix

Foreword

Much of the subject matter of environmental medicine is as old as the Hippocratic school of healers; much of it belongs to the 1970s. In the 4th century BC, weather, climate, water supply, eating habits and mode of life generally, were taken to be essential considerations for "whoever wishes to investigate medicine properly". Now, such considerations need to be supplemented by a multiplicity of others. They include gross atmospheric pollution by motor and industrial emissions, and by nuclear radiations; pollution of the land by industrial wastes and pesticides, and pollution of streams, rivers, lakes and seas by sewage, industrial effluent and oil. There is overcrowding in big cities, and the tempo and tensions of life in them are such as to cause stress. Drug taking, cigarette smoking and promiscuity are on the increase in Westernized industrialized countries and cacophony knows no bounds.

The environment is under constant interference and attack by man. Some of the changes being wrought are imperceptible and insidious; others are blatantly obvious. What is beyond doubt is that society is altering the balance of a relatively stable environmental system by its actions. The ecology of life on planet Earth is being radically changed by man. From the medical point of view some of the changes may be good, some may be harmful and others may well prove catastrophic. That there are no precise views of the impact of the changes being wrought in the balance of the great natural forces and of the new environments being created is cause for serious concern. It is palpably unwise to continue to interfere with man's habitat without, at the same time, striving to determine the real and lasting effects of such actions on man's health and general well-being. To ignore these effects could well lead to the extinction of life on this planet.

This book is intended to demonstrate the way in which environmental conditions are, or may be, causatively related to man's health or man's diseases. It is essentially inter-disciplinary in nature, with both an environmentalist and physician as editors and with contributors drawn from a variety of academic disciplines. The book begins with a group of general or introductory chapters dealing with certain selected natural and man-made aspects of the environment These are followed by contributions which demonstrate, for a selection of diseases, the importance of environmental hazards in both the so-called developing and developed countries. These chapters show how environmental factors or influences act and react on people, on communities, and on society as a whole. Insofar as man's response to environmental influences is limited by hereditary or genetic factors, a chapter on the geography of genes is included to serve as a counterpoint to those chapters more particularly concerned with external environmental conditions and with disease relationships. To conclude, and especially for those wishing to pursue further studies in environmental medicine, a chapter is included which presents information on conventional and computerized methods for the retrieval of relevant articles, books, papers and theses pertinent to enquiries in the field of environmental medicine.

From the point of view of the medical profession, this book could well be regarded as a watershed. The traditional approach to medicine has always been the steadfast assumption that the primary role of the doctor is the treatment of the individual patient. Now the practice of medicine is moving rapidly into an era where social and environmental factors in relation to disease may come to be regarded as of paramount importance.

xi

This book is replete with examples of the ways in which, in a subtle and arcane manner, the health of the patient can be affected by his environment and by his social milieu. That this approach is germane and eminently justified in the world of the 1970's there can be no doubt. As the medical profession gradually becomes increasingly disenchanted with an approach to disease dominated by high technology and massive feats of "medical engineering" it is inevitable that the pendulum will swing back to a much more detailed consideration of the basic environmental factors which operate to produce pathological conditions in man. The contributors to this book have recognized and faced up to the accelerative thrust of change in medicine. They are to be congratulated for being in tune with the tides of contemporary medical history.

GMH & JAL
September, 1973

Chapter 1

The Environment, Its Influences and Hazards to Health

G. Melvyn Howe

The term 'environment' suggests different things to different people. To the meteorologist and climatologist, for instance, it means the atmosphere, but to the environmental engineer it usually means the atmosphere in an enclosed space. 'Environment' to the ecologist is synonymous with the term 'habitat' within which plants and animals live. Historically environment has been frequently equated with sanitation. Such views on what constitutes environment are rather restricted. In the present context the term 'environment' is assumed to have a broader connotation and to refer to the totality of the external influences—natural and man-made—which impinge on man and affect his well-being. It thus embraces the life support systems and also the multiplicity of stimuli and hazards, direct and indirect, which man experiences. This, the *external* environment (nurture) stands in contrast with man's *internal* environment (nature) which relates to an individual's biological system, i.e. his genetic make-up.

LIFE SUPPORT SYSTEMS

Life support systems are those environmental conditions which are essential for human life. They include air, water, food and shelter. For instance, a supply of free oxygen is essential for respiration. This is obtained from the atmosphere. A man deprived of oxygen dies in a few minutes. It is fortunate that oxygen is abundant almost everywhere in the atmosphere and is uniformly distributed over the globe. Where oxygen does run short, as in very high mountains, then man cannot live at all, though he may pass through such regions in the pressurized cabins of modern aircraft or struggle up into them for a few hours or days—at the cost of great difficulty and exhaustion. In the lower part of the atmosphere (i.e. the troposphere) where man normally lives, pure dry air is a mixture of at least nine gases. By volume two of these, oxygen and nitrogen, make up over 99 per cent, 21 per cent consisting of oxygen and 78 per cent of nitrogen. Should the percentage of oxygen fall below 16 per cent anoxia develops. This affects bodily functions and brain centres. Life cannot be sustained if the oxygen concentration is below 6 per cent.

Water is another integral part of the life support systems. The need for water is as universal as the need for oxygen. However, in contrast to oxygen, water has a variable spatial distribution in that some regions have an excess of water, either annually or seasonally, and others a deficiency, either annually or seasonally. Furthermore, much of the water is too salty for man. *Homo sapiens* can live at most for about a week without drinking water, but in individual cases survival may be measured in days.

Food nutrients, in certain minimum amounts and proportions, are required to ensure active life and successful procreation of the species. Carbohydrates, fats and proteins form the major portion of man's diet while minerals and vitamins are present in smaller quantities. That man is closely dependent upon his environment for the provision of these requirements is all too obvious.

An equally close man–environment relationship exists in the case of shelter or housing. Shelter is necessary to provide protection from environmental extremes of heat, cold, moisture, aridity, sunshine or wind. Life or settlement would be impossible in several parts of the world were it not for adequate shelter. At the same time shelter provides the place where people care for most of their bodily needs, rear their children and keep their possessions. Important too is the fact that shelter provides a retreat where the individual may enjoy quietude and privacy, particularly where there are large numbers of people living in close proximity to one another.

HAZARDS AND INFLUENCES OF THE ENVIRONMENT

The environment provides man with the essential life support systems but it also presents him with a variety of hazards which may prejudice his health. If health is 'a state of complete physical, mental and social well-being and not merely the absence of disease or infirmity' (WHO) it represents a balanced relationship of the body and mind and complete adjustment to the total environment. Disease, on the other hand, is maladjustment or maladaptation in an environment, a reaction for the worse between man and hazards or adverse influences in his *external* environment. The response of the individual to these influences is conditioned by his genetic make-up or *internal* environment. Environmental influences or hazards may be categorized as (*a*) physical, (*b*) biological, and (*c*) human.

Physical Environment

Ever since the time of Hippocrates the weather and climate have been postulated as influencing, either favourably or unfavourably, man's well-being. The main problem is to isolate those components—solar radiation (including heat, measured in terms of temperature), air movement (wind), precipitation (rain, mist, snow, sleet, hail), moisture (humidity, fog, dew, frost), etc.—which have a direct and/or specific influence on man.

Solar radiation includes cosmic rays, gamma rays, X-rays, ultraviolet rays, luminous rays and infrared rays. Prolonged exposure to radiations may induce skin burns, cancers, genetic mutations and other biological changes. These and other effects of climate on man are discussed in Chapters 3 and 5 respectively.

Whilst water is essential to support life it can, nevertheless, provide a range of health hazards. Many pathogens responsible for disease are essentially water-borne (Table 1).

TABLE 1

Diseases associated with water-borne pathogens

Pathogen	Disease
Algae	Gastro-enteritis.
Bacteria	Cholera, dysentery, paratyphoid, typhoid.
Parasites	Malaria, tapeworm, yellow fever, schistosomiasis.
Protozoa	Dysentery.
Viruses	Infectious hepatitis, poliomyelitis.

In addition to the water-borne pathogens there are often impurities in water which may have adverse effects on health. These include pesticides, herbicides, lead, zinc, mercury, arsenic, nitrates, fluorides, selenium, molybdenum, cadmium and sodium (see Chapters 2 and 4).

Differences in the trace element contents of soils, vegetables, and atmosphere, in the character of the water supply, and in background radiation can be related to differences in the basic rock structure of countries and localities, in rock type, and in the general relief of the land. For instance, sedimentary rocks have a lower content of the radioactive elements uranium and thorium and provide less gamma-ray background than igneous rocks. Rocks, together with their overlying soils, may have anomalous trace element or micro-nutrient contents. The essential trace elements are more important in the nutrition of man than their organic micro-nutrient counterparts, the vitamins. The former cannot be synthesized as can the vitamins, but must be present in the environment within a relatively narrow range of concentration. Both trace-element deficiencies and excesses kill. Soils derive their trace elements from the soil, parent rock material, applied fertilizers, and agricultural dusts and sprays, and pass on their trace-element characteristics to vegetable matter growing on them. Vegetable matter used as food may thus repeat the trace element peculiarities of the soil and of the parent geological material. An excess of mercury, lead, cadmium and selenium, whether eaten in vegetable matter or animal foods, can seriously affect health. Deficiencies of, for example, copper, iron, manganese, zinc, iodine, fluorine, cobalt and molybdenum may give rise to nutritional problems.

Biological Environment

Man's body provides a rich ground for parasites. Several parasites live in or within man permanently, without causing any structural change or functional disturbance. On the skin there are staphylococci, in the mouth non-pathogenic strains of streptococci and in the colon the organism, *Escherichia coli*. In the atmosphere there are pathogenic bacteria and pathogenic viruses which are responsible for many human diseases. One of the intriguing features of the micro-organisms which attack man is their natural history and the ways in which they, the disease agents, are transmitted from person to person. It is here that relationships between disease agents, the diseases they cause, and the physical and human environment, are particularly evident. For example, cholera, has a two-factor complex—causative organism and host. The causative organism, *vibrio cholerae*, is introduced into the human body directly and, as far as is known, only man can be infected by it. Physical environmental factors thought to correlate with, and possibly to govern, cholera endemicity are high temperatures, low-lying lands, ponds and lakes and other bodies of water rich in organic matter and salts, and sheltered from the rays of the sun and from the rain. On the other hand, typhoid caused by the bacterium *Salmonella typhosa*, usually enters the body through the mouth in contaminated food, milk or water. Water contaminated by infected sewage provides a major means of spread. There are other infectious diseases which are vector-borne. Bubonic plague, for instance, is primarily an infectious disease of certain species of rodent. The pandemics of this disease in the past were the result of fortuitous invasions of the human body by internal pathogens of these rodents. In the Great Pestilence or Black Death of the fourteenth century the pathogen was the bacillus *Pasteurella pestis* which produced endemic infection in rats. It was transmitted from plague-infected rats to man through the bite of infected rat fleas (*Xenopsylla cheopsis*). Rodents acted as host to the plague bacillus and the flea was the carrier or vector. The malaria parasites (*Plasmodia*) are transmitted by female *Anopheles* mosquitoes. Yellow fever, a viral disease, is transmitted by the mosquito *Aedes aegypti*. This same vector also carries dengue. Mosquitoes are either known or suspected vectors for a group of viruses thought to cause encephalitis.

Each case disease is the result of a complicated natural history. Different hosts and vectors of several pathogenic viruses, pathogenic bacteria, and parasitic organisms, are differently affected by geographical conditions and controls. Furthermore, disease agents may be carried by different vectors in different parts of their range, as in the case of malaria.

Whether it be causative organism (virus, bacterium, spirochaete, rickettsia) intermediate host, or vector, each element in the disease complex has its own specific environmental requirements. Each element including man himself, is inescapably bound up with the geographical environment. Disease in any given locality is the result of a combination of geographical circumstances which bring together disease agent, vector, intermediate host, reservoir and man at the most auspicious time. Knowledge of these relationships and of each element in the complex is a prerequisite to a true understanding of infectious disease, its distribution and control. But, relationships are rarely simple or static: on the contrary they are complicated and undergo continuous change. Despite this, man's ability to modify certain environmental conditions, his use of pesticides and his ever-increasing technology, have succeeded in bringing a new perspective to the control of vector-borne diseases. Several of the viral and microbial diseases are effectively inhibited by antibiotics such as chloramphenicol and the synthetic penicillins. In consequence smallpox, plague, typhus and cholera are now virtually banished from the West and the incidence of scarlet fever, diphtheria and poliomyelitis is now very low. Serious infectious disease is now absent from most of the developed countries of the world as are the environmental conditions likely to bring about their return. However, in the present era of rapid air travel a person may acquire a serious infectious disease and travel thousands of miles in a fraction of the incubation period of the disease. Thus, he may enter a country such as Britain in the silent stage of an infection in the absence of any clinical symptoms. Such a situation, whereby a person may be at home for a week or more before becoming ill and where the diagnosis of the illness may be delayed or missed, is likely to be exacerbated with the increasing speed, size and frequency of air travel.

Human Environment

The human, or socio-cultural environment, is essentially man made and relates to the density, geographical distribution and mobility of populations, to occupations, to socio-economic status, to housing, to diets, to habits and to customs.

The size of a country's population is obviously important in that the larger the number of people the more they are at risk of exposure to disease. If the population is too large in proportion to its natural resources, the availability of capital, etc., then, accordingly, living standards will be much lower than they might be under different circumstances.

Overpopulation does not depend merely upon the total number living in a country or on the density of the population. A population density of 60 persons to the square kilometre may mean overpopulation in one area but underpopulation in another. Much depends upon the available resources and the degree of cultural development. A country is not necessarily overpopulated simply because it is incapable of providing sufficient food to support its people; such a country may be able to employ its labour force more effectively in manufacturing industry, exporting its surplus of manufactured goods in exchange for food-stuffs. Where this is not possible and a country becomes overpopulated, as is evident in, for example, Mainland China, India, Pakistan, Indonesia and parts of the Mediterranean and the West Indies, the evils of overcrowding—slums, mass unemployment, poverty and disease—are all too evident.

Population density and population distribution vary within countries. Patterns of settlement are the outcome of physical, historical and economic factors. In Egypt, for instance, there is a striking contrast between a large area of desert which is virtually uninhabited and the Nile valley and delta which have a huge concentration of people. In places there is a rural density of over 400 persons per square kilometre, and, including towns and cities, there is an overall density of 960 per square kilometre. The average figure of 25 persons per square kilometre for the whole of Egypt is thus totally misleading.

Bangladesh, with an overall population density of 433 persons per square kilometre, is the most densely populated country in the world. Taiwan comes second with 384 persons per square kilometre followed by Belgium and South Korea, each with 316 persons per square kilometre. The United Kingdom is the ninth most densely populated country in the world; England and Wales together are second after Bangladesh and Taiwan. Over 80 per cent of the population of the United Kingdom lives in towns and 7·4 million out of a total population of 55·34 million (1971), i.e. 13 per cent, live in Greater London. If the optimal population for a country is taken as 'the maximum which can be maintained indefinitely without detriment to the health of individuals from pollution or from social or nutritional stress' (Southwick), then clearly Britain is overpopulated.

People today are mobile to a remarkable degree, and in countries like Britain there is a slackening of close family ties and social relationships. Many urban dwellers have moved further away from centres of towns and their places of work and have taken up residence in suburbs, in nearby market towns and in so-called 'new towns'. Change of place of residence involves not only physical disturbance but also a social disturbance since people are obliged to create entirely new social environments for themselves. There is at the same time the inevitable journey to work. This may be long or short, but it involves extra energy and takes toll of physical and mental reserves.

Pilgrimages of people to places such as Mecca, Jerusalem or Lourdes, seasonal movement of labour, as in the USSR, Switzerland or Bolivia, the movement of traders along ancient caravan routes, or roads and railways or by sea, present hazards to health. The paths of infection by such diseases as cholera or plague have been traced along ancient trade routes by land and sea. Ships carry people who may be contaminated; ships also carry rats which may alight at ports of call. Seaports have thus become, quite naturally, secondary sources of disease dispersion. The role of air travel in the modern spread of disease has already been noted.

Occupation is one way of measuring 'socio-economic' status, with its implications in respect of family income, living standards and life styles. Furthermore, many occupations carry with them hazards to health. As early as 1775 Percival Pott drew attention to soot as a cause of scrotal cancer in chimney sweeps. Silicosis is a risk in quarrying and glass manufacture and there is an above-average incidence of pneumoconiosis among coal miners. Lead, mercury, arsenic, fluoride, asbestos, chromium and benzene are among the recognized poisonous materials used in modern industry. These and hundreds more new chemicals are being introduced into the environment each year. Man's reaction to such chemicals depends on the toxicity of the chemical, the duration of exposure, the concentration of the chemical, and the genetic susceptibility of the individual. Diseases or disorders commonly associated with atmospheric pollutants and thought to be aggravated by them, include chronic bronchitis, pneumonia, lung cancer, emphysema and asbestosis.

Housing in the West has improved considerably. In medieval Britain, for instance, the common folk lived in dark, verminous, unventilated, wattle and daub, thatched-roofed dwellings which harboured plague. Densely packed, back-to-back dwellings lacking drains, water closets or other basic facilities characterized nineteenth-century Liverpool, Glasgow, Manchester, Birmingham, Leeds, Sheffield, Bradford, Nottingham, Newcastle and other industrial towns, all of which had bad records for tuberculosis. Housing in twentieth-century Britain is of stone, brick or concrete, supplied with clean, running water and flush toilets, usually with baths and showers; the houses are now often centrally heated and/or air conditioned. The artificial interior climates of many of these houses, offices and public buildings are often quite different from the climate out of doors. Whether such artificial climates are optimal for a person's physical or mental functioning is not known. It is thought, however, that air conditioning provides relative freedom from infection for the occupants of buildings in which it is installed due both to the withdrawal of infected air and to the filtration of incoming air.

Food constitutes yet another important factor of the human environment worthy of the closest of attention in the context of human health. Man needs food as a source of energy for undertaking work, and as a source of raw material with which to perform the processes of tissue building and the perpetuation of the species, *homo sapiens*. Diets in most of the densely populated developing countries of the world are defective both in quantity and quality. Among the most serious deficiency states now prevailing in such areas are kwashiorkor, beri-beri, pellagra, rickets and goitre. At the same time poor nutrition predisposes to infections, particularly to tuberculosis and to infestations such as hookworm. In the West there has been a profound change in food and dietary habits. In particular this has involved a reduction in the amount of protein consumed and an increase in carbohydrate intake. It has been suggested that, compared with man's total evolutionary history, the relatively short time since he changed from a protein-rich to a carbohydrate-rich diet has not permitted adaptation. Sugar, in particular, is used in ever-increasing quantities, and chemicals are added to food and drink ostensibly to improve palatability and appearance. High blood lipid levels (lipid being the general term which includes fats and fat-like compounds) are thought to be of aetiological importance in coronary arterial disease.

The unrestricted use of powerful chemical pesticides based on organo-phosphorus compounds and chlorinated hydrocarbons, while revolutionizing the chemical attack against harmful insects and pests, upsets the balance of soil ecosystems and may well prove a serious source of contamination of food supplies. Dichloro-diphenyl-trichlorethane (DDT), dieldrin, aldrin and related substances persist in soils and accumulate in animal and body fat.

The modern town, the social habitat of industrialized man, is characterized by high density living, overcrowding, a polluted atmosphere and noise. Air pollution is a major social problem. Sources include motor vehicles, railway trains, aircraft, domestic heating, chemical plant, fuel-burning factories and offices, refuse disposal and thermo-electric generating stations. Studies have shown that atmospheric pollution is associated with the occurrence and worsening of many respiratory diseases including chronic bronchitis, lung cancer, emphysema and asthma. As distinct from man-made atmospheric pollution there is the atmospheric pollution from natural sources such as pollens and dust which, either singly or possibly in association with certain kinds of food or emotional stress, cause asthma.

Cigarette smoking, a social habit of long standing, has also been associated epidemiologically with respiratory and heart diseases. Carcinogenic hydrocarbons, notably 3,4-benzpyrene, have been isolated from tars of cigarette smoke and from the soot of polluted atmospheres.

Other social habits likely to prejudice health and promote disease include gluttony leading to obesity, chronic alcoholism leading to physical disease (gastritis and cirrhosis of the liver), mental deterioration (see Chapter 16) and promiscuity leading to venereal disease (see Chapter 15). Several countries now appear to be experiencing an epidemic of drug dependence among teenagers in which opiates, cocaine, amphetamine and cannabis have pride of place.

In contrast to the rather more tranquil life of past generations the tempo and tensions of contemporary Westernized urban societies are such as to lead to 'stress', believed by some to be a factor in producing coronary heart-disease, cerebro-vascular disease and some forms of cancer. How much mental illness is due to the stresses of modern life, to genetic causes, or to influences in youth in the Freudian sense, remains to be established. Certainly cacophonous noises from motor vehicles, railway trains, jet aircraft, building site machinery, pneumatic drills and a multiplicity of other man-made sources, provide a contributory environmental hazard. For example, it is thought that the acoustic discomfort associated with winds blowing against high-rise and tower blocks of flats results in nervous strain among the occupants, largely due to fear of actual physical harm.

MAN'S INTERNAL ENVIRONMENT

External environmental influences and hazards act on man, but the response in almost every case is thought to be conditioned by the genetic make-up of the individual (Chapter 19). There appears to be a hereditary predisposition on the part of some individuals to certain diseases. Illness attributable solely to inherited characteristics, such as haemophilia, is rare. On the other hand some associations between the ABO blood groups and disease have been proved beyond reasonable doubt. Thus the incidence of peptic ulcer is 40 per cent higher in persons of blood group O than in those belonging to blood groups A, B or AB. The increased risk attaching to blood group O is about 25 per cent more common in gastric ulcer. Again, it has been established that cancer of the stomach is about 25 per cent more common in persons of blood group A than in those of other groups. Pernicious anaemia is similarly associated with group A. In addition, there appears to be an association between group A and diabetes mellitus, bubonic plague, and carcinoma of the cervix, pancreas, prostate and stomach. Smallpox and pituitary adenomata have an O group association; broncho-pneumonia, infantile diarrhoea, salivary gland tumours and tumours of the ovary appear to have an A group association.

It is open to question whether the associations are truly causal in the sense that a person of blood group O is intrinsically more liable to peptic ulcer, or whether persons of other blood groups are protected against the disease. Similarly it is possible that blood group A does not carry a liability to cancer of the stomach but that other blood groups have a special protection against the disease. At present views on this subject are conflicting.

Further Reading

Arthur, D. R. (1969). *Survival: Man and his Environment*, London, English Universities Press Ltd.
Arvill, R. (1969). *Man and Environment*, London, Penguin Books.
Bach, W. (1972). *Atmospheric Pollution*, New York, McGraw-Hill Book Company.
Barr, J. (1970). *The Assaults on our Senses*, London, Methuen.
Black, J. D. (1968). *The Management and Conservation of Biological Resources*, Philadelphia.
Bresler, J. B. (Ed.) (1968). *Environments of Man*, Reading, Mass., Addison-Wesley Publishing Co.
Burton, I. (1968). 'The quality of the environment: a review. *American Geographical Review*, **58**, p. 472.
Carr, D. E. (1965), *The Breath of Life*, New York, W. W. Norton & Co. Inc.
Clements, F. W. and Rogers, J. F. (1960), *Diet in Health and Disease*, Sydney.
Economic Commision to Europe (1971), Symposium on problems relating to the environment. *United Nations*, New York.
Ehrlich, P. R. and Ehrlich, A. (1970). *Population Resources and Environment*, San Francisco, W. H. Freeman & Co.
Holdren, J. P. and Ehrlich, P. R. (Eds.) (1971). *Global Ecology*, New York, Harcourt Brace and Jovanovich Inc.

Howe, G. M. (1972). *Man, Environment and Disease*, New York, Barnes and Noble Books, and Newton Abbot (England), David and Charles.

Loraine, J. A. (1972). *The Death of Tomorrow*, London, Heinemann.

Medical Research Council (1956), *The Hazards to Man of Nuclear and Allied Radiations*, London, Cmnd. 9780, HM Stationery Office.

Mellanby, K. (1969). *Pesticides and Pollution*, London, Fontana.

Perloff, H. S. (Ed.) (1969). *The Quality of the Urban Environment*, Baltimore, John's Hopkins Press.

Powles, J. (1972). 'The Medicine of Industrial Man'. *The Ecologist*, **2**, 10, p. 24.

Purdom, P. Walton (Ed.) (1971). *Environmental Health*, New York, Academic Press.

Rodda, M. (1967). *Noise and Society*, London, Oliver and Boyd.

Royal College of Physicians (1970). *Air Pollution and Health*, London, Pitman.

Royal Commission on Environmental Pollution (1971). First Report, Cmnd. 4585, London, HM Stationery Office.

Royal Commission on Environmental Pollution (1972). Second Report. Cmnd. 4894 London, HM Stationery Office.

Ward, M. A. (Ed.) (1970). *Man and his Environment*, Proceedings of the first Banff conference on pollution (May 16–17, 1968), London, Penguin Press.

Chapter 2

Some Trace Element Concentrations in Various Environments

Harry V. Warren

INTRODUCTION

Everybody is interested in health, particularly his own. One way of thinking of health is to consider that it represents the degree to which an organism is in harmony with its environment. Thus those who rate human ecology high among their interests, may well have some useful contributions to make in a sphere of learning that intrigues so many of us, namely our health.

Unfortunately a subject as complicated and all embracing as health is far too vast to be dealt with in one chapter of one book. In this chapter an attempt will be made to indicate the role that may be played by some trace elements in determining the health, or absence of health, in animals and man. A few elements only will be selected for illustrative purposes. The author's interest in trace elements may well appear to lead to an apparent overemphasis with respect to their influence on health. However he is well aware of the importance of other factors involved in determining the well-being of an organism, but suggests at the same time that bacteria and viruses must exist in an environment and that the trace element content of that environment may in many instances determine the pathogenesis of a particular disease-producing organism.

GEOGRAPHICAL BACKGROUNDS

Living matter is made up of various combinations of those elements that occur in air, sea, and land. Life has existed on this planet for thousands of millions of years. Initially life was most intimately related to the ocean. A few hundreds of millions of years ago some forms of life became more closely associated with the land but of course, directly or indirectly, they also had relationships with both atmosphere and hydrosphere.

When man took to cultivating food crops a few thousands of years ago he was in effect setting in train a series of events the results of which have only recently become of any significance to his welfare. Early civilization, which was closely linked with increasingly stable agricultural practices, largely centred around great river deltas such as the Nile and the Tigris and Euphrates. Great rivers tend to have vast and geologically diverse hinterlands and when their deltas were flooded, as they tended to be periodically, their soils were enriched, or fertilized by the many elements carried by the river in suspension and in solution. Under these circumstances both the vegetal crops and the animals living off the herbage on the deltas tended to contain trace elements in amounts that reflected, in a general way, the trace element contents of substantial portions of the earth's crust. In a crude way animal and vegetal life established themselves in harmony with their environment. Species not in harmony with their environments tended to die out.

9

THE INDUSTRIAL REVOLUTION AND THE POPULATION EXPLOSION

About one hundred and fifty years ago several changes commenced and these changes were, in some parts of the world at least, to have profound effects on the trace element content of some foods and thus on the health of man.

These changes were brought about by two principal happenings, known as the 'Industrial Revolution' and the 'Population Explosion'. These two changes were interrelated and had marked effects on one another. The Industrial Revolution changed the occupational patterns and modes of life for numerous working people but essentially involved the use of vastly increased amounts of energy, largely made possible by the combustion of great quantities of fossil fuels, initially coal but later on petroleum, natural gas, and lignite. Concomitantly with this increased use of energy came what is commonly referred to as the population explosion. This population explosion involved not only the production of much greater quantities of food but also the crowding of more and more people into urban communities. The Industrial Revolution was also accompanied by a much greater use of many metals whether considered absolutely or on a per capita basis.

Thus insofar as trace elements are concerned, a virtual revolution has taken place during the past century and a half, and this revolution has been particularly marked during the past fifty years. During the past fifty years the world's population has doubled, the per capita consumption of the base metals, copper, lead and zinc, has tripled and furthermore populations have increasingly tended to move from rural into urban areas where now many cities have population densities ranging from five to forty thousand persons per square mile.

CHANGING PATTERNS OF FOOD SUPPLY

Thanks to the internal combustion engine, man has been able to grow foodstuffs and to transport those foodstuffs from areas previously economically inaccessible. Many of these previously inaccessible areas are composed of soils that are not naturally recharged with nutrient elements as were those delta soils on which man had for so long depended for his food crops. In order to recharge these soils, to a large extent continental in nature, it was essential to introduce increasing amounts of fertilizing materials. With increasing demands for food the supplies of organic fertilizers were not forthcoming, and more and more agriculturists were forced to use large amounts of mineral fertilizers. This in turn has led to the introduction of a new set of problems.

Increasing urbanization has in itself led to a further change in the trace element content of many diets. The refining, packaging, and freezing of many foods—all practices which in themselves involve the use of increasing amounts of energy—have altered materially the trace element concentrations in foodstuffs between what are found in original farm products and what are found in finished products as eaten by consumers.

It is proposed to show how various factors are tending to alter the trace element contents of the food that is eaten, the water drunk, and the air that is breathed. Except in a few instances knowledge is not available to assess the medical significance of the varying trace element intakes suggested by the data here presented. Nevertheless, it is generally accepted that trace elements in general do have some relationship to health, and it is hoped that the empirical data made available in this contribution will be integrated into the more sophisticated studies that undoubtedly will be made by medical men, nutritionists, and epidemiologists.

REPORTED 'NORMAL' CONCENTRATIONS OF SELECTED TRACE ELEMENTS IN THE EARTH'S CRUST AND IN SOILS

Table 1 gives in condensed form relevant data concerning the trace elements to be discussed.

TABLE 1

Average contents and range in contents of selected trace elements in lithosphere and in soils as reported by various authors
(in ppm)

Element	Lithosphere Goldschmidt 1954	Shacklette *et al.* 1971 (USA)	Vinogradov 1959 (Global)	Swaine 1955 (Global)	Present report 1971 (Canada and Gt. Britain)
		Average	Average	Range	Average
Manganese	1,000	560	850	200–3,000	800
Nickel	100	20	40	5–500	30
Zinc	80	54	50	10–300	90
Copper	70	25	20	2–100	30
Cobalt	40	10	8	1–40	10
Lead	16	20	10	2–200	20
Molybdenum	2·3	—	2	·2–5	1·5
Cadmium	0·18	—	0·5	<1	0·5

It must be admitted that the data presented for Canada and Great Britain are not statistically based. In preliminary investigations the amount of contamination, particularly in Great Britain, became appreciated only as work progressed. Samples known to be contaminated were eliminated from data used to arrive at what might be considered normal. Nevertheless, too many subjective judgements were involved for these data to be accepted as anything other than tentative suggestions.

In the light of the data given in Table 1, the trace element contents of the following soils serve to illustrate the degree to which certain urban soils have been contaminated.

TABLE 2

Trace element contents of a few representative urban soils
(in ppm)

Location	Mn	Ni	Zn	Cu	Co	Pb	Mo	Cd
Leeds, Yorkshire	2,140	49	355	165	21	520	6	·8
Birmingham, Warwickshire	710	50	635	230	13	345	3	3·4
Coventry, Warwickshire	1,170	66	715	120	23	455	6	3·6
Los Angeles, California	230	11	165	22	3	975	1	·5
San Diego, California	1,105	36	750	68	15	1,820	1·5	2·6
Vancouver, British Columbia	740	27	395	53	13	510	1·5	1·0
Toronto, Ontario	440	19	220	77	7	485	1	2·0
Liverpool, Lancashire	215	26	150	76	5	220	4	1·0
Bradford, Yorkshire	580	51	425	90	15	560	11	·8

TRACE ELEMENT CONTENTS OF SOME URBAN SOILS

Urban communities tend to have distinctive patterns in the trace element content of their soils. The age of a community, the occupations of its inhabitants, the climate, the density of population, and a host of other factors combine to alter markedly the trace element concentrations found in the soils of urban areas. Table 2 provides examples taken from well-known cities.

The above examples were selected for illustrative purposes only but they could be repeated again and again. If such samples are representative then the degree of lead and zinc contamination in some urban areas may be readily appreciated.

TRACE ELEMENT CONTENTS OF SOME URBAN DUSTS

Because of the obvious contamination of urban soils it is natural to expect the particulate matter in the air of cities to show evidence of contamination. In Table 3 a few examples of the trace element contents of city dust are reported.

TABLE 3

Trace element contents of a few urban dusts
(in ppm)

Location	Element							
	Mn	Ni	Zn	Cu	Co	Pb	Mo	Cd
Toronto, Ontario	585	22	320	120	5	6,780	1·5	15
Los Angeles, California	537	83	850	138	8	2,750	4	29
New York (heavy traffic area) New York	—	—	—	—	—	20,000	—	—
Vancouver (west end), British Columbia	760	270	1,680	625	25	6,450	—	6
Vancouver 'A' (industrial), British Columbia	—	—	5,400	320	—	12,000	—	—
Vancouver 'B' (industrial), British Columbia	2,036	37	2,240	304	25	6,010	—	17

— = values not known.

The limited number of results reported in the above table suggest that collections of urban dust might well repay investigation, and if the dust samples did little else they might at least indicate the kinds of metal contamination that should be further investigated.

INDUSTRIAL CONTAMINATION

As anticipated, soils near some industrial plants contain abnormally high concentrations of some trace elements. Some such concentrations are recorded in Table 4.

TABLE 4

Trace element contents of a few soils selected from areas close to industrial plants
(in ppm)

Location	Element							
	Mn	Ni	Zn	Cu	Co	Pb	Mo	Cd
Richmond, British Columbia	360	48	845	132	21	2,060	—	<1
Vancouver, British Columbia	665	28	525	214	18	480	—	<1
Vancouver, British Columbia	4,190	11	1,095	745	8	685	—	1·1
Toronto, Ontario	585	22	320	120	5	6,780	1·5	15
Toronto, Ontario	920	66	2,095	460	18	58,900	11·3	29
Noranda, Quebec	845	49	565	1,285	16	245	3	15
Sudbury, Ontario	160	650	106	370	25	47	1·5	0·6
Trail, British Columbia	515	18	665	42	6	625	1·0	14
Teeside, Yorkshire	785	32	260	65	11	365	2·0	1·1

The above are representative soil samples, they contained neither the highest nor the lowest values encountered in each area. The object in presenting them is to indicate that as far as metal contamination is concerned there is little to choose between urbanized and industrialized localities.

UNEXPLAINED ANOMALOUS SOILS

Metal imbalance is sometimes found in areas where it is totally unexpected. The following examples are from predominantly agricultural areas.

TABLE 5

Trace element contents of a few soils where anomalous conditions were not anticipated
(in ppm)

Location	Element							
	Mn	Ni	Zn	Cu	Co	Pb	Mo	Cd
St. Peter, Jersey	425	12	180	35	6	310	1·0	0·5
Chard, Somerset	400	24	300	45	5	155	2·0	0·3
Sherborne, Dorset	750	50	315	60	18	600	2·5	ND
Aldershot, Dorset	550	25	230	41	15	240	2·0	0·4
Eastling, Kent	1,135	54	405	62	25	635	4·0	ND
Wye, Kent	830	30	250	47	14	830	2·0	ND

ND = not detected.

TRACE ELEMENT CONCENTRATIONS IN MINING AREAS

Soils from mining areas are apt to be high in one or more trace elements even though the mines in that particular area have not been active for decades. Table 6 provides illustrative data.

TABLE 6

Trace element contents of a few soils selected from former or presently operating mining areas
(in ppm)

Location	Element							
	Mn	Ni	Zn	Cu	Co	Pb	Mo	Cd
Morwellham (Tamar valley), Devonshire	1,800	54	385	705	24	440	2·5	0·8
Charterhouse, Somersetshire	1,595	53	1,040	39	18	4,540	3·0	2·1
Bonahaven, Argyllshire	1,020	4	205	24	2	86	2·0	<1·0
Riondel, British Columbia	840	25	340	34	10	300	2·0	1·0
Machynlleth, Montgomery-shire	1,545	106	1,695	295	42	9,010	17·0	ND
Bethesda, Caernarvon	1,625	43	625	115	20	425	8·0	0·9

Tables 2, 4, 5 and 6 suffice to demonstrate that the soils in which food supplies are grown are in many cases contaminated by one or more metals. Fortunately, by no means all of the metal present in a soil finds its way into the agricultural products growing on that soil. Whether or not metals are incorporated into the food derived from a particular soil depends on many factors including the texture, pH, and organic content of the soil. Work done in the University of British Columbia has shown impressively that much of the lead extracted from a soil may find its way into the tops and outside peelings of potatoes rather than into the heart. Similarly beet tops have been found on many occasions to contain greater concentrations of such metals as lead than do beet roots.

Agriculturists are aware that the trace element content of vegetal matter varies at different periods in the growing season. Thus some plants can be eaten with impunity at specific times of the year yet may prove lethal at other times.

RELATIONSHIP BETWEEN TRACE ELEMENTS IN WET, DRY, AND ASHED MATERIALS

One of the problems facing anyone involved in health-trace element problems is that findings are presented by different academic disciplines in various ways. Geologists usually report their biogeochemical results in terms of ash whereas nutritionists think in terms of 'wet' or natural weight. Other workers summarize their findings in parts per million of dry weight.

It would seem desirable to use wet weights when dealing with health matters but doing so presents certain elementary problems. Unless adequately equipped the problem of transporting samples two or three hundred miles from the field to a laboratory where wet and dry weights can conveniently be determined, can present difficulties. Furthermore, the weather, climate and the actual time of collecting the samples may on occasion materially affect wet and dry relationships.

Although the writer's experience is too limited to make anything but the most tentative of hypotheses, it is suggested that the following conversion factors be used until other, possibly more acceptable ones become available.

TABLE 7

Wet, oven dry and ash relationships for some vegetables

Vegetable	Dry weight as percentage of wet weight		Ash weight as percentage of dry weight	
	Between 75% and 85% fall between	Working average	Between 75% and 90% fall between	Working average
Lettuce	4% and 10%	6·5%	15% and 20%	1·0%
Cabbage	4% and 10%	7·0%	8% and 11%	9·3%
Potato	10% and 20%	19·5%	4% and 6%	4·7%
Bean (except broad-)	7% and 14%	11·0%	5% and 9%	7·3%
Carrot	8% and 14%	11·0%	5% and 8%	6·8%
Beet	8% and 14%	11·5%	7% and 10%	8·5%

'NORMAL' TRACE ELEMENT CONTENT OF SOME VEGETABLES

It is of interest to consider the trace element content of vegetables after they have been prepared by the prudent housewife, i.e. after the vegetables have been washed, scraped or peeled as circumstances dictate, or as in the case of such vegetables as lettuce or cabbage, stripped of their outer leaves.

There is as yet only limited information concerning the trace element content of vegetables grown under widely differing circumstances. Nevertheless standards of safety are being set up in some countries. On the basis of such data and of the writer's experience with several collections of food and fodder in the United Kingdom and Canada the following 'normals' for six widely-consumed vegetables are suggested as the basis for a working hypothesis (Table 8).

ABNORMAL OR ANOMALOUS CONCENTRATIONS OF TRACE ELEMENTS IN VEGETABLES

Using the data presented in Tables 7, 8 and 9 it is possible to assess the degree of contamination to which people may be exposed. In practice only a small proportion of the world's inhabitants are exposed to extremes of metal contamination. Nevertheless, there are several places in the world where diets are extremely limited and where food supplies are available from restricted sources. Furthermore, little is known about the long term implications of continued small, but above-normal, intakes of such elements as lead, zinc, cadmium, or mercury.

At present little is known about the form in which these trace elements are present in vegetables much less the degree to which they are digested, absorbed, and passed through the human system. In short it is not yet possible to assess the full significance of the trace element variations that have been found to exist in foods and which are presented in the following tables.

TABLE 8

Metal content of some random collections of vegetable samples and
suggested normals for some Canadian and British vegetables

	(In parts per million of ash)					
	Mn	Zn	Cu	Pb	Mo	Co
A. *World-wide vegetation* (320–3 000 samples)	3,300	1,220	165	41	13	7
B. *Average herbs* (26 plants studied)	2,100	630	120	41	NR	NR
C. *Native vegetation* New Mexico (101 samples)	520	460	65	20	NR	NR
Maryland (38 samples)	3,440	300	150	100±	NR	NR

	In ash				In wet			
	Zn	Cu	Pb	Mo	Zn	Cu	Pb	Mo
D. *British and Canadian vegetables* (50–200 samples)								
Lettuce	400	60	20	5	4·9	0·74	0·25	0·06
Cabbage	300	40	16	30	1·9	0·26	0·10	0·20
Potato	320	100	40	16	2·9	0·92	0·40	0·15
Bean (except broad-)	450	70	30	60	3·6	0·56	0·24	0·48
Carrot	450	70	30	30	3·4	0·52	0·22	0·22
Beet	420	80	20	4	4·1	0·78	0·20	0·04

NR = not reported.

TABLE 9

Abnormal trace element contents of some potatoes
(in ppm wet weight)

Location	Element			
	Zn	Cu	Pb	Mo
Noranda	4·9	2·4	1·7	0·02
Sudbury	4·1	1·7	0·7	0·02
Riondel	6·9	0·4	0·5	0·09
Trail	8·0	1·9	1·0	0·30
Oyama	4·2	0·4	1·4	0·25
Vancouver	7·4	2·4	0·6	0·11

All the above samples were of potatoes and they all came from Canada but similar illustrative material could equally well have been selected from any of a dozen or more other vegetables and from any one of thirty or forty localities.

It may thus be safely concluded that trace element imbalances in soils and in vegetables may result from either geological and geographical causes, for which man is not responsible, or from industrial and urban contamination, for which man must be held accountable.

FERTILIZERS AS SOURCES OF CONTAMINATION

Fertilizers, including mineral fertilizers such as triple superphosphates and such products as milorganite, and even some seaweeds constitute a further source of contamination. Limestone and some smelting slags, both of which are widely used to increase productivity, may also on occasion introduce unwanted amounts of some elements into soils and thus into food supplies.

The following data, all taken from *The Trace Element Content of Fertilizers* (1962) by D. J. Swaine, provide some indication of the extent of the contamination.

Table 10 data are not intended to suggest that all fertilizers contain harmful ingredients. However they do suggest that in his efforts to increase the productivity of the soil man must be careful not to introduce harmful elements. Organic fertilizers such as milorganite, bone meal and seaweed are as liable to introduce such elements as lead, cadmium, and arsenic into the soil as are inorganic fertilizers such as limestone, phosphate rock and superphosphate.

VARIABLE RELATIONSHIPS BETWEEN TRACE ELEMENTS IN SOILS AND VEGETABLES

Cereals and vegetables do not always take up the amount of harmful elements in a soil as might be feared. In several instances either the pH of the soil, its organic content, or its clay fraction prevents a significant portion of the harmful element from being incorporated in the vegetal matter. Bearing this in mind it is of interest to examine the trace element contents of some soils and of the potatoes growing on them. Trace element contents of the vegetables are expressed in parts per million of ash and of soil as parts per million of aquaregia perchloric acid extractable metal.

TABLE 10

Abnormal trace element contents of some fertilizers

Element	Fertilizer	ppm	Country
Lead	Concentrated sludge	12,000–61,000	Germany
	Dried sewage sludge	6,700	USA
	Digested sludge	1,900	USA
	Milorganite	2,000	USA
	Seaweeds	<300–400	Ireland
	Limestones	5–100	USA
	Factory lime	0–600	Sweden
	Flue dust	1,000	England
	Blast furnace dust	650–62,900	Japan
Cadmium	Milorganite	about 2,000	USA
	Phosphate rock	<100–100	USA
	Superphosphate	50–170	Australia
Arsenic	Brown coal fly ash	about 1,000	Germany
	Seaweeds	2–58	Canada
	Milorganite	150	USA
Molybdenum	Lignite ash	1,500	USA
	Activated sewage sludge	6–45	USA
	Bone meal (steamed)	<100–200	USA and England

It would not be difficult to present a wealth of further data to show how various vegetables and cereals take up widely different amounts of trace elements from a particular soil. However, to indicate vegetal-soil relationships it would seem adequate to select potatoes from different parts of the British Isles and present the zinc, copper, lead, and molybdenum contents of their ash and of the soil in which they were growing. This will illustrate the complexity of the relationships between soils and plants.

The trace element content of a wet, oven dried, or ashed vegetable may be compared with the trace element content of the soil in which that vegetable grew. From experience it has been found that ash-soil relationships are satisfactory. These relationships are used in the Tables that follow.

TRACE ELEMENTS IN THE ATMOSPHERE

Even if it is assumed that each person inhales twenty cubic metres of air each day it is safe to assume that under normal conditions the quantity of trace elements absorbed from air is much less than that absorbed from food. Schroeder (1970, p. 799) reports that 'estimates for absorption of particulate matter consider that 25 per cent of inhaled material is exhaled and 50 per cent deposited in the upper respiratory passages and subsequently swallowed. The other 25 per cent, if soluble, is absorbed from the lung; if it is insoluble it accumulates. These relative amounts depend on particle size. Thus, the intake of a metal from air is about 75 per cent of that in air, more or less'.

TABLE 11

The copper content of the ash of some potatoes and the soils on which they are growing
(in ppm)

Location	Ash	Soil	Copper in ash, copper in soil
Birmingham (a), Warwickshire	270	85	3·2
Bonahaven, Argyllshire	205	45	4·6
Bradford, Yorkshire	200	85	2·4
Birmingham (b), Warwickshire	195	540	0·4
Doddington, Kent	195	27	7·2
Plas Newydd, Anglesey	135	24	5·6
Leeds, Yorkshire	125	215	5·2
Grassington, Yorkshire	105	42	2·5
Chinley (a), Derbyshire	105	81	1·3
Chinley (b), Derbyshire	95	105	0·9
Sherborne, Dorsetshire	85	60	1·4
Teeside, Yorkshire	85	35	2·4
Charterhouse, Somerset	65	44	1·5
Teeside, Yorkshire	65	62	1·0
Bridgend, Argyllshire	60	43	1·4
Priddy, Somerset	50	65	0·8
Crutch, Caernarvon	40	62	0·6
Aberystwyth, Cardiganshire	40	40	1·0
Machynlleth, Montgomeryshire	40	263	0·15
Easton Royal, Wiltshire	30	24	1·2

TABLE 12

The zinc content of the ash of some potatoes and the soils on which they were growing
(in ppm)

Location	Ash	Soil	Zinc in ash, zinc in soil
Machynlleth, Montgomeryshire	1,490	1,825	0·8
Bradford, Yorkshire	1,245	360	3·5
Birmingham, Warwickshire	1,130	1,210	0·9
Birmingham, Warwickshire	1,000	445	2·2
Bonahaven, Argyllshire	940	270	3·5
Doddington, Kent	465	150	3·1
Leeds, Yorkshire	455	430	1·1
Charterhouse, Somerset	360	730	0·5
Teeside, Yorkshire	345	250	1·4
Sherborne, Dorsetshire	340	310	1·1
Aberystwyth, Cardiganshire	305	200	1·5
Plas Newydd, Anglesey	300	105	2·9
Chinley, Derbyshire	285	330	0·9
Chinley, Derbyshire	280	290	1·0
Easton Royal, Wiltshire	270	360	0·7
Crutch, Caernarvon	250	275	0·9
Teeside, Yorkshire	245	350	0·7
Priddy, Somerset	240	440	0·5
Grassington, Yorkshire	225	755	0·3
Bridgend, Argyllshire	180	135	1·3

TABLE 13

The lead content of the ash of some potatoes and the soils on which they were growing
(in ppm)

Location	Ash	Soil	Lead in ash, lead in soil
Leeds, Yorkshire	705	525	1·3
Teeside, Yorkshire	425	140	3·0
Plas Newydd, Anglesey	375	71	5·3
Teeside, Yorkshire	336	370	1·0
Sherborne, Dorsetshire	325	600	0·5
Crutch, Caernarvon	220	285	0·8
Birmingham, Warwickshire	200	700	0·3
Chinley, Derbyshire	175	345	0·5
Grassington, Yorkshire	160	2,315	0·07
Machynlleth, Montgomeryshire	140	9,075	0·02
Bradford, Yorkshire	110	335	0·3
Birmingham, Warwickshire	105	205	0·5
Easton Royal, Wiltshire	95	175	0·5
Aberystwyth, Cardiganshire	60	165	0·4
Chinley, Derbyshire	50	260	0·2
Bonahaven, Argyllshire	45	85	0·5
Charterhouse, Somerset	25	4,105	0·01
Doddington, Kent	20	70	0·3
Bridgend, Argyllshire	<18	155	<0·12
Priddy, Somerset	12	710	0·07

TABLE 14

The molybdenum content of the ash of some potatoes and the
soils on which they were growing
(in ppm)

Location	Ash	Soil	Molybdenum in ash, molybdenum in soil
Chinley, Derbyshire	18	3·6	5·0
Teeside, Yorkshire	12	2	6·0
Leeds, Yorkshire	11	10	1·1
Birmingham, Warwickshire	10	5·4	1·8
Priddy, Somerset	9	3	3·0
Birmingham, Warwickshire	7	2	3·5
Easton Royal, Wiltshire	4	<1	>4·0
Grassington, Yorkshire	3	4	0·75
Plas Newydd, Anglesey	3	13	0·2
Machynlleth, Montgomeryshire	3	13	0·2
Doddington, Kent	2	1	0·5
Bradford, Yorkshire	2	12	0·2
Chinley, Derbyshire	2	2	1·0
Crutch, Caernarvon	2	6	0·3
Aberystwyth, Cardiganshire	2	3	0·7
Charterhouse, Somerset	2	3	0·7
Teeside, Yorkshire	2	1·7	1·2
Sherborne, Dorsetshire	2	2·5	0·8
Bonahaven, Argyllshire	<21	2	—
Bridgend, Argyllshire	<18	<2	—

Most of our atmospheric pollution comes from burning coal and petroleum. There are however indirect sources of metal contamination. Materials used as fertilizers, pesticides and herbicides, find their way via soil into dust and thence under special circumstances into the atmosphere. Lead, derived largely from burning additives to gasoline, is possibly one of the best documented of all pollutant metals.

At the close of his article, Schroeder (*op cit.*, p. 805) concludes: 'Only three (metals) represent real or potential hazards to human health; nickel, cadmium, and lead'. It must be appreciated, however, that Schroeder was discussing air pollution only.

In the limited investigations conducted by the writer it has been noted that there are high concentrations of molybdenum, manganese, zinc and mercury in certain soils. The origin of some of these metal concentrations is not clear, but they may well have been introduced into the soil at least in part by atmospheric fallout. At all events they often find their way into the food supply.

Because so many measurements have been made on the lead contents of the atmosphere it is of interest to note how it may contribute to man's total intake of lead. In rural areas the lead content of the atmosphere may be expected to run from 0·002 to 0·15 micrograms per cubic metre and in urban areas from 0·1 to 2·3 micrograms per cubic metre. In crowded urban areas with much traffic a content of 10 micrograms per cubic metre may frequently be encountered. Thus a person living in an urban area might inhale from between 2 and 46 micrograms a day and perhaps as much as 200 micrograms if exposed to heavy traffic for a period of twenty-four hours. In contrast a person eating 200 grams of potato containing 6 parts per million of lead (wet weight) ingests 1.2 milligrams of lead. Fortunately, only a small percentage of the metal would be absorbed. Perhaps it should be noted that a normal intake of lead is assumed to be from 0·3 to 0·4 milligrams a day.

TRACE ELEMENTS IN WATER

No attempt is made in this paper to discuss the intake of trace metals from water by humans. The World Health Organization has set water standards for the various elements. If these are met, as they probably are in most instances, one can drink water in most developed areas in perfect safety.

In areas where there is an appreciable rise and fall in the water table and where there are concentrations of heavy metals, important variations in the trace element content of the groundwater may occur from time to time. It appears that the greatest opportunities for trace elements to enter groundwater occur either when the water table is rising through the zone of oxidization, or when rain comes after a long dry spell, or as is particularly common during spring time when the water table withdraws through the zone of oxidization. This suggests that in many areas the random sampling of water supplies is meaningless and indeed may be most misleading.

That the water supplies of many communities are frequently contaminated by the method adopted to transport water from reservoir to user is widely recognized. The degree of contamination depends on several factors, in particular the acidity of the water being transported. Where the pH of the water is suitable the pipes have lasted for several decades, but with peaty water lead pipes have been ruined in less than twenty years. With almost neutral water combined with modern plumbing it is possible to find that tap water may contain significant amounts of such metals as iron, copper, and zinc which have been picked up between the reservoir and the point of delivery.

TRACE ELEMENTS IN HUMAN DIETS

The writer does not propose to discuss the dietary, nutritional or medical aspects of trace elements. Nevertheless, it is relevant to note that of the trace elements discussed in this article many are known to be essential for human health. It is when such trace elements are present in anomalous amounts that health problems may arise. Consequently if the significance of the different concentrations of specific trace elements in various vegetables is to be fully appreciated it is essential to know what is considered to be the normal intake of an average, 70 kilogram man.

The writer has drawn freely on two major sources for the broad and generalized statements which follow. These sources consist primarily of several articles published by Henry A. Schroeder (1961, 1961 (*a*), 1967, 1968, 1970) and a reference book entitled '*Trace Elements in Human and Animal Nutrition*' by E. J. Underwood. These and other references are referred to in the Bibliography of this Chapter.

With the data of Table 15 in mind readers may be able to deduce the significance of the anomalies which are reported elsewhere in this paper. Much has still to be learnt about the form or forms in which the above elements are present in vegetables, the interactions that these elements have with one another and the disposition of these elements after ingestion.

TABLE 15

Average daily intake of some trace elements by 'Normal' man

Element	Common range	Possible average
Iron	15–22 milligrams	18 milligrams
Zinc	10–15 milligrams	12 milligrams
Manganese	2–4 milligrams	3 milligrams
Copper	1–3 milligrams	2 milligrams
Nickel	300–500 micrograms	400 micrograms
Lead	300–500 micrograms	350 micrograms
Molybdenum	200–600 micrograms	330 micrograms
Cobalt	140–580 micrograms	310 micrograms
Cadmium	200–500 micrograms	300 micrograms

Nevertheless, with the above data as guide lines it should be possible to assess the potential significance of any anomalous concentrations of trace elements which may be reported from time to time and to decide whether or not industrial, urban and or other forms of contamination pose a potential health hazard.

BIOLOGICAL CONCENTRATION OF TRACE ELEMENTS

The question is frequently asked, 'Is it safe to eat sea food?' This question has been activated by public reaction to the recent finding of mercury in some sea foods and cadmium in some vegetables. Rivers discharge mercury and cadmium into the sea every year together with a host of other elements. It is inevitable that some of these elements are taken up by phytoplankton and zooplankton and thence into higher forms of life. Scallops, oysters, and mussels pass large amounts of sea water through their digestive systems. Table 16, after Thompson (1971), illustrates the degree of trace element concentration performed by those commonly eaten shellfish.

It does not follow that seafood will concentrate all the above elements to the extent shown; the Table merely indicates what may happen under particular circumstances.

TABLE 16

Concentration factors for some metals in commonly eaten shellfish*

Element	Shellfish		
	Scallop	Oyster	Mussel
Iron	291,500	68,200	196,000
Manganese	55,500	4,000	13,500
Zinc	28,000	110,300	9,100
Nickel	12,000	4,000	14,000
Copper	3,000	13,700	3,000
Lead	5,300	3,300	4,000
Molybdenum	90	30	60
Cadmium	2,260,000	318,000	100,000

* The ratio of the element concentration in the organism and the element concentration in sea water.

The ability to concentrate trace elements is not confined to shellfish. Species of plants and trees have been found to be capable of concentrating some elements occurring in a soil by one and even two orders of magnitude over that possessed by neighbouring plants growing on the same kind of soil. Mint has the ability to collect lead, beans attract molybdenum. Douglas Fir is avid for arsenic. The affinity of members of the astragalus family to absorb selenium has long been reported.

SUMMARY AND CONCLUSIONS

The geology and geography of the earth's crust is such as to ensure that the trace element contents of soils and of the vegetal matter growing on these soils will vary greatly.

The Industrial Revolution and the Population Explosion have accentuated the uneven distribution of these elements in the soils and food supplies of the world.

It is probable that in certain specific instances natural and man-made imbalances may be related to human health or morbidity.

Some so-called geochemical anomalies should be examined systematically in relation to possible local and regional patterns of human disease. Such examinations have proved fruitful in relation to animals and may eventually shed light on human disease.

Acknowledgments

This paper is based on work made possible by a Canada Council Grant (Killam Award) given to further studies of environmental problems, and in particular those involving pollution. The Donner Canadian Foundation had previously provided funds for the purchase of atomic absorption equipment. Grants from the Geological Survey of Canada, the National Research Council of Canada (Grant No. A 1805), and the Defence Research Board of Canada (Grant No. 7510–16) facilitated the acquisition of basic data.

The Royal College of General Practitioners in Great Britain, acting with the enthusiastic support of Dr R. J. F. H. Pinsent, and the College of Family Physicians in Canada, spearheaded by Dr W. A. Falk, have freely given assistance and advice.

Dr and Mrs L. E. Lefevere and Mr E. Wilks collected the soil and vegetable samples from various localities in Great Britain and North America, respectively.

Several colleagues in the Faculties of Medicine and Agriculture in the University of British Columbia have participated with advice and criticism, and especially Drs R. E. Delavault and K. Fletcher of the Department of Geology, who paid particular attention to analytical procedures. Mr A. S. Dhillon and Miss Anne Baxter carried out the analytical work and coped well with the problems inherent in dealing with such a diversity of samples.

Bibliography

Cannon, Helen L. and Anderson, Barbara M. (1971). *The Geochemists Involvement in Pollution Problems*. The Geological Society of America Memoir 123.

Goldschmidt, V. M. (1954). *Geochemistry*. Oxford, The Clarendon Press.

Schroeder, H. A. (1961). Trace metals and man. *The Spex Speaker*, VI, No. 2, June, pp. 1–4.

Schroeder, H. A. and Balassa, J. J. (1961). Abnormal trace metals in man: lead. *J. of Chronic Diseases*, **14,** No. 4, October, pp. 408–425.

Schroeder, H. A., Nason, A. P. and Tipton, Isabel H. (1967). Essential trace metals in man: cobalt. *J. of Chronic Diseases*, **20,** pp. 869–890.

Schroeder, H. A. and Tipton, Isabel H. (1968). The human body burden of lead. *Arch. Environ. Health*, **17,** December, pp. 965–978.

Schroeder, H. A., Balassa, J. J. and Tipton, Isabel H. (1970). Essential trace metals in man: molybdenum. *J. of Chronic Diseases*, **23,** pp. 481–499.

Schroeder, H. A. (1970). A sensible look at air pollution by metals. *Arch. Environ. Health*, **21,** December, pp. 798–806.

Shacklette, H. T., Hamilton, J. C., Boerngen, Josephine G. and Bowles, Jessie M. (1971). *Elemental Composition of Surficial Materials in the Conterminous United States*. US Geol. Survey Prof., Paper 574D.

Swaine, D. J. (1955). *The Trace Element Content of Soils*. [Commonwealth Agricultural Series], Farnham Royal, Bucks, Commonwealth Bureau of Soils.

Swaine, D. J. (1962). *The Trace Element Content of Fertilizers*. Technical Communication No. 52. Commonwealth Bureau of Soils, Farnham Royal, Bucks, England.

Thompson, G. (1971). Spectroscopy in oceanography. *The Spex Speaker*, XVI, No. 2, June p. 4.

Underwood, E. J. (1962). *Trace Elements in Human and Animal Nutrition*, 2nd Edn., New York, Academic Press Inc.,

Vinogradov, A. P. (1959). *The Geochemistry of Rare and Dispersed Chemical Elements on Soil*, 2nd Edn. [Translated from Russian by Consultants Bureau, Inc.], New York, Plenum Publishing Corporation.

Warren, H. V. (1965). Medical geology and geography. *Science*, **148,** No. 3669, April, pp. 534–539.

Chapter 3
Radiation and Health Hazards

R. D. E. Rumsey

INTRODUCTION

After the first demonstration of X-rays by Wilhelm Roentgen in 1895, Henri Becquerel discovered in 1896 that pitchblende ore was capable of darkening photographic film. Becquerel showed that the ore contained small amounts of uranium and thorium salts which emitted energy in the form of penetrating radiations similar to X-rays. Continuing this work, Pierre and Marie Curie demonstrated that even smaller quantities of a third radioactive element, radium, were also present in the ore and that radium evolved a gas, known as radon, which is itself radioactive. As pitchblende is relatively widely distributed throughout the earth's crust, it became immediately apparent that man's environment contained a hitherto un-recognized factor which subsequently became known as 'radioactivity'.

Legend has it that Becquerel carried a tube of radium around with him in his waistcoat pocket for demonstration purposes; while the radium did not harm the material of his waist-coat, it nevertheless produced a burn on his body. This initial demonstration that radiation was capable of harming living tissue, in this case causing skin erythema, also suggested that the property might be adapted and controlled for the treatment of malignant tumours. Later, when radiologists and chemists who had been working with uranium, thorium and radium began to contract diseases of the blood and certain types of cancer, it was realized that radio-activity produced a generalized deleterious effect on the human body. The detonations of atomic bombs at the end of the Second World War and the subsequent testing of more powerful devices, resulting in radioactive contamination of the global environment, served to demonstrate the lack of knowledge about the effects of radiation, not only in individuals but in populations. Since that time much research has been carried out and certainly much has been achieved. The picture, however, is very far from complete, and the state of our knowledge remains unsatisfactory. As our depth of understanding increases so does our respect for the hazards of radioactivity; this is demonstrated by the reduction in safety levels of radiation by a factor of 30 in as many years.

THE NATURE OF RADIOACTIVITY

Radioactivity is energy liberated by the fragmentation of large atoms into smaller ones. This may occur spontaneously as with atoms such as uranium, thorium or radium, or it may be produced synthetically, a process known as nuclear fission, resulting in the release of vast amounts of energy in the form of heat, light, noise and radiation.

Since the early discoveries by Roentgen, Becquerel and the Curies, it has become recognized that the elements in our environment exist as a mixture composed of isotopes of the same matter. By far the greater proportion of the element consists of the stable isotope, but the residue is composed of one or more unstable isotopes which fragment or 'decay' spon-taneously with the emission of radioactivity to form different atoms. For example, the element potassium which occurs in abundance throughout the earth's crust as well as in all living organisms, exists as a mixture of two isotopes, potassium 39 and potassium 40. Potassium 39 is the stable isotope and constitutes the greater proportion by far of the naturally occurring element. However, potassium 40, the radioactive isotope, occupies 0·0119 per cent of naturally occurring potassium. Potassium 40 decays to give the stable isotope of calcium (calcium 40) in the process emitting β and γ radiations. Radioactive isotopes are produced artificially by bombardment of a stable element in a nuclear reactor.

The half life of radioisotopes, or radioactive nuclides as they are also called, is a measure of the rate of radioactive decay of that isotope and is defined as the time required for one half of the total atoms of a nuclide to disintegrate. Half-lives vary immensely, for example the half-life for the disintegration of potassium 40 to calcium 40 is $1·3 \times 10^9$ years and the decay of polonium 212 to lead 208 is 3×10^{-7} seconds. Obviously radionuclides with longer half-lives constitute the greater potential environmental hazard.

Radioactive emissions from nuclides are of differing types; briefly they fall into four main categories.

    (i) Alpha (α) radiation consists of high energy particles identical with helium nuclei. Alpha particles have a low penetrative power corresponding to less than 0·1 millimetres in tissue and a few centimetres in air.

    (ii) Beta (β) radiation consists of electrons or positrons of varying energy and a penetrative power of up to a few millimetres in living tissue.

    (iii) Gamma (γ) rays and X-rays are similar in nature to one another and possess higher penetrative powers. Unlike α and β radiations which are composed of particles, γ and X-rays are electromagnetic waves similar to light rays but of shorter wave length. Gamma rays have shorter wave lengths than X-rays and consequently possess greater powers of penetration; indeed, they can travel distances of up to one mile through air and may only be stopped by heavy shielding.

    (iv) Neutrons are unchanged particles with a wide range of energy and penetrative power.

The unit of radioactivity is the 'curie' which is based on the rate of disintegration of radium atoms, and is defined as that amount of material in which $3·7 \times 10^{10}$ disintegrations occur per second. One curie is such a colossal amount of radioactivity that radiation in biology is more often measured in microcuries (μc), one millionth of a curie, or even in micromicrocuries ($\mu\mu$c), one millionth of a microcurie. X-rays are, however, measured in 'roentgens' (r) or milli-roentgens (mr) which are one thousandth of a roentgen.

RADIOBIOLOGY

A proportion of the constituent atoms of substances in the path of radiation become charged particles known as ions. Ions have greater chemical activity than neutral atoms, and it is thought likely that the harmful effects of ionizing radiations on living tissues are mediated by their influence on the normal chemical processes of the cell.

The amount of radiation actually absorbed by living tissue is measured in rads, 1 rad being equivalent to the energy absorbed when 1 roentgen is received by 1 gramme of 'soft tissue'. As different types of radiation produce different biological effects, a second unit of absorbed dose, the 'rem' or roentgen equivalent for man is defined to compensate for this variation. In the case of β, γ and X-irradiation, one rem is equivalent to one rad. Alpha particles are estimated to have a 'relative biological effectiveness' (RBE) of ten times that of X-rays, so that 1 rem for α particles is equivalent to 10 rads.

Different living organisms as well as different cell types within the same organism possess varying degrees of radiosensitivity. For instance, dosages of some thousands of rems are required to impair the reproductive capacity of certain bacteria, whereas the growth of certain human cells is inhibited by doses of as little as 100 rems. In man the lethal dose of radiation to the whole body varies from 200 to 1,000 rems; dogs are thought to be more radiosensitive and rabbits less so, with mice, rats, hamsters and monkeys, along with man, being intermediate. In the human body the most sensitive tissues include the gonads, the lymphatic glands, the haemopoietic cells of the bone marrow and the small intestinal mucosa; on the other hand, a tissue such as muscle is less sensitive. Young cells and children's tissues are more radiosensitive than are those of adults. Absorbed dosages of radiation are frequently quoted in terms of a particular organ of high sensitivity, for example the 'gonad dose' or the 'bone marrow dose'. However, certain other individual organs are capable of withstanding doses of radiation which would be ultimately lethal if applied to the whole body.

The biological effects of radiation depend not only upon the intensity of the radiation but also on the time for which it is applied. Thus, dosages of 1,000 rems applied for 1 minute produce an effect similar to doses of 100 rems applied for 10 minutes. Natural radiation exposes the human body at a constant rate of approximately 0·1 rems annually. A radiological examination may produce a similar annual exposure but at a far higher dose rate, the actual exposure occupying only fractions of a second. However it is generally considered that for the same dose of radiation, the lower the dose rate the smaller the biological hazard.

SOURCES OF ENVIRONMENTAL RADIATION

Before considering its effects on health, it is important to discuss the sources of environmental radiation, both natural and man-made, and to compare the dosages received by man from each source.

Radiation is assessed as internal or external depending on whether the source is inside or outside the body. Internal radiation arises following the inhalation or ingestion of radioactive substances. If such substances are not excreted but instead are incorporated into the tissues of the body, then the source rests permanently in direct contact with living cells causing the maximum potential damage from all types of radiation—α, β and γ. Sources external to the body are less potentially hazardous, particularly if they contain a high proportion of α or β particle irradiation which have considerably lower penetrative power than do γ rays.

By far the greatest dosage of radiation to which man and all the other species which have evolved on this planet are, and always have been, subjected arises from the natural environment. These sources will be discussed first in an endeavour to place in perspective the sources of radiation which are man-made.

Natural Radiation

The total dosage of natural radiation is approximately 100 m rads per year derived from the various sources which are summarized in Table 17.

Terrestrial Radiation

Terrestrial radiation emanates from radioactive elements such as thorium, uranium and radium which occur in the soil and in the earth's crust. These elements are still present in nature because they possess half-lives similar to the geological age of the planet. Since soil and rock compositions vary greatly, it is not surprising that the absorbed dosage of gamma rays from this source varies from one part of the world to another. Acid igneous rocks

TABLE 17

Sources of natural radiation

Source	Mean Gonadal dose-rate (m rads/year)
External radiation:	
Terrestrial radiation	47
Cosmic radiation	28
Atmospheric radiation	2
Internal radiation:	
Potassium 40	19
Carbon 14	1·6
From diet	<1

and those with a high silicon content have been found to contain high proportions of radioactive elements. Some examples of the variation encountered throughout the world from this source are given in Table 18.

TABLE 18

Variation in terrestrial radiation throughout the World

Region	Annual gonadal dose (m. rads)
London	24–29
Leeds	48
Cornwall (granite regions)	50–300
Edinburgh	36
Aberdeen	57
Kerala, Southern India	800–4,000
Espirito Santo ⎱ Brazil Guarapary ⎰	300–1,600

(From 1958 UNSCEAR Report and Spiers, 1960)

Man has modified the effects of terrestrial radiation by living in houses often constructed from the local stone. On the one hand this may shield him from external environmental radiations but on the other hand, it may increase his radiation dosage if the building materials themselves are rich in radioactive ore-bearing rock.

Cosmic Radiation

Cosmic radiation reaches this planet from the depths of space. Fortunately, the majority of this energy is removed by filtration in the upper atmosphere. The dosage from cosmic rays varies by approximately 14 per cent with latitude but to a far greater degree with altitude above sea level; this is illustrated in Table 19.

TABLE 19

Annual doses from cosmic radiation
at different altitudes above sea-level

Altitude (m)	Annual dose (m rads) (approx.)
0	30
1,000	35
2,000	45
3,000	75
4,000	100
5,000	170

(From Lindell & Dobson, 1961)

Cosmic radiation is a well documented hazard of space travel; it is also a factor to be considered in the health of aircrew of supersonic transports of the immediate future. It is thought that cosmic rays produce carbon 14 in the upper atmosphere; this falls eventually to earth as carbon dioxide where it takes part in the natural carbon cycle. By this mechanism naturally produced carbon 14, as well as that produced by nuclear explosion, is incorporated into all living tissue producing internal β radiation in man.

Atmospheric Radiation

External atmospheric radiation arises from the radioactive gases radon and thoron present in minute amounts in the atmosphere. These gases are inhaled by man and in a small way constitute internal radiation of the lungs where their effect appears to be localized.

Internal Radiation

Approximately one fifth of all natural radiation dosage arises from radioisotopes present in living tissue. The most important of these is potassium 40 with carbon 14 contributing to a lesser degree. The element potassium occupies between 0·12 per cent and 0·36 per cent of total body weight; 0·0119 per cent of this fraction is radioactive potassium 40. Even this minute proportion of the total constituent mass of the body is capable of delivering a dose of 19 m rads to human tissue per year.

Recently it has been ascertained that the human diet contains extremely varied amounts of radiation mainly in the form of α-emitters such as radium and thorium. The radiation content of certain foods has been found to be particularly high as indeed have some sources of drinking water; examples of these are shown in Table 20.

Man-made Radiation

Sources of man-made radiation fall into two categories, firstly those that affect the total population of the planet or at any rate significant populations in any given country, and secondly those that affect small proportions of a population; generally this second group of people is potentially at risk by virtue of their occupation.

Fall-out from Nuclear Explosions

Detonations of nuclear 'devices' in the atmosphere produce worldwide radiation pollution. Following the explosion a certain proportion of the radiation generated falls locally in

TABLE 20

Radiation content of some human foods and drinking water

Food	Concentration of α activity $\mu\mu c/kg$
Fruit and vegetables	1–23
Milk and milk products	1–15
Meats and fish	1–130
Breakfast cereals	9–58
Bread and flour	9–147
Nuts	5–17,000

Water	Concentration of α activity $\mu\mu c/litre$
Ocean	0·01–0·07
Water from surface drainage ⎫	0·04
Water from deep bore-holes ⎬ UK	0·40
Some Cornish tap water ⎪	4·0
Some Spa waters ⎭	40·0
Springs in Boulder, Colorado, USA	300,000
Springs in Joachimstal, Czechoslovakia	500,000

(From Mayneord, 1960, and Lindell & Dobson, 1961)

particulate form. However, a large fraction of the total is ejected with great energy into the stratosphere, particularly if the explosion is in excess of 1 megaton. This fraction of radiation is consequently distributed throughout the upper atmosphere of the planet whence over a period of time, it is deposited on the earth through rainfall. The deposition of radionuclides is not uniform over the surface of the earth; indeed, a larger proportion falls over the northern temperate zones which include Europe, the USSR and North America.

Table 21 illustrates the total doses in the two hemispheres of radiation in millirads as a result of all nuclear explosions carried out before 1968.

TABLE 21

Variation in nuclear fall-out between the hemispheres

Organ dose	Whole world	Northern Temperate Zone	Southern Temperate Zone
Dose to gonads	80 m rads	110 m rads	33 m rads
Dose to bone marrow	140 m rads	170 m rads	51 m rads
Dose to cells lining bone surface	220 m rads	240 m rads	66 m rads

(From 1969 UNSCEAR Report)

It has been estimated that more than half of the total dose to man from all external sources liberated by nuclear explosions in the past has now been delivered. However, sources resulting in internal radiation are more hazardous to the population; of these radionuclides the most dangerous are considered to be strontium 90, caesium 137, and carbon 14, all of which have long half-lives. Other products of nuclear fall out such as iodine 131, which is a short lived isotope, and medium life isotopes such as strontium 89, zirconium 95, ruthenium 103, ruthenium 106 and cerium 144 are also liberated in considerable quantities. Such radionuclides are lesser risks, particularly since the mass testing of nuclear bombs has ceased.

The test ban treaty of 1958 does not signal any reduction in the hazard to health from this source of environmental pollution; indeed, it has been estimated that of the total dose of strontium 90 to man resulting from nuclear tests, only one-eighth was delivered by 1967.

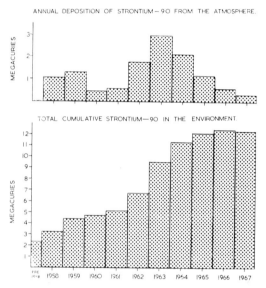

Fig. 1   Disposition and accumulation of strontium 90 in the environment. (Adapted from 1969 UNSCEAR Report.)

Similarly between two-thirds and three-quarters of the dose of caesium 137 was delivered by that time and only one-tenth of the total dose of carbon 14 will have been delivered by the end of this century. The reasons for this are two-fold; firstly, the half-lives of all three isotopes are long, that of strontium 90, 28 years, of caesium 137, 30 years and of carbon 14, 5,640 years. Secondly, strontium and caesium, once deposited, are assimilated into plants from which, after variable amounts of time, they pass into the human body either directly or indirectly as milk and its products.

Figure 1 compares the annual deposition of strontium 90 with the accumulated amount of the isotope in the environment over the years 1958–1967. It may be seen that the maximum years of deposition, 1962, 63 and 64, occurred several years after the last large series of nuclear tests in 1958 and that the cumulative content of the isotope in the environment had not reached maximum levels until several years later.

Once ingested, caesium 137 and carbon 14 are distributed evenly throughout the body where they constitute whole body radiation. The effect of caesium 137 is ameliorated by the fact that it is excreted from the body within a few months. On the other hand, the effects of strontium 90, are much more hazardous. The body is unable to distinguish between the elements strontium and calcium, so that both are incorporated almost permanently into bone. As a result radioisotopes of strontium irradiate the blood-producing cells of the bone marrow as well as the entire human skeleton.

Nuclear Power

Although at the present time it might be fair to say that the radioactive effluent from nuclear power stations and other reactors constitutes no hazard to public health, the potential risks, however remote, of accidents at such establishments and the transporting and storage of high energy waste products for hundreds of years merit attention. Certainly the proliferation of nuclear power stations during the latter third of the present century will increase the risk of such an environmental catastrophe. Accidents at nuclear plants involving any hazard to the general public have so far been exceedingly rare. The best documented of such incidents occurred at Windscale, Cumberland in October 1957. The accident resulted in the release of a total of 33,000 curies of radioactivity composed mainly of iodine 131, tellurium 132 and caesium 137 into the surrounding environment. Radioactive fall-out was measured throughout England and Wales as well as a large part of the mainland of Europe and Scandinavia. Estimations of the maximum doses received by those living within 25 miles of Windscale varied between 1 and 16 rads, the higher doses being absorbed by the thyroid glands of children. Adults living in the Leeds area were estimated to have received 0·2 rads to the thyroid and those in London, 40 millirads.

Another accident occurred in 1966 in an experimental breeder reactor on the shores of Lake Erie in the USA. At one stage during the emergency that ensued the evacuation of the neighbouring metropolis of Detroit was considered. This incident illustrates one of the major prerequisites in the siting of nuclear power stations, namely that they should be placed as far as is possible from large centres of population.

Medical Radiology

X-rays have been used for diagnostic and therapeutic purposes for the greater part of this century. However, in recent years, largely as a result of state welfare schemes, the use of X-rays for medical purposes has proliferated to such an extent that a sizeable fraction of the total population is now being irradiated from this source. This potential hazard only occurs in highly developed nations in which routine X-ray diagnostic and therapeutic facilities are readily available. Individuals undergoing routine X-ray procedures are not considered at risk. Nevertheless, if the doses absorbed from each radiological procedure are integrated for the total number of persons irradiated, a genetic dose to the whole population may be estimated.

In Britain the medical use of X-rays both inside and outside the National Health Service was thoroughly investigated by a Ministry of Health Committee appointed in 1956 and headed by Lord Adrian. This Committee, which submitted its main report in 1960, estimated that a total of 21 million X-ray examinations were carried out in the United Kingdom in the year 1957. This figure included 13 million examinations carried out in Health Service Hospitals. The Adrian report further estimated the total annual genetic dose to the population from this source to be 19·3 milliroentgens per person. The composition of this total dose is illustrated in Table 22.

It may be seen from the table that the greatest source of irradiation is general diagnostic radiology (14·09 mr), of which 13·2 mr was estimated to account for diagnostic radiology carried out in National Health Service Hospitals. The Adrian Committee did not believe that the levels of radiation dosage indicated any need for a restriction of radiological practice but recommended many improvements in radiological technique which, if implemented, might reduce the total annual genetic dose from 19·3 mr to a figure as low as 6 mr. In 1964 the British Institute of Radiology stated that technical improvements suggested in the Adrian Report had undoubtedly reduced patient dosage in diagnostic radiology; however, on the other hand, the total annual number of radiological examinations in 1971 had increased by two-thirds since 1957. Similar computations in other developed countries indicate considerable variability in the dosage delivered to the total population from this source; indeed, doses in the range of 150–240 m rads per year have been estimated for those resident in Australia and the USA.

TABLE 22

Annual genetic dose from the medical uses of irradiation (1957)

Use	Dose (m roentgens) per person
General diagnostic radiology	14·09
Mass miniature radiography	0·01
Dental radiography	0·01
External radiotherapy {non-malignant diseases	4·47
{malignant diseases	0·52
Internal radiotherapy {non-malignant diseases	0·13
{malignant dieases	0·05
Total annual genetic dose	19·3
Natural background level	100

(*Courtesy of Ministry of Health, 1960*)

Surveys carried out in the USA (Penfil & Brown, 1968) have indicated that by far the greatest component of the genetically significant dose is received by young men between the ages of 15 and 29 undergoing radiological examinations of the pelvis and abdomen. It is, however, possible to achieve a marked reduction in the genetic dose by using such examinations for this particular age group sparingly and also by ensuring that the lowest gonadal dose possible for the particular procedure is administered (ICRP publication 16).

Similarly it has been possible to calculate mean bone marrow doses to human populations, and such surveys have indicated that the magnitude of this dose is similar to the genetic dose. Approximately one-third of the total bone marrow dose is derived from chest radiology and one-quarter from examinations of the gastro-intestinal tract. (ICRP publication 16).

Occupational Exposure and Maximum Permissible Doses

The increasing use of radioactivity in industry and medicine combined with the proliferation of nuclear power has resulted in many more people becoming exposed to radiations in their occupation. The dose of radiation received by workers in these industries is today strictly monitored, usually by means of radiation-sensitive film badges designed to conform to international standards of radiation safety. In Britain the blood of workers subjected to occupational radiation is also monitored routinely in order to detect early signs of possible radiation damage. It was noted by the Medical Research Council in 1960 that over-dosages of radiation occurred very rarely indeed, and that the estimated contribution from occupational radiation to the genetically effective dose to the total population of the country was about 0·5 per cent of the natural background, or some sixty times less than the maximum permissible dosage to the general population.

Standards of radiation safety are expressed as maximum permissible doses which are agreed upon by the International Commission on Radiological Protection (ICRP). ICRP recommendations on maximum permissible doses are universally accepted. However, in Britain they are reinforced by the recommendations of the Medical Research Council's Committee on Protection against Ionizing Radiations.

Since it is not possible to state with certainty that any dose of radiation is harmless, ICRP recommendations are based on practical considerations which attempt to balance the undoubted advantages of using radioactivity against its hazards. Thus, maximum permissible doses for occupational exposure are higher than those for the general public, a proportion of which are children who are considered to be more sensitive to radiation. Moreover, it is not possible to carry out monitoring schemes for all members of the public.

The present maximum permissible dose to the gonads or bone marrow for adults who are occupationally exposed to radiation is 5 rems in a year; the similar dose limit for members of the public is 0·5 rems per year. The genetic dose to the total population is considered to be even more critical and has been recommended to be no more than 5 rems over a single generation time of 30 years, plus the lowest practicable contribution from medical radiology over the natural background level. This dose corresponds to 170 m rem per year or approximately a 200 per cent increase over natural background, taking into consideration medical radiology in this country. In effect, no significant proportion of any community or occupation at present receives anything approaching the maximum permissible dose.

Other Sources

Other sources of environmental radiation include X-ray fluoroscopy for shoe fitting, radio-isotopes present in the luminous dials of watches and clocks and television receivers. The design and construction of equipment used in X-ray fluoroscopy for shoe fitting has been improved to the extent that at present no hazard is considered to exist. Absorbed dosages of radiation from the luminous dials of watches and clocks have fallen to negligible levels following the introduction of isotopes with emissions of low penetrative powers. The risks to the population from the almost universal use of television receivers is also considered negligible.

THE EFFECTS OF RADIATION ON HEALTH

The effects of ionizing radiations on health are generally classified under two headings—somatic disease and genetic damage.

Somatic Disease

Of the somatic diseases or conditions caused by radiation, cancer is the most important. The induction of malignant disease is not spontaneous, but the appearance of the first symptoms can vary from two years to thirty years after the original exposure. Leukaemia, which is generally classified as a cancerous condition for radiobiological purposes, has a relatively short time lapse between exposure and the onset of the disease. For this reason, it was originally believed that leukaemia was the only malignant disease caused by radiation damage. However, it is now known that virtually all types of cancer are capable of being induced by radiation.

Before adequate radiation protection was available, several occupations carried risks of malignant disease for those engaged in them; for example, painters using radioactive luminous paint suffered from a higher risk of cancer of the bone than did the general population. Lung cancer was at one time very common in miners employed in radioactive ore mines. It has also been shown that radiologists are particularly prone to leukaemia and to various types of skin cancer.

Radiation induced disease is a serious side effect of radiotherapy, *i.e.* in the treatment of some otherwise incurable conditions with large doses of X-rays or radioisotopes. This fact was initially demonstrated in Britain by a Medical Research Council team on the basis of a retrospective study involving 14,000 patients treated by radiotherapy for ankylosing spondylitis. It was noted that these patients were nine times more likely to develop leukaemia and twenty-six times more likely to develop aplastic anaemia than were the general population. The survivors of the atomic bomb explosions in Japan have been observed subsequently by the Atomic Bomb Casualty Commission, whose continuing reports represent our best source of information regarding health hazards associated with ionizing radiations. These studies demonstrated increased incidences of several diseases including leukaemia, cancer of the lung and cancer of the thyroid gland, particularly in children. (Maki, Ishimaru, Kate and Wakabayashi, 1968).

Radiation also induces non-malignant somatic damage such as cataracts in the eyes, impaired fertility, bone marrow atrophy and a reduction in the cellular defence mechanisms against infection. It is also known to reduce the life span and to impede growth and development following previous irradiation *in utero*. However, the doses of radiation necessary to induce such changes are very large indeed compared with any dosage which might be termed environmental. Indeed, the survivors of Hiroshima and Nagasaki are estimated to have received several hundred rads, and the dosage administered to ankylosing spondylitics was often greater than 1,000 rads (Court-Brown & Doll, 1965). However, on the basis of such evidence and by extrapolation from certain animal studies, it is possible to estimate the correlation between the incidence of disease and dose rate for different tissues of the body. The highest sensitivity is exhibited by bone marrow and thyroid for both of which it is estimated that a dose of one rad increases the incidence of disease by 'one to several' cases per 100,000 of the population. Tissues with median sensitivities are lymph nodes and reticular tissue followed by pharynx, bronchus, pancreas, stomach and large intestine. Low sensitivity tissues include larynx, oesophagus, small intestine, skin and bone for which it is estimated that a dose of one rad increases the incidence of disease by one case per million. It must be emphasized that these figures, estimated by the ICRP (ICRP publication 15), 'are only intended to serve as a working hypothesis and will need revision as more information is accumulated'.

Genetic Damage

Even less is known regarding the ability of ionizing radiations to induce genetic damage in the human. However, this category of disease, by its threat to the authenticity of heredity, must pose serious questions as to the wisdom of ever permitting increases in environmental levels of radiation, however small these may be. The dose levels of radiation producing genetic damage are considered to be lower than those inducing malignant somatic disease. The effects of genetic damage are likely to be irreversible and cumulative. Furthermore, there is no indication of damage until the advent of the next generation thirty years later.

Although the mechanism whereby radiation induces genetic damage is unknown, the effect on the cell nucleus is two-fold. Firstly, radiation may directly damage and thus modify chromosomal structure and secondly, radiation may increase the gene mutation rate. In successive generations both these genetic changes may be manifest in two ways—firstly a general reduction in intelligence, emotional stability, physique, viability and other attributes and secondly, specific visible detriments resulting in disease states.

Genetic deteriorations are extremely difficult to detect in human populations. Moreover, it is thought that these characteristics may depend on the summational influence of a number of genes so that increased mutation rates would be less likely to produce a general reduction in abilities but instead a greater variation about the same mean. Specific detrimental traits of differing degrees of severity account for between 2 and 4 per cent of all live and still births. Such genetic changes are determined by the action of a single specific gene or by gross chromosomal abnormalities. These changes vary in severity from small alterations in the pigmentation of the iris to serious defects such as Down's syndrome, haemophilia, achondroplasia and phenylketonuria. The incidence of such diseases would show immediate but varying increases following acceleration of the genetic mutation rate. The majority of all harmful mutants are eventually eliminated from a population because they are less efficient at one or more phases of the reproductive cycle. For example, deaths occurring before implantation are thought to be the most important means of eliminating such effects; sterility in the adult is another.

Although there have been no reported increases in the incidence of such hereditary diseases at the present time, a small but significant shift in the ratio of male to female children born in Japan after 1945 has been noted. This might be interpreted as an indication of genetic damage. However, much uncertainty still exists about the eventual form which genetic damage might take. Furthermore, the dose rates capable of producing such charges in the human are largely speculative. Present recommendations are that no one should receive a dose in excess of 5 rems over a period of 30 years, particularly in the first 30 years following conception.

THE ASSESSMENT OF HAZARD

The need for safe maximum permissible dose levels is of paramount importance if the predictions for the future use of nuclear energy and all other forms of radioactivity are to be fulfilled. Safe dose levels are estimated, however, by extrapolation from observations on the effects on humans of much higher doses of radiation. Moreover, these higher doses are never continuous, varying from a single dose resulting from a nuclear bomb detonation to a small series of exposures arising from radiotherapy courses for ankylosing spondylitis. Experiments on various laboratory species provide valuable information on the relationships between radiation dose and cellular damage at lower dose rates. Nevertheless, the science of hazard evaluation in radiobiology is still imprecise mainly because the dose-response relationships at low levels of radiation exposure in the human remain unclear.

The relationship between biological effect and larger doses of radiation between 100 and 1,000 rads is generally agreed to be linear or almost so. At smaller doses the shape of the curve is controversial and depends upon the presence or absence of a threshold radiation dose below which there is zero biological effect (Fig. 2). However, it is becoming apparent that all the dose response relationships illustrated in Figure 2 are not only possible but likely. Most tissues have the ability to repair damage done to them by radiation and other forms of injury by replacing dead cells. This process would accelerate with increasing radiation dosage until the cell renewal mechanism is 'swamped' at a threshold level of radiation dosage. On the other hand, cells might be injured sublethally and might not be replaced; in such circumstances even the smaller doses of radiation might be expected to have biological consequences. In reality, radiation would probably induce both general patterns of damage with one type predominating, depending upon the capabilities and fragility of the cells involved.

Figure 2 illustrates three broad categories of dose-response relationships between low radiation dosage and biological effect. Type (*a*) has no threshold and might be predicted on the basis of subcellular considerations. Type (*b*) possesses a well defined threshold; an example is the relationship between whole body radiation of mammals and the biological effect as expressed as a lethal dose. Type (*c*) possesses a less well defined threshold due to the

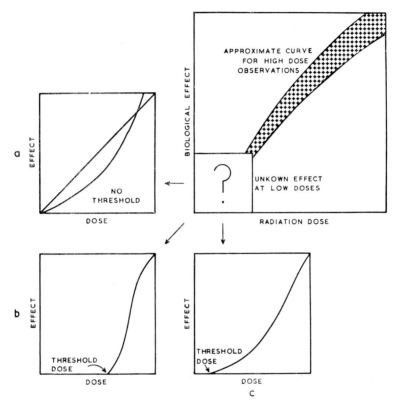

Fig. 2   Possible dose-response relationships for the harmful biological effects of radiation. (Adapted from Lindell & Dobson, 1961 and ICRP publication 14.)

wide range of individual sensitivities to radiation. Table 23 attempts to categorize the harmful effects of radiation in terms of their probable dose-response relationship.

From Table 23 it csn be seen that malignant changes and genetic damage may be potentially hazardous at even the lowest of doses, since there appears to be no threshold dose of radiation for either of these biological effects.

In the light of these theoretical considerations one might expect to detect disease susceptibilities in certain parts of the world subjected to high levels of environmental irradiation. However, it is notoriously difficult to construct adequate control groups into the basic experimental design of epidemiological studies of this nature. Preliminary surveys that have been carried out nevertheless suggest that the inhabitants of Guarapary in Brazil and certain parts of New York State, USA, where terrestrial radiation is very high, show a greater incidence of congenital abnormalities than do people living in different environments.

Mortality statistics for the disease of leukaemia have risen steadily over the past decades in all developed nations. Although much of this increase might be explained by improved standards of diagnosis, environmental radiation has still not been eliminated as the possible causative agent in this condition.

TABLE 23

Probable dose-response curves for different harmful effects of radiation

Biological effect	Type of dose response curve
Malignancy	(a)
Eye cataract	(b)
Impaired fertility	(c)
Bone-marrow insufficiency	(b)
Retarded growth and development	(b)
Genetic damage	(a)

It is most unlikely that it will be possible to define totally safe dosages of radiation in the near future, and it is possible that the human race will find it necessary to withstand a small degree of radiation pollution in order to reap the advantages of nuclear power and other radioactive appurtenances which will be used increasingly for the remainder of this century and beyond.

References

British Institute of Radiology (1964). Memorandum on implementation of second report of the Adrian Committee on radiological hazards to patients. *Br. J. Radiol.*, **37**, 559–561.

Court-Brown, W. M. and Doll, R. (1965). Mortality from cancer and other causes after radio-therapy for ankylosing spondylitis. *Br. Med. J.*, **2**, 1327–1332.

Eisenbud, M. (1963). *Environmental Radioactivity*. New York, McGraw-Hill.

International Commission on Radiological Protection (1966). London, Pergamon Press, ICRP Publication.

International Commission on Radiological Protection (1969). *Radiosensitivity and Spatial Distribution of Dose*, London, Pergamon Press, ICRP publication 14.

International Commission on Radiological Protection (1970). *Protection Against Ionising Radiation from External Sources*, London, Pergamon Press, ICRP publication 15.

International Commission on Radiological Protection (1970). *Protection of the Patient in X-ray Diagnosis*. London, Pergamon Press, ICRP publication 16.

Lindell, B. and Dobson, R. L. (1961). *Ionising Radiation and Health*. World Health Organisation Public Health Paper 6. Geneva.

Maki, H., Ishimaru, T., Kate, H. and Wakabayashi, T. (1968). *Carcinogenesis in Atomic Bomb Survivors*. Technical Report 24–68. Atomic Bomb Casualty Commission.

Mayneord, W. V. (1960). Naturally occurring alpha activity. Appendix E in *Hazards to Man of Nuclear and Allied Radiation*, M.R.C. Report. Cmnd. 1225. London, HMSO.

Medical Research Council (1960). *The Hazards to Man of Nuclear and Allied Radiation*. Cmnd. 1225. London, HMSO.

Ministry of Health (1960). Radiological hazards to patients. Second report of the Adrian Committee. London, HMSO.

Penfil, R. L. and Brown, M. L. (1968). Genetically significant dose to the United States population from diagnostic medical roentgenology. *Radiology*, **90**, 209–216.

Rogers, J. C. (1971). Ionising radiation. In *Environmental Health*, pp. 351–377, edited by P. W. Purdom, London, Academic Press.

Spiers, F. W. (1960). Gamma Ray dose rates to human tissues from Natural External Sources in Great Britain. Appendix D in *Hazards to Man of Nuclear and Allied Radiation*. M.R.C. Report. Cmnd. 1225. London, HMSO.

United Nations Scientific Committee on the Effects of Atomic Radiation (UNSCEAR) (1958). Report to the General Assembly, Official Records, 13th Session, Supplement No. 17 (A/3838). New York, United Nations.

United Nations Scientific Committee on the Effects of Atomic Radiation (UNSCEAR) (1964). Report to the General Assembly, Official Records, 19th Session, Supplement No. 14 (A/5814). New York, United Nations.

United Nations Scientific Committee on the Effects of Atomic Radiation (UNSCEAR) (1966). Report to the General Assembly, Official Records, 21st Session, Supplement No. 14 (A/6314). New York, United Nations.

United Nations Scientific Committee on the Effects of Atomic Radiation (UNSCEAR) (1969). Report to the General Assembly, Official Records, 24th Session, Supplement No. 13 (A/7613). New York, United Nations.

World Health Organization (1962). *Radiation Hazards in Perspective*. Geneva. WHO, Technical Report Series 248.

Chapter 4

Water in Relation to Human Disease

Geoffrey Ffrench

Man's dependence on water is evolutionarily related to his body tissue composition comprising, in the adult, some seventy per cent of water, most of it intracellular. For a 70 kg. man this means fifty litres of water weighing 50 kg. With the current world population of approximately 3,800 million, 1·75 billion litres of water are being cycled through human bodies to maintain their homeostasis: this is equivalent to 650 times the volume of water in Lake Victoria Nyanza.

Early man's intimate reliance upon water was tempered with an ability to withstand considerable fluctuations in his internal and external water resources. This developed as a result of the unconscionably slow progress of the evolution from the fully aquatic organisms to the modern well-adapted land animals, some of which can dispense with water in its common natural state, as, for instance, the ass in the arid deserts. Homer Smith (1951) provided the explanation of the physiological adaptation which was required for man's internal system of water and mineral conservation, when he traced the comparative development of the filtration and tubular reabsorption mechanisms of the kidney.

But man himself had to take a number of different paths to achieve this, as evidenced by the extraordinary physical appendage of the female bushmen of the Kalahari desert of South Africa; this is the steatopygous buttock which is thought to be a water resource organ, through the oxidation of fat. There are parallel examples among animals of adaptation to water deprivation, as for instance the camel's remarkable ability to store water in the blood and extra-cellular tissues by means of a special micro-molecular serum protein which acts like a demand pump and storage tank, expanding and contracting the plasma volume as necessary (Perk, 1964).

This close reliance on water has inspired religious faith and paradoxically a mistrust amounting to actual dread and terror of an otherwise life-giving substance (Frazer, 1963). This has come down to us in the form of the mystical concepts of the power of water for good and evil in such diverse examples as the heathen rites of Midsummer Day, the Christian baptism, the bathing festivals of the Hindoos, the burial rites of the Sikhs and many other practices such as that of the Maoris jumping into a river and releasing a stalk of fern, so allowing it to float away carrying with it their sins! This ritualistic attitude to water is still possessed by many people, primitive and otherwise as illustrated by the devil which dwells in the great sunken crater Lake Bosumptwe in Ashanti, Ghana. The fishermen take their catch from simple log floats, fearing to anger the spirit by defiling the water with anything more man-made. There is such a 'devil' in the British culture, in the belief of the Loch Ness Monster! There are many primitive folk who believe that to look into a pool of water will result in death because a malignant spirit will drag the reflection or soul under the water and devour it. This widespread belief extends across time and the world from the peoples of ancient India and Greece to the Zulus of Africa and natives of the New Guinea highlands, who abhor the use of water on their bodies despite its abundance.

The harm that can come to man from water forms a basic creed from the days of the Flood, apparently of a gigantic scale, causing a world disaster perhaps some 10,000 years ago when the land mass, now occupied by the Atlantic Ocean, disappeared beneath the waves. This gave rise to cataclysmic changes in the remaining continents in the form of climatic and flood phenomena which created much of the land form as we know it today, particularly the deserts and inland seas (see Fig. 3).

Fig. 3   The world from Apollo 12, showing the desert areas of the eastern hemisphere. (*The Daily Telegraph Magazine.*)

What then is this thing Water which has so mystified man but which he cannot do without? Basically it is a mineral compound, hydrogen dioxide, incompressible in its liquid state, but occurring also in solid and gaseous forms. It is formed by the bonding of two hydrogen atoms to one of oxygen, resulting in a dipolar molecule with positively and negatively charged poles (Fig. 4). This characteristic gives it the facility to undertake a number of functions such as separation into H^+ and OH^- ions, available for chemical reactions in nature and illustrated by the formation of carbonic acid in rain water from atmospheric carbon dioxide.

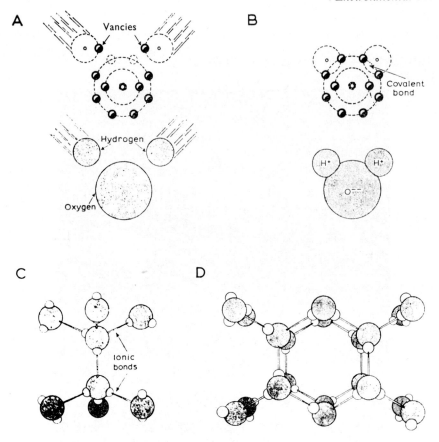

A. The joining of two hydrogen with one oxygen atom.
B. The covalent bond in the water molecule.
C. The ionic bonding of water molecules.
D. The structure of ice.

Fig. 4   The chemical structure of water. (From Chorley, R. J. (Ed.), *Water, Earth, and Man* (1969).

Water molecules can orientate themselves in an electrical field resulting in a very effective solvent action when in contact with molecules of other chemicals: this is achieved because the atoms of the latter are simply held together as a result of the opposition of the electrical charge of the ions. An example of this is the solution of natural salt, Na Cl, in water. The negative and positive ends of the molecule of water become attached respectively to the Na$^+$ and Cl$^-$ ions thus neutralizing their charges and causing the least disturbance to separate them.

   The ionic binding of hydrogen atoms of different water molecules results in its surface tension and capillarity, important properties in the biological application of water, in particular water-borne infections and parasitic disease (Fig. 5). In their usual state at room temperature, 17°C (62·5°F), the bound water molecules are interspersed with unbound molecules free to rotate and become agitated, which gives water its properties of flow and lubrication, both of which enter critically into bio-engineering. With changes of temperature, either up or down, viscosity is respectively decreased to the point of gaseous formation (vapour) or increased to the point of solidification with lowered density as the hexagonal crystalline structure of ice develops, accounting for the floatation of solid water in liquid water (Fig. 4). The reasons for the failure of ice to sterilize water containing pathogenic bacteria are not clear, but perhaps it allows expansion and contraction of bacterial membranes without destruction by intra-cellular crystallization. This failure to kill bacteria is by no means universal, and the speed and mechanism of freezing and subsequent thawing which regulate intra-cellular crystallization appear to be important to the survival of bacteria, and indeed other organisms including fungi and protozoa.

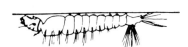

LARVAE

Larve have no long breathing tube or siphon; they rest just under surface of water and lie parallel with it.

Larvae have distinct breathing tube or siphon on the eighth segment of the abdomen; they hang from surface film by this siphon, except in *Mansonia*, which obtains air from aquatic plants.

Fig. 5 Mosquito larvae hanging from surface film of water. (From Chandler, A. C. (1955), *Introduction to Parisitology*).

When a medium containing water and electrolytes freezes, water separates out as ice, demonstrated by the low salinity of sea-ice compared to sea-water. In biological media, whether extra-cellular or intra-cellular, the result of ice-crystal formation is to leave a greater concentration of electrolytes in the remaining medium, which could have a denaturing effect on proteins, whether enzyme systems or DNA and RNA. There is some evidence to show that where bacteria are exposed to an increased concentration of the electrolytes of their usual medium, again intra- or extra-cellular, they are more susceptible to injury and death: the practical implications may only be surmised.

Water and energy are necessary and continually interacting components of the earth's layers except for the core; changes in state between liquid and solid water account for a significant proportion of the earth's energy in the form of heat. The incompressibility of water has provided a means of storing energy and using it in concentrated or controlled form, the essence of hydraulic engineering. The flow of water has been harnessed to water-wheels, locks, water-hammers, clocks and other mechanical devices to achieve a direct conversion of energy, used by man for many thousands of years for primitive agriculture and now a great source of electrical power: recently the flow properties of sea-water have been utilized for the same purpose by damming tidal waters. The discovery of steam-power and its rather inefficient control by Thomas Newcommen (1663–1729) to provide the energy for the water-pumps of Britain's coal mines in the eighteenth century led to the great surge forward in modern man's progress which lifted him out of a comparatively static condition into the exponential development of our times which threatens, by the very overproduction of man himself, to end his existence prematurely; as some would have it, the 'disease of humanity'.

The uses of water are almost limitless, whether as a transport medium or as a depository and vehicle for waste products of all kinds, natural and man-made; the introduction of renal dialysis has emphasized its importance in the former. Heated natural waters have been used since antiquity for their healing and medicinal qualities and are now harnessed to provide central heating of houses as well as continuing to be a tourist attraction in the Western USA and New Zealand. Those in Siberia do not have the same appeal as yet!

The multiple use of natural and artificially impounded water in rivers, canals, lakes or ponds is coming more and more into our concept of modern living. It can provide at one and

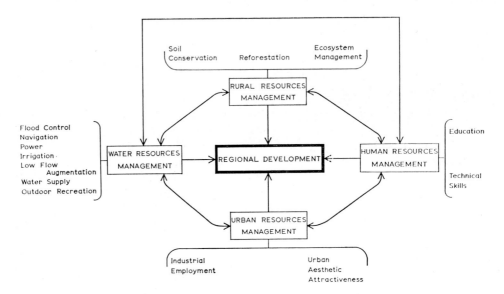

Fig. 6   Complex of resource management projects. From Chorley, R. J. (Ed.), *Water, Earth, and Man* (1969).

the same time great sources of power, storage facilities for direct human consumption and agricultural irrigation (Fig. 6). Increasingly important in the modern world, these water stores are a means whereby fish can be bred and harvested (e.g. Volta and Kariba dams) and the population, both local and tourist, can use them for social as well as for more practical utilitarian and aesthetic reasons. (Fig. 7).

Until the sudden upward surge in the world's population which began, apparently for the first time, in the seventeenth century, man and water had complemented each other without either exerting disproportionate demands (Fig. 8). This has now changed and water has become increasingly costly and sometimes hard to come by, requiring expensive and technically difficult means to achieve it, exemplified by modern distillation and desalinization of sea-water and the seeding of clouds by chemicals to produce rain. Water, by its relative unavailability or its potential for providing a medium for agents harmful to man, usually in the form of disease vectors or toxic solutes, has contributed to the stability of the world's population: perhaps the most expressive illustration has been the waves of water-borne cholera which have almost perennially swept across the world from its traditional home in the south east of Asia. (Fig. 9).

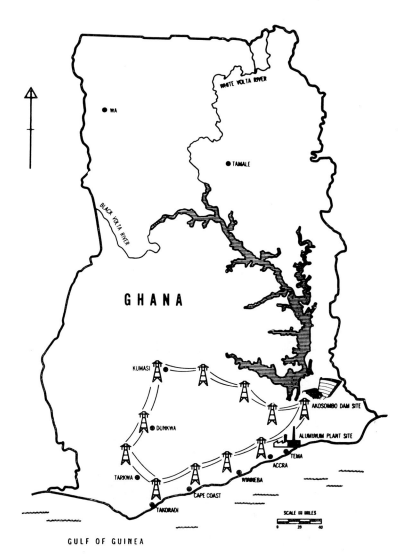

Fig. 7   The Volta River Scheme showing damsite, aluminium plant site and transmission lines. (Hughes, J. P., 'Health aspects of the Volta River Project in Ghana'. *Proc. 5th Conf. Ind. Coun. Trop. Health*, Boston, 1964.)

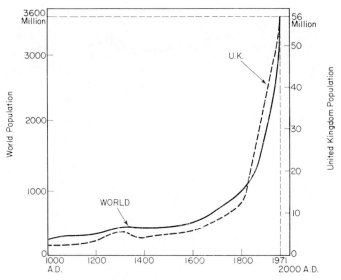

Fig. 8   World and UK population graphs.

CHOLERA SPREAD IN THE WORLD FROM   1865—75   to   1960—71

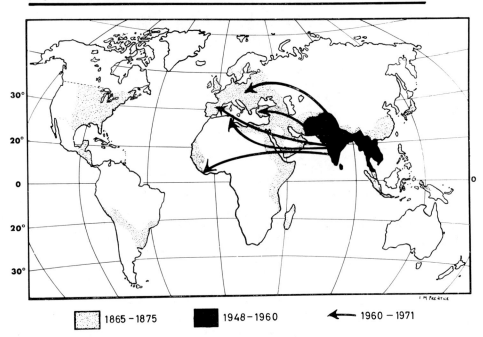

Fig. 9   World distribution of cholera.

Let us therefore examine this capacity for water to harbour man's enemies, a capacity which apparently does not diminish but threatens once again to exert a modifying effect upon human population growth.

One of the curious facts of history is that nearly all the early civilizations developed and flourished in dry climates on the edge of deserts. Their use of the available water required the most careful and ingenious application of hardly won knowledge; undoubtedly the problems themselves stimulated the people's ability to advance. Man learnt early in his existence that life could be sustained in a climate where water deprivation was the common experience and we continue to see examples of this adaptation. While the physiological needs for water do not greatly differ, perhaps surprisingly, among fully acclimatized people living in hot or cold, wet or dry climates, their capacity to maintain a suitable micro-climate, conserve their resources and estimate their needs are the identifiable characteristics of widely differing people, be they Eskimos whose water problem for much of the year is that it is in solid form, or the Arab of the desert whose scant supplies are impossibly brackish to our sophisticated tastes.

Coincidental with physiological demands for water there developed the civilized use of water for many different purposes; unfortunately these needs were, more often than not, satisfied by the same water source, leading to the appalling consequences which water-borne disease was to reap until modern times. Even yet in some modern cities the foul water containing human excreta mixes with the rain and flood waters in sewers, overloading of which can lead to pollution of houses and communities with this potentially dangerous natural effluent.

Water conservation and supply can now be described as a 'growth industry' (Venner, 1972) and the close interdependence between water supply, river management and sewage treatment in this country has become clear, although previously there was little integration of studies, personnel or the necessary financing. Only by these means shall we be able to meet the demands of today and of the future. Undoubtedly it has been the remarkable control of infectious pollution which contributed so greatly to the surge forward in population growth in Europe from the middle of the nineteenth century onwards.

While human life is, with carbohydrates, fats, proteins, vitamins and trace elements, dependent on a regular intake of water, this water can either improve, though not maintain, poison or deprive the body by the presence or deficiency of inorganic and organic constituents, some in solution, some in suspension or emulsion and some floating on the surface or free-swimming. We shall take each of these in turn and review their capacity to increase or reduce the quantum of disease with which man has been burdened since the dawn of his life.

DEPRIVATION OF TOTAL WATER

Death from insufficient water intake can be precipitous within a matter of hours or insidious over many days, dependent upon whether the reduction is total or partial, and on the metabolic requirements during the time of restriction, the latter being dictated by muscular activity and the external environment, both cold and heat.

Men in the deserts and mariners at sea have perished in this manner since the beginning of time. It has been claimed that man has been able to adapt physiologically to a reduced intake, but the evidence for this is still lacking. What man has done is to conserve and identify sources of water which, to our civilized eyes, do not appear to exist. It was Dr Alain Bombard (1956) in his craft L'Heretique, so aptly named, who provided the evidence which convinced a sceptical world that man could live at sea without fresh water for many days, in his case sixty-five. He did this by a careful assessment of his needs, and the estimated sources of supply of physiologically acceptable water and food from the inhabitants of the sea water surrounding him (Table 24). Perhaps it was the sheer heresy of Bombard's story more than anything else which has eventually made the average man aware that usable water exists in various forms and media which are not at once apparent. But nearly a century earlier the remarkable Francis Galton (1872) had provided our predecessors with a compendium of ways and means of replenishing drinking water and referred to the field-lore that was already possessed by people inhabiting apparently waterless places, or which can be developed by newcomers. A great deal of sound information is given, ranging from the value of the pericardial fluid of the turtle, often found floating in tropical seas, to succulent roots beneath the barren earth. At the same time that Bombard was surviving at sea the picture was being clarified on land through the descriptions of those remarkable survivors from the Stone Age, the Bushmen of the Kalahari desert and the aborigines of Central Australia (Van der Post, 1961; Thompson, 1962).

TABLE 24

Principal constituents of fish

Percentages by weight

Fish	Water %	Proteins %	Fats %
Ray	75·80–82·20	18·20–24·20	0·10–1·60
Basking shark	68·00	15·20	16·00
Dolphin	77·00–78·89	17·25–19·00	1·00–3·31
Rays Dream	78·90	18·42	0·34
Sardine	78·34	16·30–21·00	2·00–12·00
Anchovy	76·19	21·92	1·11
Bonito	67·50–69·17	18·53–24·00	7·00–12·46
Bass	77·00–79·94	18·53–19·96	0·84–2·50
Mullet	75·60	19·50	3·90
Mackerel	68·84–74·27	17·59–23·10	5·14–8·36
Tunny	58·50	27·00	13·00
Fish roe	48·80–78·31	11·50–54·90	1·16–16·20

(Bombard, A., 1956.)

Body requirements for water intake will vary depending upon the loss of water in sweat, urine, stools and exhaled air. While water is the subject of our discussion its electrolyte contents must be remembered for they have perhaps the ultimate say in the biochemical exhaustion of a water-deprived cell. Most of the significant studies in water needs have taken place since the Second World War, although before that, practical knowledge had come from the work of Haldane (1905), Moss (1524) and others in deep mines and steel foundries. From experience gained subsequently it is now clear that the optimum water needs of a fighting force in the desert, such as the 8th army and the Africa Korps in North Africa in 1941–43, far exceed the allowances actually made and no doubt contributed to the problems of both armies in the hot desert. Britain's experience in the Aden Protectorate and later in the Arab States bordering the Persian Gulf has confirmed laboratory workers' results and set new standards for all to follow (Leithead and Lind, 1964; Macpherson, 1960). Perhaps the most acute and glaring military example of water deficiency leading to appalling catastrophe was the systematic destruction of the Egyptian army's water-carriers from the air by the Israeli air force while it was crossing the Sinai desert in the short lived war of June 1967. The story of the subsequent sufferings of many thousands of Egyptian soldiers has never been released but for those who suffered or witnessed it, it was a terrible experience.

Under conditions of physical and heat stress, fifteen litres or more of water a day may be needed to maintain fluid balance: thirst is the initial and sometimes the only early symptom of water deprivation, but there are situations where even thirst may not be experienced until a severe deficiency has occurred. This may happen when there is an accompanying relatively greater loss of salt than water. With thirst there soon comes general weakness, the skin is

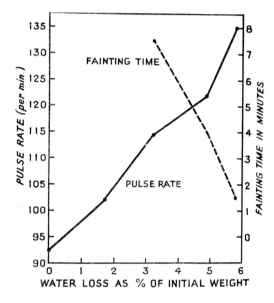

Fig. 10   Water loss as percentage of weight. (Adolph *et al*, 1947.)

dry and inelastic and the eyes are typically sunken: the urine becomes concentrated, the plasma volume is depleted leading eventually to heart and kidney failure.

A healthy man can lose perhaps one per cent of total body water before thirst symptoms arise, but beyond that the build-up of discomfort is rapid (Adolph, 1947): after ten per cent the mind becomes clouded, there is inability to drink because of the dry mouth and throat and total cardio-vascular collapse and death is imminent (Fig. 10). Meanwhile there has occurred rapid protein breakdown of the tissues, particularly when exacerbated by exercise contributing to a build-up of nitrogenous waste products: renal failure may be precipitated by the deposition in, and blockage of the renal tubules by myoglobin. (Schrier *et al*, 1947).

It should be remembered that discussion of body water deprivation must always take into consideration the electrolyte constituents, of which the most important is sodium chloride: it is beyond the present brief to enlarge upon this.

The sensitivity of young infants and children to water and electrolyte loss is well-known among clinicians but often overlooked by others.[6] Acute water deprivation shows an accentuated response to that of the adult, but chronic restriction may result in a characteristic clinical picture called 'Thirst Fever' (Shaker, 1966). (Fig. 11).

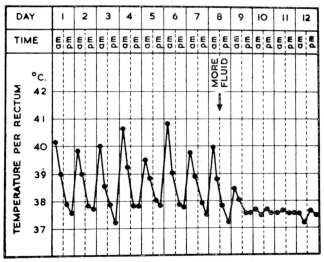

Part of temperature chart of Case 10, before and after administration of more fluids.

Fig. 11   Temperature chart of Thirst Fever. (Shaker, 1966.)

DEPLETION OF ESSENTIAL CONSTITUENTS

The optimal requirements of physiologically active substances in drinking water are, in contrast to potentially destructive elements, surprisingly poorly documented. Their study has been overlooked because essential nutrients can usually be found in the food. However there may be significant nutritional deficiencies dependent on the quantity and quality of the diet. Oxygen deficiency in water is destructive to fish life and affects the food chain of human communities who rely on fish for their protein sources. Oxygen deficiency of water comes into the category of pollution and will not be further discussed here. In 1850 Chatin[15] showed that if iodine was insufficient in the air, soil or water in certain regions, goitre was frequently present among those relying for their food and water on local supplies: the understanding of endemic goitre as a result of McCarrison's work (1917) has now become one of our medical geographical leitmotivs. Goitre continues to exist in isolated areas which are otherwise slow to develop (*see* Ward, 1966) (Table 25). That iodine deficiency is not the sole cause of endemic goitre must be considered elsewhere, but its identification was one of the preludes to what has become a fascinating path of detection, no longer limited to water alone but involving genetic factors and goitrogenic agents.

The more debatable deficiency of fluorine (as fluoride) has been superseded by the now well-established practice of adding fluorine (as sodium fluoride) to the drinking water for its protective effects against dental caries during the development of the teeth: the argument over the ethics of this continues. The threshold value of fluorides has been generally accepted to be 1 part per million: control of this concentration is important because of the known adverse effects of relatively high concentrations, in excess of 2 parts per million, in causing dental mottling and severe bone and joint disorders.

Relative deficiency of the hardness of water, which some have recently sought to incriminate in the production of chronic renal and myocardial disease, should be reviewed. The deficiency is referring to the temporary and permanent hardness, due respectively to magnesium and calcium carbonates which can be thrown out of solution by boiling and sulphates of calcium and magnesium, including at times chlorides and nitrates, which remain in solution. Both

TABLE 25

Incidence of goitre November 1965: villages of Upper Lunana

	Examined	Goitre	Percentage
Chozo:			
Male	41	8	19%
Female	38	19	50%
Total	79	27	34·5%
Dyotta:			
Male	11	4	36%
Female	15	11	73%
Total	26	15	54%
Thanza:			
Male	32	14	14%
Female	42	30	71%
Total	74	44	58%
Tyonchho:			
Male	42	25	59%
Female	44	26	59%
Total	86	51	59%

Iodine deficiency in Bhutan. (Ward, 1966.)

temporary and permanent hardness reflect a geographical relevance, not necessarily over a wide scale, for changes in hardness of water can be found within very narrow geographical limits; for instance, the soft water obtained from gravels and sand at Tunbridge Wells, Kent, showed a total hardness of 40 ppm., a very soft water, and that of Tonbridge, 5 miles distant, of 260 ppm., a moderately hard water. Again the 700 feet deep well chalk waters of Lombard Street, London E.C. and similar 150 feet wells at Leadenhall Street, E.C. gave 75 and 740 ppm. respectively (Taylor, 1949). It is of some interest, that, generally speaking, in those water sources which have been tested, surface, lake and reservoir, rivers and streams, springs, shallow wells, boulder clay waters, gravels and sands, deep wells, chalk water, clay waters, sandstone waters, limestone and coal measures, the total hardness is significantly higher in most areas south and east of a line drawn from the Humber to the Mersey, then to the Severn and thence perpendicularly to the South Coast near Weymouth. While Morris *et al* (1961) some ten years ago declared that there was then 'no indication that local water hardness is indicating some important social environmental cause of death from cardiovascular disease' they did point out that the deaths per 100,000 from cardiovascular disease bore a reciprocal relation to the hardness of local water, those with a total hardness of less than 50 ppm, a very soft water, returning a death rate of 664 per 100,000 falling to 543 when hardness exceeded 250 ppm. They interposed, however, an important corollary that could have an indirect bearing on these figures and also related to the generally higher death rates in the north and other parts of the country. During the rapid expansion of industry in the latter part of the 18th century, sources of soft water were regarded, for good technical reasons, as being more satisfactory for steam raising and other industrial purposes. The subsequent development of the industrial communities together with the accompanying well-known adverse public health and social features may have contributed and indeed continue to contribute to the excessive mortality. Seven years later, the same group of workers (Crawford *et al*, 1968) could still find 'no acceptable explanation . . . for associations . . . between water hardness and mortality'. The most recent research in this field, again by the same authors (Crawford *et al*, 1971), has been able further to relate cardiovascular mortality, in distinction to general mortality, to changes in water hardness over a period of thirty years, higher rates being observed where water was softened and lower rates when it became harder. They were quick to point out that 'in the present study, the water hardness changes occurred at various times before or during the years of our mortality date, and we emphasize that our findings do not warrant any conclusion about the time factor'. There has remained a striking similarity, both in the USA and the UK of some association between water hardness and mortality, which is firm evidence of a 'water factor' in all cardiovascular disease: this bears an analogy to the higher infant mortality associated with soft water which makes a mineral factor suspect. Subsequent work has been following these leads, not only in relation to the plumbo-solvency of water but to other minerals, in particular calcium and magnesium exchange at cell surfaces which could relate to myocardial sensitivity. The same authors go on to say 'It may be that we should be thinking of water as a general factor in mortality operating through the cardiovascular system and therefore more noticeable in cardiovascular (and bronchitis) death rates —the mechanism may be in heart failure' (Crawford *et al*, 1968). Since recent attention has been drawn to the qualities required of water supplied to renal dialysis units to avoid metabolic accidents (Reed & Tolley, 1970, 1971), the relevance of these comments cannot be overstressed.

While justifiable concern has arisen regarding the plumbo-solvency of soft waters (Wilson, 1966), the actual role of lead is not clear although it has been suggested that 'it could be replacing calcium and magnesium in important relationships' (Crawford & Crawford, 1969). The significance of lead and other mineral pollutants of water will be discussed in a later section.

The problem of providing both a potable drinking water and one free from injurious effects either by depletion or excess of mineral and other constituents poses an increasingly complex problem for water engineers and public health authorities because of the restricted nature of total supplies and the need for recycling. Careful microbiological and chemical monitoring of all elements which must find their way into the water supplies would be the ideal but at present the authorities must rely to a great extent upon legal restraints and safeguards against pollution. Two methods of potable water production which require careful reintroduction of minerals are the desalinization of sea water by distillation and deionization, the first a process requiring a relatively cheap fuel source and distilling under a vacuum at 80°C, the other being an electrolytic process. The final product has to be reconstituted to taste and quality by the addition of either natural waters or measured chemicals or more usually a combination of the two. While these methods have now been in use for some twenty-five years in those countries without a natural potable water supply, their introduction into areas of more traditional water resources is now being seriously considered. The break-point is the cost effectiveness using available fuel supplies: in Britain's case this means natural gas or oil. These developments may well stimulate further research into deficiency or excess of water constituents because of the need to supply an artificially produced and physiologically active but harmless water supply.

CHEMICAL WATER POLLUTION

As this contribution is being written the air vibrates with the clamour of reports of serious chemical and mineral pollution in streams and rivers, bays and estuaries and the sea itself, either from the systematic disposal of industrial effluent or the accidental release of material in transit. The sudden explosive pollution of the Rhine by the insecticide Endosulphan, the organic mercurial poisoning in Minamata Bay, Japan, and the Torrey Canyon disaster off the Cornish coast stand out as recent dreadful warnings. Over the centuries such disasters have always occurred, though on a smaller scale and have entered into the myths and legends of our history, such as the plague of St Anthony's fire due to the pollution of rye, blighted by the fungus ergot.

The sight, smell, taste and even the feel of water impinge on one's aesthetic sensibility according to current needs and desires. It is recorded of David Livingstone that at the end of a long and waterless march he knelt to drink at a pool of foul water, the customary drinking and soiling place of the local game. To him it was the most delectable fluid imaginable: yet it was from that time that he suffered the bloody dysentery which eventually led to his death. Anyone who has visited Venice during the humid season will agree that the odorific qualities of its waters compete, sometimes successfully, against the beauty of its setting. Human and industrial effluent has been pouring into Venice's canals since its foundation over a thousand years ago: yet it remains a place of delight. It is this predominance of the expedient, the habitual and the aesthetic needs over the judgement, wisdom and the common weal which has perpetuated our disregard of the obvious until it is almost too late.

Much nearer home, in fact still in many homes today, remains the problem of lead in drinking water deriving from naturally or artificially polluted sources in the catchment area, or a more sinister source, from the solution of lead into water during its passage through domestic lead pipes. A large proportion of underground and household piping in this country is made of lead, which, under the influence of acid plumbo-solvent soft and some acid hard waters, can lead to clinical, but more importantly to sub-clinical and unrecognized intoxication. Unexpected or unrecognized electrolytic action affecting lead pipes, either by their being connected to metal of another kind, usually copper, or having an earthing wire attached, will result in significant quantities of lead being released into the drinking water. The habit of drinking a cup of water in the morning taken from pipes in which the water has lain overnight is particularly dangerous in the circumstances outlined above. In modern building techniques, lead pipes are proscribed where the pH of the water supply is less than 7·8.

There are several studies implicating lead contamination of water at its source: perhaps the most suggestive one as far as this country is concerned, was by Allen-Price (1966) from West Devonshire. He reviewed local death records over twenty years, and showed that there was a remarkable sporadic distribution of cancers in that area of Devonshire which seemed to be inextricably related to the water supplies. He postulated that there might be a 'cancer-provoking ingredient in these water supplies which could be isolated by micro-chemical examination'. Variations in the lead content of the different local water supplies could be correlated with the local incidence of cancers and subsequently significant amounts of radon 222 in some of the same water supplies were found (Warren *et al*, 1966).

Lead affords perhaps the best model because a fair amount is known concerning its competitiveness with other ions in the enzymatic processes of human metabolism: it should however be pointed out that there is now sufficient evidence to implicate a number of other elements in the metabolic disturbances of man and his domestic animals. (Table 26). Many of these elements are incorporated in essential metallo-dependent enzymes in a variety of body organs.

TABLE 26

Symptoms and signs associated with excess or deficiency of some metallic ions

Symptom or sign	Pb	Hg	Cd	F	As	Zn	Cu	Se	Te	Co	Cr	Ni
Abortion	●			●								
Abdominal pain	●	●		●	●							
Alopecia	●			●	●							
Anaemia	●	●		●	●	●						
Anorexia	●	●			●					●		
Arthritis					●							
Bronzing of skin					●			●				
Carcinogenesis	●				●						●	●
Cirrhosis							●					
Delay in wound healing						●						
Dental mottling				●								
Dermatitis		●			●			●			●	●
Fatigue	●	●			●							
Foetal damage	●											
Garlic breath								●	●			
Headache	●	●	●	●	●							
Hepato-splenomegaly							●					
Hyperkeratosis					●							
Hypogonadism						●						
Metallic taste	●	●	●									
Myocarditis										●		●
Myopathy	●											
Nausea and vomiting	●	●	●		●		●			●		●
Nephritis	●	●	●	●								
Osteomalacia			●									
Peripheral neuritis	●				●							
Polycythaemia										●		
Pulmonary effects		●	●	●		●	●	●			●	●
Rhinitis										●	●	
Salivation		●										
Skeletal deformity			●	●								
Thirst		●										
Thyroid disorder										●		
Tremor		●			●							
Vertigo								●				
Weight loss	●	●			●							

Enzymes are proteins, complex interlocking amino-acids, possessing specific spatial orientation of chemical constituents. This allows the enzyme molecule to fit that of the substrate upon which it acts. It may consist of several components—iso-enzymes—each of which still act on the same substrate. Foreign, toxic substrates do not possess the precise spatial requirements which natural foodstuffs possess: thus only incomplete metabolic degradation occurs.

Enzymes have additional requirements in that they need co-factors or activators, metals and/or vitamins. Anything that will inactivate them will render the enzyme inert, e.g. beryllium, cyanide, lead, mercury, arsenic: competition or more correctly molecular substitution is the mechanism invoked.

Accumulation of an enzyme's product will inhibit its activity, a natural feed-back system. Therefore:

(1) Enzymes combine with a toxic substance:
(2) Inhibition may occur as a result, the products of reaction giving rise to blocking, resulting in:
(3) Slowing of vital functions and cell death.

Importance is now attached to this trace metal aspect of epidemiology: West (1971) will shortly publish a geochemical analysis of strata throughout Britain which may bear a significant relation to the findings reported by Howe (1970) in his National Atlas of Disease Mortality.

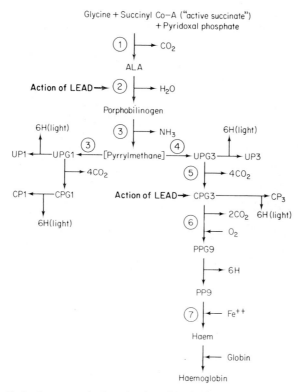

Fig. 12 Cycle of protoporphyrin and action of lead. (de Bruin, A., 'Certain effects of lead upon the animal organism'. *Arch. Env. Health*, **23**, 249–264, 1971.)

To return to the case of lead, it is known to interfere with the succinic acid pathway in the liver, in the metabolism of glycine involving acetyl co-enzyme A. The pathway is deviated to that of laevulinic acid leading to an excess of delta (Δ/δ) amino-laevulinic acid: excessive amounts of this intermediate metabolite are excreted in the urine and can be utilized as a reliable and sensitive monitoring test of lead absorption. Lead interferes elsewhere with the manufacture of haemoglobin through its inhibition of iron entering into the protoporphyrin cycle (Fig. 12). Porphyrins build up in the red blood cells and may be the cause of the degradation of ribo-nucleic acid which gives the stippling effect. The red cell life is shortened and initially reticulocytosis occurs following initial marrow stimulation; however marrow depression and anaemia are the usual sequelae. Porphyrins also exert a toxic effect on nervous tissue and muscle cells leading to well-known clinical pictures. The interference with porphyrin metabolism may also be recognized by urinary estimations of coproporphyrin III. Both delta amino-laevulinic acid and the latter correlate well with blood lead levels.

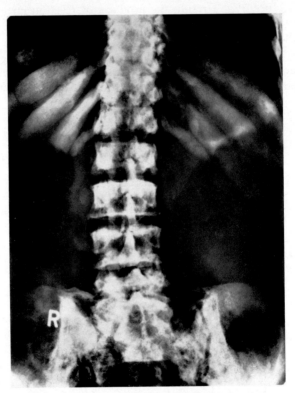

Fig. 13   X-ray of bone fluorosis.
(*Courtesy of Ilford Ltd., London*)

The story of lead serves as a reminder of what may happen when water of unbalanced natural constitution is all that is provided, but another severe effect of chronic intoxication from a natural source is seen in areas where fluorine as fluoride occurs naturally in waters in concentrations above 5 ppm. This leads to bizarre sclerosis of bones due to deposition of insoluble calcium fluoride or of a more complex salt (Fig. 13): there follows fixation of joints, crippling arthritis, and sometimes paraplegia, far beyond the mild effects of the mottling of permanent teeth noted when concentrations are in excess of 2 ppm., a relatively common phenomenon found in this country and elsewhere. The fluorine exerts a direct local action on enamel-forming cells leading, during the first 9 years of life, before the development of permanent teeth, to a malformation of the cementing substance between the enamel rods on the outermost part of the surface of the enamel. Where fluorides are added to water deficient in them, up to a maximum of 1 ppm., care must be taken to adjust the concentration

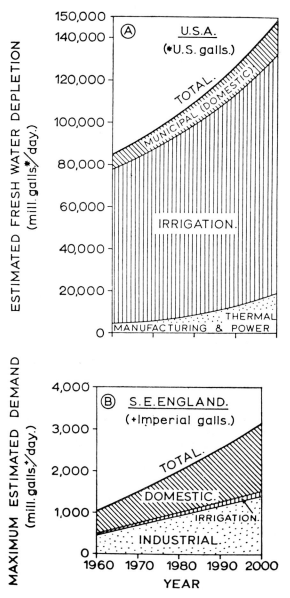

Fig. 14   Estimated fresh water depletion in USA and UK. From Chorley, R. (Ed.), *Water, Earth, and Man* (1969.)

so that excessive fluoride absorption is avoided during high fluid intakes in very hot weather: this is particularly applicable to tropical countries.

Suggestions have recently been made that the increasing difficulty of providing chemically pure water warrants the serious consideration of dual supplies, one of suitable potability for drinking, the other of a lesser standard for domestic, industrial and agricultural purposes (Reed & Tolley, 1969). The consumption of water for actual cooking and drinking is estimated to be only a small percentage, certainly less than 5, of all water consumed in developed countries: industrial and irrigation uses for water are now dominant. As an example it is estimated that by the end of this century the largest single use of water will be for cooling purposes by the thermal electricity generating industry. (Fig. 14). While this suggestion of dual water supply for already developed areas raises substantial problems, the principal is being applied in those arid areas of the world where water is in very short supply, whether it is the known sweet water supply of the desert Arab and the brackish water holes for camels (Figs. 15 & 16), or the separately piped reconstituted drinking water alongside a piped brackish water for his urban brothers who obtain it from the distillation of seawater (Ffrench & Hill, 1970).

Fig. 15   Sweet water of a desert oasis.

We have discussed the pressing question of chemical pollution of domestic water supplies in developed countries, and we should remember that, compared to problems facing man across the globe, it is a minute fraction of the total water pollution picture, over-shadowed by the effects of micro-organisms, about whose mode of action we now know far more, but yet are still unable to control completely.

Fig. 16   Brackish water of the desert.

BACTERIAL POLLUTION

The sources of micro-organism pollution of water are two-fold; one is as the result of man and animals using it for bathing and working in involving it in their cycle of excretion and ingestion, the other is through the agency of one or more vertebrate and invertebrate intermediate hosts and vectors. Throughout his history man has been the victim of waterborne infection either by enteric organisms such as the dysentry and salmonella bacilli, the cholera vibrio and numerous protozoa, or by worm infestation through intermediate hosts such as fish, snails or crustacea. This threat to his life and health was accepted as inevitable until the development of microbiology in the second quarter of the nineteenth century. Previous to this epidemiology, not yet developed into a science, had identified water as a source of illness and death but it was the work of Snow (1855) which finally related specific infection to local water supplies during the London cholera epidemic of 1854–55 and fixed the guilt. Pasteur's work of bacterial identification and culture, followed by the other giants of nineteenth century medicine, enabled measures to be taken to combat the pandemics of typhoid and cholera which recurred so frequently at that time. Subsequent development of vaccines further fortified the public health measures and by the time the South African War was fought (1899–1902), means were available, though not always accepted or correctly applied, to parry the threat to all armies in the field, of greater losses from insect and waterborn sickness than from wounds. Regrettably the opportunity was missed.

The desperate search for water, as desperate now as the quest for oil, calls for an increasing proportion of the industrial countries' efforts. These have ranged from the conversion, by the Dutch of the Zuider Zee from sea to fresh-water lake, to the Americans with far-seeing plans to entrap all the waters of North America (Fig. 17). Britain is considering barrages across Morecambe Bay in Lancashire and the Wash. In the industrial development of newer countries the need for power has led them, with the help of world funds, to build great dams

Fig. 17   The North American Water and Power Alliance proposals.
(Courtesy of the Ralph M. Parsons Company, Los Angeles and New York)

along the course of rivers which, besides providing the power for the turbines, have created inland seas and lakes which ultimately will be used for water-consumption, fish-breeding, and the other uses discussed earlier. These magnificent conservation projects have inevitably led to problems of biological pollution even though the possibility may have been anticipated. The most prevalent guilty plant agents are the water-hyacinth (*Eichhornia crassipes*) and various forms of algae plants, which had already an evil reputation in great natural stretches of water. Unfortunately one of the most reliable biological controls of water hyacinths is the very snail harbouring the intermediate stage of the world-wide schistosome worms which have retarded human development wherever they are found (Fig. 18).

'During the rainy season the people of the village obtain their drinking water from three sources: (*a*) rain-water collected from the roof drains off houses roofed with corrugated-iron sheets, the water being stored in earthen pots; (*b*) a well, which in the dry season contains only a little water; (*c*) temporary streams situated about 400 yards west of the village.

In the dry season the well and three ponds in the dried-up stream bed provide the village water. The ponds are shallow and the people wade into them to draw water. The well is not much used because it contains little water' (Schrier *et al*, 1967). The village referred to is Akufo in Western Nigeria but could be one of thousands in Africa and elsewhere.

The pollution of drinking water, either in rivers, wells or tanks by urine and faeces, however indirectly, will lead to the recovery of bacteria originating from the human or animal colon: this remains the keystone of all bacterial water testing. Above a certain permissible level— taking into consideration the character and source of the water—faecal contamination is judged to have occurred and with it the danger of pathogenic organisms, whether of the ubiquitous salmonella and dysentery groups or the dreaded *cholera vibrio* and its analogue, the *El Tor vibrio*. With all of these the principle of the carrier state operates, which in these days of jet air travel, even more than the religious and refugee migrations, remains a sword of Damocles poised over our heads. It is accepted that water is only one of the pathways to

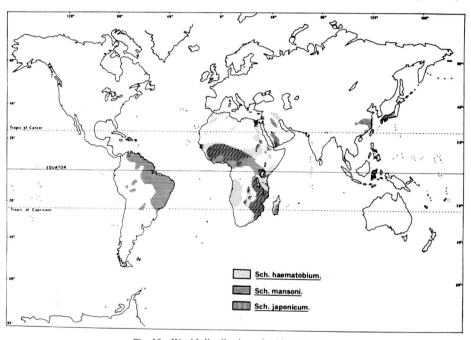

Fig. 18   World distribution of schistosomiasis.
(*Courtesy of the Wellcome Museum of Medical Science, London*)

infection by these and lesser breeds of organisms. Once the danger is recognized steps can be taken to control the use of water for drinking and cooking, treat it so that it can be safely drunk, and attempt to find and seal the source from which it is infected. It is often mistakenly thought that rapid filtration of water will sterilize it: filtration is indeed used for sterilization in certain circumstances such as settling and filtration beds, but rapid filtration should never be relied upon for sterilization. Again, the chance of infection by a pathogenic organism is dose-related and restriction of intake may help to build up immunity. The action of ultra-violet light from the sun on the surface of water while effective in water of low organic content should not be relied upon. Boiling or chemical sterilization by free iodine or chlorine is the method of choice, bearing in mind that the presence of organic matter will fix the halogen ions and render them unavailable for their purpose. The methods used to achieve the object will differ according to circumstances, and the appropriate sources should be consulted (Ward, 1966).

The history of Public Health is replete with examples of local and widespread outbreaks of water-borne disease and perhaps the remarkable thing is how relatively few people were clinically affected, considering that all must have been exposed. The development of true immunity and the multiple social and medical factors which influence it must be at least a partial explanation. The spectre of bacterial warfare and the terrible accidents which have followed the ingestion of exotoxin from the spore-bearing bacillus *Clostridium botulinum*, should be a warning that only careful planning, civil discipline, routine water sterilization and vaccination can be expected to prevent catastrophe, particularly when dealing with typhoid and cholera outbreaks.

It is a mistaken idea of many travellers that carbonated 'pop' drinks, because of their carbon dioxide gas and high sugar content, are not potential sources of infection: this is just not true as anyone who has had the responsibility of testing such drinks bacteriologically in tropical countries will agree. Neither is sea water a sterilizing medium; in fact it may be a very potent source of infection, particularly by salmonella organisms carried by seagulls and other sea-living animals including oysters and fish which may also harbour the *Vibrio parahaemolyticus*, a not uncommon cause of illness in Asia. Heavily saline solutions are no barrier to bacterial growth as the writer experienced when the strong saline waters of the Persian Gulf had to be tested for contamination due to the drowning of some thousands of imported sheep when the vessel carrying them overturned within a quarter of a mile of the sea-water intake of a major water-distillation plant (Ffrench & Hill, 1970).

VIRAL POLLUTION

Within recent years, particularly following the poliomyelitis outbreaks of the nineteen forties, the hazard of viral contamination of drinking and swimming water has been appreciated. The poliomyelitis, infective hepatitis, measles, foot and mouth disease of cattle, echo, reo and adenoviruses are all potential contaminants, more especially of bathing places. A factor of importance in their control is the knowledge that virus infection can be spread by these means: it requires the successful chlorination and filtration of such water sources. While viruses are very susceptible to sunlight, it again is unlikely that this can play any useful part, and the use of swimming baths may have to be proscribed during a virus epidemic. The role of sewage-polluted water in virus infections is not clear, but the possibility of the danger certainly exists, particularly in relation to poliomyelitis and infective hepatitis, either directly by ingestion of such water or through the agency of arthropods transferring infection to food or directly through their bites: the latter suggestion is only speculation. The role of inter-mediate hosts in virus infections is still emerging and such agents as arthropods, worm larvae and ova may all be involved to a greater extent than is at present realized: water may be the natural medium of the intermediate hosts.

SPIROCHAETAL POLLUTION

The borderline between bacteria and protozoa is occupied by the spirochaetal organisms, a very large diffuse group, most being harmless saprophytes. But some are parasitic and actively pathogenic, wreaking havoc with human life and health. The genus *Leptospira* requires the medium of water for the passage of infection, whether of the pathogenic variety or not. Those leptospires which infect man normally inhabit an animal reservoir and it is the excreta of the animal host which is the link. Providing the water is not acid, the organisms are capable of surviving for several weeks. Human infection takes place by the organism penetrating the mucous membranes of mouth, nose and eye or abrasions in the skin. Many are the local names given to the illnesses resulting, such as 'slime fever', 'marsh fever', 'swamp fever', 'swineherd's disease', '7-day fever', 'Fort Bragg Fever', 'Pseudo-dengue' and so on. They are due to different species of leptospires whose effects range from quite mild indisposition through generalized aches and pains, skin rashes and meningism, to the severe and fatal Weil's disease, a hepatitis and nephritis with haemorrhages throughout the skin and body caused by *Leptospira icterohaemorrhagiae*, whose natural host is the brown rat, *Rattus norvegicus*.

PROTOZOAL POLLUTION

For practical purposes the discussion will be restricted to the free-living *Entamoeba histolytica* and the more recently recognized opportunist agents *Naegleria* and *Hartmanella*, both free-living amoebae. *E. histolytica*, in the trophozoite and cystic form is a pathogen and commensal respectively of the human lower small intestine, appendix, caecum, colon and occasionally other organs, in particular the liver: even the skin may be the site. It may also be found in the gut of primates. Sewage-contaminated water is thus an important source of human infection and indeed was the route taken in the tragic Chicago outbreak at the World Fair of 1933: defective plumbing caused over 1,000 cases and 58 deaths. It should be emphasized that *E. histolytica* can prosper outside the tropics and requires only an absence or breakdown of good public hygiene. Endemic amoebiasis still exists in many temperate areas and Chandler (1955) recorded a 60 per cent incidence in the Kola peninsula in which lies the sea-port of Murmansk, situated entirely within the Arctic circle, between 74° and 70°N and 32° and 42°E. It is the cystic form of *E. histolytica* which survives outside the body and reinfects: the cysts may live up to ten days in water at room temperature, but if cooled, survive for several weeks.

Amoebae of stagnant pools have served many generations of school children, since the day in 1676 when Anthony Van Leeuwenhoek first identified them with his new invention, the microscope. That they also provide a very useful biological control of bacteria is not generally appreciated. Within recent years human infection with certain of these organisms of the genera *Naegleria* and *Hartmanella* (*Acanthamoeba*) (Symmers, 1969) has been reported on several occasions from as widely separated areas as Czeckoslovakia, the USA, Australia and the UK. The late Aldo Castellani had suspected the pathogenicity of one particular organism, subsequently named after him *Hartmanella castellani* (*see* Douglas, 1930). In the following thirty-five years much work was done in defining the pathogenicity of these organisms (Cuthbertson *et al*, 1961; Fowler & Carter, 1965). The important point is that the free-living organisms are found in a variety of water habitats, slow moving streams, ponds, garden puddles and of course the inevitable sewage. While the rarity of the severe effects is not in doubt, the possibility of less severe consequences particularly in young children has been discussed by Symmers (1969).

Strictly speaking, while the human malarial parasites *Plasmodium vivax*, *P. malariae*, *P. falciparum* and *P. ovale* are not water-borne organisms, their vector the anopheles mosquito is a water-breeder, and thus from time immemorial stretches of water have had an evil reputation long before the malarial cycle in the mosquito and warm-blooded host was defined. Around the Pontine marshes of Rome and in the fens of East Anglia, in fact throughout most of the settled temperate and tropical world, the (malarial) ague was a frequent illness. The flooded areas surrounding the newly established city of Calcutta in the eighteenth century perhaps reflects this evil reputation pithily: '. . . three miles to the northeast is a salt water lake that overflows in September and October and then prodigious numbers of fish resort thither, but in November and December, when the floods are dissipated, those fish are left dry and with their putrefaction affect the air with thick stinking vapours, which the Northeast winds bring with them to Fort William that they cause there a yearly mortality (Gupta, 1916). The fish were incidental; it was the mosquito which killed.

The mosquito larvae and pupae develop in almost any water except the open sea, even in the salty flats of bays and estuaries. I can do no better than quote Chandler (1955) to describe the intimate relation of mosquitoes to water which explains their adaptability and their geographic range: 'Mosquito larvae, unless suspended from the surface film by means of the breathing tube, have a tendency to sink . . . some species are habitual bottom feeders: others feed at the surface; some live on microscopic organisms, others on dead organic matter: and still others attack and devour aquatic animals, including young mosquito larvae of their own and other species. Soluble and colloidal substances in water can also be utilized'. It is only by the control of breeding at the water sites, and of transmission by inhibiting the development of the parasite in the human body, that this terrible scourge can be overcome. Unhappily we are still a long way from that goal.

WATER-BORNE WORM INFESTATION

While the larval stages of the human hook-worms, *Ankylostoma duodenale* and *Necator americanus*, and the minute human nematode *Strongyloides stercoralis* require moist soil for their development, as do the soil nematodes, it is the trematodes or flukes which, perhaps above all other human parasites, have created and continue to cause the greatest havoc. The liver fluke *Fasciola hepatica* and lung flukes such as *Clonorchis sinensis* and *Paragonimus westermanni* have relatively restricted geographical distribution but the human blood flukes, *Schistosoma haematobium*, *S. mansoni*, and *S. japonicum* are ubiquitous (Fig. 18). All flukes require one or more intermediate hosts in the form of fish, water-inhabiting crustaceans, snails, or bivalve molluscs. Except for the blood flukes they all have a further encystment stage, usually on water vegetation: it is the combination of certain species of snails or molluscs with the specific vegetation and the animal or human host which creates the cycle. The prevention and control of fluke infestations continue to challenge the engineers, the chemists and the doctors. It requires the control of snails and molluscs and the education of the people to deposit their excreta elsewhere than in water in which they work or play and thus avoid infection by skin or mucous membrane-penetrating larvae. In some areas they will need to be persuaded not to eat raw crustaceans. The immensity of the task becomes clear, when it is realized that centuries of habit will have to be eradicated, for raw meat eating continues to be prevalent throughout the world, whether in North Americans and Europeans (pork), in English (beef and oysters), in South East Asians (crab and crayfish), in Philipinos (goats' intestines) or Australian aboriginals and Arabs (lizards). The habit is therefore still universal and forms a strong link in the chain of human parasitism.

Among the human cestodes, the flat or tapeworms, the only one of significance which requires a watery medium is the fish-tapeworm *Dibothriocephalus latus*, having the distinction of being the largest tapeworm found in man and known to reach up to 30 feet. It requires an intermediate host, a species of fresh-water crustacean (Cyclops), its natural host being fish, mainly of the pike family, where the free worm-like larvae live in the muscles: these fish are eaten raw by the Baltic people and in Japan, Canada, Siberia and Chile. The worm is of particular medical interest as it causes a pernicious anaemia type of clinical picture through

Vitamin B_{12} competition in the human gut: the geographical location of this is of interest, for despite the world-wide prevalence of infestation, the characteristic anaemia is not found universally among the people harbouring the worm, suggesting additional factors, one of which may be the location of the parasite in the host's intestine; the more distal the worm the less likely the interference with Vitamin B_{12} absorption (Rausch *et al*, 1967).

Of the nematodes or round worms, in addition to the already mentioned hook-worm and strongyloides, the other one of water importance is the well-known Guinea or Medina worm

Fig. 19   The Guinea of Medina Worm, Dracunculus medinensis, coiled on a stick after withdrawal from the skin. (Henschen, F. (1966). *The History of Disease*, London.)

found in most tropical areas. The life-cycle requires the female to discharge the larva on to the skin after burrowing to the surface. It is then washed off into the water of open wells or pools. The larvae, ingested by Cyclops, are swallowed and migrate through the host's tissues. This is a very common condition among many tropical peoples in both dry and wet climates without a piped water supply and extends far back into history. Its very name, *Dracunculus medinensis*, which means the little dragon of Medina, signifies its effect upon the senses and it was very probably the 'fiery serpents' which were the scourge of the Israelites beside the Red Sea. It has been postulated that the 'staff and serpent' of Aesculapius, which for centuries has been the symbol of healing, may be none other than the Medina worm coiled around a stick, which still remains the traditional manner of withdrawing the worm little by little each day from under the skin (Fig. 19).

Analogous with the indirect importance of water to malaria, is the disease known as 'river blindness' resulting from the inflammatory effects of the tissue larval stage of the insect-borne adult filarial worm *Onchocerca volvulus*, which lives in the subcutaneous and connective tissues of the human host (Fig. 20). The insect vectors are the devilish little blood-sucking blackflies or buffalo gnats, *Simulium damnosum, S. narvi* (Africa) and

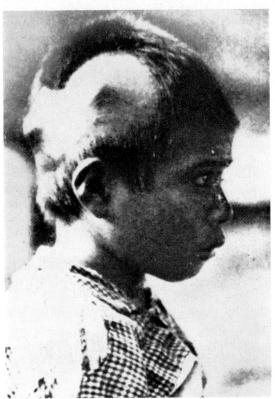

Fig. 20   Nodules, caused by the adult worm of *Onchocerca volvulus*, under the skin of the scalp of a Guatemalan child. (Strong *et al*, 1934.) *Onchocerciasis*, Harvard University Press).

S. othraceum (Central America) which require well-oxygenated fast-running fresh water for their breeding (see Fig. 21): it is the classic riverine disease and it is concentrated particularly at points along a river or stream suitable for damming: hence its potential danger to construction workers in these days of giant dams and power stations. This concentration of human beings, often non-immune to the local diseases, has provided a virgin soil for the spread of insect-borne disease, the oustanding example being the ill-fated attempt by the French engineer de Lesseps to construct a Panama Canal after his initial success with the Suez Canal: the environmental conditions were utterly different. This concentration of non-immune construction workers led the late George Macdonald (1957) to describe an epidemiological pattern as 'malaria of the tropical aggregation of workers'.

   The other important riverine disease, albeit having a more remote link with water, is the protozoal infection by species of Trypanosomes, carried by the biting tsetse flies of the genus Glossina. These flies breed and concentrate in the shade of low bushes and trees near the banks or rivers and water holes in Africa. It is here that they become infected by animals and humans coming to drink or collect water and here that they infect new victims (Fig. 22).

Fig. 21 Fresh water breeding area of *Simulium damnosum.* (*Courtesy of Wellcome Museum of Medical Science, London*)

Approximate ranges of tsetse flies.

≡ . . . range of entire genus *Glossina*

\\\ . . . range of *Glossina palpalis*

/// . . . range of *Glossina morsitans*

Fig. 22 Distribution of African trypanosomiasis. (Chandler 1955.)

THE SEA

With these short descriptions of the significance of water in relation to human disease it will
be realized that I have mostly been discussing sources and sites of water other than the sea,
which has, as far as infectious disease goes, a better record, except for that delectable bottom
feeder of estuaries and effluents, the oyster, a not infrequent source of typhoid infection. But
the sea holds within it many other hazards to man, despite all the beauty to which Jacques
Cousteau has introduced us; it continues to present challenges to our use of it both in our
work and play. These may take the form of fish, man-eating, biting and stinging (Ffrench,

Fig. 23   Pterois volitans, the Chicken or Lion Fish. This specimen caused a severe illness in a skin-diver.
(Ffrench, 1967.)

1967) (Fig. 23), muscle-paralysing snakes (Reid, 1967), stinging jelly-fish, poisonous bivalve
molluscs and irritating vegetation from a species of *Coralline alcyonidium* which our fishermen in the North Sea sometimes experience, known as Dogger Bank itch! (Newhouse, 1966)
Man has long been seeking a mechanical means of exploring beneath the surface of the sea
since the days of Archimedes: Leonardo da Vinci foretold the construction of the submarine
'with such precision of scientific and mechanical detail' (McCurdy, 1938). Until the successful practical development of the submarine during the American Civil War, underwater
activities were restricted to the extraordinary powers of submersion developed by divers for
pearls, sponges and animal and plant food in seas as far apart as the Persian Gulf, the Bay
of Bengal, the coasts of Japan, Korea and Australia. In the case of Korea and Japan it was
the women who perfected the remarkable diving techniques which have so intrigued the
physiologists (Tokyo Symposium, 1965). With the development of the air pump and rubber
life-line, painfully and gradually the experience of working at and returning from great
depths was collected at the cost of many lives. Engineering and construction needs stimulated
the development of the system of man working in caissons under water at increased atmospheric pressure, both in fresh and sea water; this brought with it the realization that here
indeed there were serious hazards to health particularly in the element of the return to
surface atmospheric pressure. The acute and chronic effects of decompression occupy an

important field of their own in the health of workers, whether they be divers, caisson workers, tunnellers under rivers or submariners. A newer aspect of their importance, is the need for widespread education arising from the adoption by many thousands of people of the sport of sub-aqua diving using various forms of aqua-lung: tragedies arise with monotonous frequency each summer. The basic element in the understanding of decompression is the knowledge that the external water pressure increases exponentially with the depth (Table 27).

TABLE 27

Chart of water depth and pressure

Depth, in feet	Pressure, in atmospheres absolute
Sea-level	1
100	4
250	8·55
500	16·15
1 000	31·3

(Miles, 1966.)

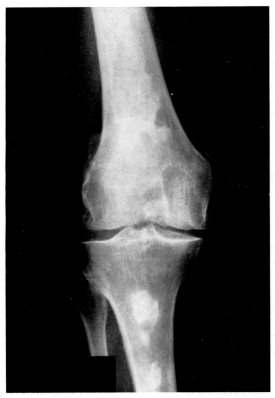

Fig. 24   Bone disease in a caisson worker. (*Courtesy of Dr Frank Pygott.*)

Provided air or other breathing mixtures such as oxygen and helium can be supplied at correct concentration and pressure, the human can continue to breathe and carry out modified work at depths down to 600 feet (Miles, 1966). However, if the rate of ascent to the surface is not carefully controlled, the gases dissolved in the blood, particularly nitrogen, will come out of solution, and be driven by the circulation to lodge in the tissues forming, in effect, gas emboli. These can cause immediate effects in such sensitive tissue as the nervous system and the lung, or delayed effects, from hours to several years, in the bone adjacent to joints (Fig. 24).

The entry of man into realms for which he was never designed has created for him new delights, new sources of wealth such as oil from the sea-bed, but also serious disabling and sometimes fatal diseases.

CONCLUSION

On this note the story of water's part in human disease must end, with the plea that it is increasingly necessary for people in all walks of life to know as much as possible about water so as to enlarge production and conserve it, if it is to continue to be a benefit rather than a curse to us and our descendants, because of its ill-considered use as a drain for the waste products fouling our environment and threatening our ecology.

References

Adolph, E. F., *et al.* (1947). *Physiology of Man in the Desert*. New York, Hafner.

Allen-Price, E. D. (1966). Uneven distribution of cancer in West Devon. *Lancet*, **1**, 1,235–1,238.

Bombard, A. (1956). *The Bombard Story*. London. Harmondsworth, Penguin Books.

Chandler, A. C. (1955). *Introduction to Parasitology*. New York, Wiley.

Chatin, A. (1850). Recherches sur L'iode des Eaux Douces: de la Présence de ce Corps dans les Plantes et les animaux Terrestes. *Comp. rend. Acad. Sci.*, **31**, 280.

Chloremis, K., Danelaton, C., Maounis, F., Basti, B. and Lapatsanis, P. (1959). Paper chromatography for amino-acids in thirst fever. *Helvet Paediat. Acta*, **14**, 44.

Crawford, M. D. and Crawford, T. (1969). Lead-content of bones in a soft- and hard-water area. *Lancet*, **1**, 699–701.

Crawford, M. D., Gardiner, M. J. and Morris, J. N. (1968). Mortality and hardness of local water supplies. *Lancet*, **1**, 827–831.

Crawford, M. D., Gardiner, M. J. and Morris, J. N. (1971). Changes in water hardness and local death-rates. *Lancet*, **2**, 327–329.

Culbertson, C. G., Overton, W. M. and Reveal, M. A. (1961). Pathogenic acanthamoeba (hartmanella). *Amer. Jour. Clin. Path.*, **35**, 195.

Douglas, M. (1930). *Jour. Trop. Med. and Hyg.*, **42**, 191.

Ffrench, G. E. (1967). A case of lion fish sting. *Jour. Trop. Med. Hyg.*, **70**, 42.

Ffrench, G. E. and Hill, A. G. (1970). *Kuwait: A Geomedical Study*. Berlin, Springer-Verlag.

Fowler, M. and Carter, R. F. (1965). Acute pyogenic meningitis probably due to acanthamoeba sp: a preliminary report. *Brit. Med. J.*, **2**, 740.

Frazer, J. G. (1963). *The Golden Bough*. Abridged edition. London, Macmillan.

Galton, F. (1872). *Art of Travel*. Newton Abbot, Devon. David and Charles (Reprint, 1971).

Gilles, H. M. (1964). *Akufo: An Environmental Study of a Nigerian Village Community*. Ibadan.

Gupta, J. N. (1916). *India in the Seventeenth Century*. Calcutta, University of Calcutta.

Haldane, J. S. (1905). The influence of high air temperature. *Jour. Hyg.*, **5**, 494.

Howe, G. M. (1970). *National Atlas for Disease Mortality*. Edinburgh, 2nd Edn., Nelson.

Leithead, C. S. and Lind, A. R. (1964). *Heat Stress and Heat Disorders*. London.

Macdonald, G. (1957). *The Epidemology and Control of Malaria*. London.

Macpherson, R. K. (1960). *Physiological Responses to Hot Environments*. London.

McCarrison, R. (1917). *The Thyroid Gland in Health and Disease*. London, 286 pp.

McCurdy, E. (1938). *The Notebooks of Leonardo da Vinci*. London, Vol. 1, Jonathan Cape.

Miles, S. B. (1966). *Underwater Medicine*. London, Staples Press.

Morris, J. N., Crawford, M. D. and Heady, J. A. (1961). Hardness of local water supplies and mortality from cardio-vascular disease. *Lancet*, **1**, 860–862.

Moss, K. N. (1924–5). The mechanical efficiency of the human body during work in high temperatures. *Trans. Inst. Min. Engrs.*, **68**, 377.

Newhouse, M. L. (1966). Dogger Bank itch: a survey of trawlermen. *Brit.Med. J.*, **1**, 1,142–1,145.

Perk, K. (1964). Fluid equilibrium in the camel and its desert hardiness. *Research Rev. Medical News*. 10 Jan.

Rausch, R. L., Scott, E. M. and Ransch, V. R. (1967). Helminths in Eskimos in western Alaska, with particular reference to diphyllobothrium infection and anaemia. *Trans. Roy. Soc. Trop. Med. Hyg.*, **61**, 351–356.

Reed, C. D. and Tolley, J. A. (1970). Letter to the Editor. *Lancet*, **1**, 897.

Reed, C. D. and Tolley, J. A. (1971) Letter to the Editor. *Biomedical Engineering*, **88**, (Feb.).

Reed, C. D. and Tolley, J. A. (1969). Toxic substances in drinking-water. *Zivilisationkrankheiten*. Nr. 6.

Reid, H. A. (1957). Snakebite in the tropics. *Brit. Med. J.*, **2**, 359.

Schrier, R. W., Henderson, H. S., Tisher, G. C. and Tanner, R. L. (1967). Nephropathy associated with heat stress and exercise. *Ann. Int. Med.*, **67**, 356–376.

Shaker, Y. (1966). Thirst fever in infants in Kuwait. *Brit. Med. J.*, **1**, 586–588.

Smith A. U. (1961). *Biological Effects of Freezing and Supercooling*. London, Williams and Wilkins.

Smith, W. (1951). *The Kidney: Structure and Function in Health and Disease*. New York.

Snow, J. (1855). *On the Mode of Communication of Cholera*. London.

Symmers, W. St. C. (1969). Primary amoebic meningo-encephalitis in Britain. *Brit. Med. J.*, **4**, 449–454.

(1965). *Physiology of Breath-holding Diving and the Ama of Japan*. Symposium at Tokyo, Japan, 31 Aug.–1 Sept. Washington DC.

Taylor, E. W. (1949). *The Examination of Waters and Water Supplies* (Thresh, Beale, Suckling). London, 6th Edn.

44. Thompson, D. F. (1962). The Bindibu Expedition I, II, III. *Geogr. J.*, CXXVIII, 144, 143–157, 262–278.

Van der Post, L. (1961). *The Lost World of the Kalahari*. London, Hogarth Press.

Venner, D. A. (1972). Water: a growth industry. *The Daily Telegraph*. 31 Jan.

Ward, M. (1966). Some geographical and medical observations in North Bhutan. *Geog. J.*, **132**, 591–502.

Warren, H. V., Delavault, R. E. and Cross, C. H. (1966). Editorial. *Arch. Environ. Health*, **13**, 412–413.

West, T. S. (1971). Personal Communication.

Wilson, A. T. (1966). Effects of abnormal lead content of water supplies in maternity patients. *Scot. Med. J.*, **11**, 73–82.

Chapter 5

The Relationship of Weather and Climate to Health and Disease

S. W. Tromp

INTRODUCTION

The relationships between weather, climate and disease constitute the subject matter of the interdisciplinary science of **biometeorology** and in particular that aspect known as pathological biometeorology. Biometeorology according to the International Society of Biometeorology (1970), is 'the study of the direct and indirect effects (of an irregular, fluctuating or rhythmic nature) of the physical, chemical and physicochemical micro- and macro-environments, of both the earth's atmosphere and of similar extra-terrestrial environments, on physico-chemical systems in general and on living organisms (plants, animals and man) in particular'. It has six principal branches: (1) Plant (*Phytological*) biometeorology; (2) Animal (*Zoological*) biometeorology; (3) Human biometeorology, both of healthy man (*Physiological* biometeorology) and diseased subjects (*Pathological* biometeorology); (4) *Cosmic* biometeorology; (5) *Space* biometeorology and (6) *Palaeo*-biometeorology. Definitions of each of these aspects are given by Tromp (1963). Pathological biometeorology, the subject of the present paper, studies the influence of weather and climate on the various physiological and pathological phenomena associated with the diseases of man, the period of outbreak, intensity and geographical distribution. Obviously a thorough knowledge of the basic phenomena of meteorology is required.

BASIC METEOROLOGICAL CONCEPTS

Biometeorology makes use of a number of concepts. **Weather** comprises the day-to-day changes in meteorological conditions including temperature, rainfall, snow, etc., whereas **climate** relates to average weather conditions based on 30 or more years of observations. **Macroclimate** relates to the climatic conditions of the atmosphere above specified areas, whereas **microclimate** is a term used to designate the climatic conditions directly surrounding the living organism in a limited niche, e.g., in a room, factory or mine. As the differences between the two types of climate are usually considerable, a correlative study between them may lead to erroneous results since it is the microclimate which is the main determining factor for the health of man. Even so, there is usually a reasonably close interrelationship between the macro and microclimate of a region.

In 1919 V. and J. Bjerknes, two Norwegian meteorologists, introduced the concept of **air masses.** The term air mass is used to denote large mass of air with roughly the same physical and chemical properties. Depending on its source area, meteorologists distinguish between polar, tropical and other types of air mass. If two air masses meet, e.g. polar and tropical, they often do not mix readily but remain separated by a boundary plane known as a **frontal surface.** The line of intersection of this surface with the surface of the earth is called a **front.** If warm air moves towards and over an area with cold air there is a **warm front**; if cold air moves toward and usually beneath an area of warm air there is a **cold front.** If a cold front overtakes a warm front the warm front is pushed aloft. When the warm air is lifted off the earth's surface the resulting phenomenon is known as an **occlusion.** Usually temperature falls in the atmosphere with increasing height. Sometimes, however, at a certain level temperatures increase with height; this phenomenon is known as an **inversion.** Inversions are especially common during periods, particularly in winter, when atmospheric pressure is high and there is little air turbulence. They usually cause dense fog. If an inversion is sufficiently strong to prevent the smoke from industries and/or domestic chimneys from dispersing, **smog** (smoke-fog) is likely to form. It is a dense fog, polluted with soot particles and various volatile chemical substances; it is particularly harmful to bronchitic patients. Over 4,000 deaths were attributed to the London smog of 1952. One other important concept in biometeorology is **meteorotropism,** i.e., the sensitivity of the living organism to changing meteorological environments.

From a biometeorological point of view it is necessary to distinguish between three fundamentally different climates (mountain climate, forest climate and maritime climate) each with its own specific physico-chemical characteristics and with associated biological effects on the living organism.

MOUNTAIN CLIMATES

Particularly above 1,500 m (4,500 ft) altitude the climates are characterized by:

(*a*) **Reduced partial oxygen pressure,** i.e. a reduction of the partial pressure of oxygen in the atmosphere. It represents 21 per cent of all gases in the atmosphere, being only 159 mm mercury at sea level against a total barometric pressure at 760 mm. At 2,000 m altitude the oxygen pressure is reduced to about 125 mm. As a result the oxygen binding substance in the human blood, haemoglobin, will be less saturated with oxygen at high altitude (90 per cent at 2,000 m against 96 per cent at sea level).

(*b*) **Differences in quality and intensity of solar radiation.** This is particularly true for the invisible ultraviolet part of the solar spectrum with wavelengths of 290–380 nm (**or** mμ) or 2,900–3,800 Å (1Å $= 10^{-8}$ cm $= 0.1$ mμ). Three kinds of UV are distinguished: UV–A ($\lambda = 315$–380 nm); UV–B ($\lambda = 290$–315 nm) or Dorno Radiation (named after the biometeorologist Dorno); UV–C (< 290 nm). In UV–B and in particular the interval 297–302 nm, there are strong biological (e.g. anti-rachitic) effects, UV–C is observed only at altitudes in excess of 2,000 m. The absorption of UV in the skin protein increases considerably at wavelengths below 295 nm.

(*c*) The average daily **temperatures** are generally lower; there is usually less atmospheric **turbulence** and less **water vapour.** The heat loss from the earth (terrestial radiation) is in consequence greater than at lower altitudes. The **ozone** content is often higher; the number of large ions is reduced (from $\pm 10^4$ to 10^2), but the number of small ions is increased. There is also a reduction in the dust content of the atmosphere and in air pollution in general (this applies to both allergens and chemical pollutants).

FOREST CLIMATES

Forest climates are characterized by the following physico-chemical factors:

(1) Due to the **photochemical action** of **sunlight** on the chlorophyll bodies in the cytoplasm of plant cells (particularly that part of the spectrum with wavelengths between 490 and 760 m, i.e. green to red) carbon dioxide in the surrounding air, together with water vapour is converted into carbohydrates and oxygen (i.e. assimilation or photosynthesis of plants). In consequence the air around trees during the day becomes progressively poorer in CO_2 and

richer in oxygen. At night plants absorb oxygen and break down the carbohydrates formed during the day into CO_2 and water. Most plants produce 5–10 times their volume in CO_2 in 24 hrs. A temperature rise of 10°C may double or triple CO_2 production. Therefore the oxygen-CO_2 ratio of the air in a forest fluctuates during the day depending on such factors as environmental temperatures, hours of sunshine (intensity and solar spectrum) and speed of the wind.

(2) The **moisture content** of the air is increased, causing a reduction in the number of small ions and an increase in the number of large ions. **Electrical conductivity** of the air is reduced; as are wind speeds and general air movement, the amount of direct or reflected **solar radiation** reaching the ground, and chemical air pollution through absorption of chemical **pollutants.** On the other hand the amount of organic pollutants may increase. The general **cooling** during clear nights takes place less rapidly than in open fields. This prevents the formation of ground mist. The electrostatic **potential gradient** and the daily fluctuations in forests are very small.

MARITIME CLIMATES

In temperate lands a typical maritime climate is one in which the difference in average temperature of summer and winter (i.e. the range of temperature) is less than 15°C. A typical continental climate has differences in excess of 20°C. Some of the most striking physico-chemical characteristics of maritime climate include the following:

(1) **Considerable turbulence** due to the usually high to strong winds along the coast. This causes appreciable cooling of the human body. Atmospheric **cooling** or **wind chill** is a very important physiological stimulus. It is a function of temperature and wind speed and may be calculated, as for example, through the formula of Bedford and associates:

$$h = \Delta t \, (9 \cdot 0 + 10 \cdot 9 \sqrt{\overline{V}} - V)$$ where h = the amount of heat loss in Kcal/m²/hr

Δt = difference in temperature between subject and environment in °C

V = wind speed in m/sec

In cold polar areas the formula of Wilson is probably more applicable:

$$H = (\sqrt{100V} + 10 \cdot 45 - V)(33 - t)$$ in which H = heat loss in Kcal/m²/min

t = ambient temperature in °C

V = wind speed in m/sec

(2) **High intensity of reflected solar energy.** The amount of solar energy reflected by water may be twice that from grass, particularly with respect to the shorter ultraviolet rays. Sand reflects even four times more strongly than grass. Strong reflection of the infrared heat waves gives rise to excessive heating of the skin on days when there is little air movement.

(3) **Differences in temperature** (both diurnal and seasonal) are smaller near the coast than inland. Winter temperatures are relatively high, summer temperatures relatively low and there are fewer frosty days.

(4) **Humidity** is greater near the coast, but owing to the increase of air movement foggy conditions tend to clear up more rapidly than inland.

(5) **Trace elements.** There are higher concentrations of ozone, iodine, magnesium chloride and sodium chloride above the sea.

(6) **Air pollutants**. Dust, pollen, and chemicals, are generally rare when there are sea breezes.

(7) **A different electrical field in the atmosphere above the sea.** The number of large ions is usually less above the sea than on land, and the total number of ions is usually considerably smaller than inland.

PRINCIPAL CENTRES OF THE HUMAN BODY REGISTERING METEOROLOGICAL STIMULI, AND THE PHYSIOLOGICAL MECHANISMS INVOLVED

One of the principal structures in the brain through which changes in weather and climate affect the body is the **hypothalamus,** the principal heat regulatory centre. The principal reasons for the assumption that the hypothalamus has a regulating influence with respect to weather are:

(1) The same meteorological factor or group of factors seem to affect a number of entirely different physiological and pathological processes in the human body.

(2) The hypothalamus exerts control over the pituitary gland which in turn controls most of the hormonal processes in the body.

Registration of changes in environmental temperature probably takes place as a result of minor alterations in the physico-chemical state of the blood, circulating through the rich network of hypothalamic capillaries. At the same time, the hypothalamic nuclei receive information from the thermal receptors in the skin. Vogt* and associates showed that nor-adrenalin, adrenalin and 5-hydroxytryptamine (5–HT) occur in very small concentrations in the hypothalamus; Domer and colleagues found that excess of free 5–HT or catecholamines on one side of the anterior hypothalamus seemed to be a factor determining the level of body temperature.

Several methods have been suggested to measure the total efficiency of the thermoregulatory mechanism in the human body. A simple method (Tromp and Sargent, 1964) is the measurement of the temperature of the left handpalm with a thermocouple. After cooling in water to 10°C the rewarming is measured every 15 seconds. In a well thermoregulated subject the rewarming curve reaches its initial temperature level in about 6 minutes, whereas in poorly thermoregulated subjects, e.g. those suffering from rheumatism or asthma, the initial level is only reached after 20 or more minutes. Apart from the thermoregulatory mechanism several other meteorotropic mechanisms have been discovered. They can be broadly grouped according to the part of the body through which the stimuli are registered. Table 28 presents a summary of present knowledge of the various physiological mechanisms responsible for meteorotropic effects.

THE EFFECT OF METEOROLOGICAL STIMULI ON NORMAL PHYSIOLOGICAL PROCESSES IN HEALTHY SUBJECTS

Biometeorological studies during the past fifty years have revealed a considerable number of physiological changes resulting from thermal stresses, radiation and other meteorological stimuli.

Table 29 contains a compilation of the reported changes in normal physiological processes in man. It shows clearly that practically every part of the body is affected by changes in the meteorological environment. Six important factors may change the outcome of a meteorological stimulus. They are the antecedent conditions experienced by the subject, the phase of the biological rhythms during the time affects the stimulus organs of the body, the degree of acclimatization, meteorotropic typology (see below), the age of the subject and the sex of the subject.

In 1931 Wilder introduced a physiological principle, the **'law of the initial value'.** This states that the result of a stimulus depends on *antecedent conditions*: the higher the initial value of a physiological process, the less the tendency for the initiated process to increase. Thus the higher the excitability of the autonomic nervous system resulting from weather

---

* For full references to this and subsequent authors *see* Tromp, S. W., *Medical Biometeorology*, Amsterdam, Elsevier Publ. Co., 1963.

Table 28

Principal Centres of the Human Body registering Meteorological Stimuli*

I: Skin	II: Lungs and throat (Respiratory tract)
1. Thermal effects: through conduction, convection or infrared radiation (780–>7,000 mμ wavelength) recorded by thermoreceptors in skin and hypothalamus, effects counterbalanced by vasodilatation or constriction or sweating. According to Hardy, the human skin registers radiation heat of 15×10^{-5} g. cal/sq cm/sec. causing a rise of skin temperature of 0·003°C. Effects on: (a) *Hormonal functions:* of pituitary (antidiuretic, thyrotrophic and gonadotrophic hormones), thyroid and adrenal gland (17-oxosteroids) and pancreas (insulin production and blood sugar-level). (b) *Blood:* changes in albumin and globulin levels and blood cell composition. (c) *Electrolyte balance;* pH AND OTHER SUBSTANCES IN URINE. Cold stress decreases excretion in urine of chloride, sodium, urea, hexosamines; causes rise in pH. (d) *Liver function:* cold stress increases the liver enzymes (SGOT, SGPT and SLDH), respiration of liver cells etc. (e) Other physiological processes affected by (a)–(d). 2. UV radiation effects: particularly UV-B (290–315 mμ), interval 297–302 mμ strong anti-rachitic effect; UV-C (<290 mμ) particularly occurring above 2,000 m. Absorption of UV by skin proteins increases below 295 mμ. Effects are: (a) Melanin oxidation. (b) Increased vitamin d and histamine in the skin. (c) Increased gastric acid secretion. (d) Blood: increased hemoglobin, Ca, Mg and phosphate level. (e) Increased protein metabolism. (f) Hormones: thyroid (hyperthyroidism) and adrenal gland, gonadotrophic functions. (g) Direct lethal effects on bacteria (indirect effects on man).	1. Temperature and humidity affecting mucous membrane: Dry air: (a) Drying and decreased elasticity of muscous membranes (leading to microfissures) and decreased ciliary activity (inefficient removal of dust); (b) decreased mucous (and antibody) production; (c) decreased blood flow and warming-up of inhaled air. Cooling: (a) Decreased permeability of membrane. (b) Constriction of blood capillaries. Both temperature and humidity affect survival and penetration of bacteria and viruses Gram-negative bacteria increase with high humidity, gram-positive and influenza virus with low humidity and low wind speed. 2. Ionization: Surplus of positive ions in air, particularly CO_2 mol (conc. of 10^4–10^5 ions/sq cm/sec), causes decreased ciliary activities in trachea (from 1,400 to 1,100/min) decreased mucous transport, dry throat. headaches; surplus of negative ions (particularly O_2 mol) causes increased ciliary activity (from 1,100–1,700/min). Possible cause (acc. to Krueger): Effect of CO_2^+ on cell enzymes, releasing secrotonin, causing contraction of smooth muscles in trachea. 3. Acidity of the air. (a) pH > 8·0: increased permeability, inflation of mucous membrane cells, decreased ciliary activity. (b) pH < 7·0; shrinkage of cells. (c) pH < 5·0: increased incidence of bronchitis. 4. Decreased partial oxygen pressure. In mountains or low-pressure climatic chambers from 159 mmHg (sea-level) to 125 mm (2,000 m), causing reduced oxygen saturation of blood from 96% (sea-level), to 85% (3,000 m). 30% increase of UV radiation.

* For full references, see Tromp (1963).

TABLE 28—continued

Principal Centres of the Human Body registering Meteorological Stimuli

I: SKIN (contd.)	II: LUNGS AND THROAT (Respiratory tract) (contd.)

3. Changes in acidity of the skin
 	
(*a*) by aerosols;
(*b*) by factors affecting sweat production and evaporation.
	High altitude effects above 1,500 m:
	
	(*a*) increased lung ventilation;
	(*b*) increased adrenal activity;
	(*c*) increased heart and pulse rate;
	(*d*) changes in composition of blood cells (e.g. increased Hb);
	(*e*) increased peripheral blood circulation;
	(*f*) increased thermoregulatory efficiency;
	(*g*) greater balance between the autonomic nervous systems;
	(*h*) increase of the fibrinogen-content of blood;
	(*i*) reduced stomach acid production;
	(*j*) lowered resistance to toxic drugs.
	
	Clinical situations: asthma, bronchitis, whooping-cough, rhinitis, rheumatoid arthritis, peripheral blood circulation problems, certain forms of migraine and eczema, phantom pains.
	
	5. Increased partial oxygen pressure. Treatment of gas gangrene, CO poisoning, emphysema, myocardial infarction.
	
	6. Trace elements.
	
	(*a*) Ozone: In low concentration killing bacteria, in high concentration irritating mucous membranes and increasing incidence of infections; changes in red blood cells (spherocyte formation) after inhalation of 0·25 p.m. 3 weeks inhalation during 7 hrs/day cause increased neonatal death in mice.
	(*b*) Salt (NaCl): Fine aerosols with $\frac{1}{4}-\frac{1}{2}\%$ hypotonic salt solution ($<0.9\%$ NaCl) cause swelling of mucous membranes. Hypertonic solutions cause shrinkage.
	
	7. Air pollution.
	
	(*a*) Gases: (SO_2, CO_2, benzpyrene, etc.)
	(*b*) Particles:
	   (1) Organic (pollen, spores), causing allergic reactions.
	   (2) Inorganic: mineral dust (silicosis, etc).
	(*c*) Aerosols: Strong increase of physico-chemical action due to increased action surface.

TABLE 28—continued

Principal Centres of the Human Body registering Meteorological Stimuli

III: NOSE	IV: EYES	V: DIRECT EFFECTS ON THE NERVOUS SYSTEM
1. Direct stimulation of the olfactory nerve by volatile substances and corresponding stimulation of part of rhinencephalon, affecting emotions, behaviour, hormonal and cardio-vasomotor processes, intestinal function, etc.	1. Direct overstimulation of eyes by direct solar radiation (particularly in spring and early autumn) causing acute conjunctivitis.	1. Electrostatic (pulsating) and electromagnetic fields, affecting plants, bees, golden hamsters, nerves, reaction speed, organic and inorganic colloidal fluids.
2. Acidity of nasal mucosa. Normally pH 5·5–6·5; it affects membrane permeability. pH decreases with rest, inhalation of warm air, diluted acid, aerosol sprays etc. pH rises during rhinitis and common cold; it furthers development of gram-negative bacteria in nasal mucosa.	2. Cold stress, particularly combined with strong atmospheric turbulence triggers acute GLAUCOMA.	2. Microseismic effects (amplitude 1–$20\ \mu$ and frequency 6–8/sec) on eel (Deelder).
3. See effects described under 'Respiratory tract'.	3. Changing acidity of the eye membranes by aerosols, thermal stress, wind etc., causing Blepharitis, Conjunctivitis.	3. Direct olfactory stimulation.
	4. Retinal detachments. In the Netherlands, particularly in summer (max. June), min. in winter. In Switzerland max. March–May. min. Nov.–Jan.	4. Rohracher effect. Natural mechanical vibrations of the living organism, with frequencies 6–12/sec, ampl. 1–$5\ m\mu$. Frequency increases to 14/sec during hyperthyroidism, fever; fibrations missing in poikilothermic animals.
	5. Light-flickering during strong sunlight with certain frequencies cause epileptic attacks.	5. Cosmic ray effects. Recent studies suggest direct biological effects at a cellular level due to cosmic rays.
	6. BENOIT-MILINE EFFECT. Influence of sunlight, through eye-nerves, on pituitary, hypothalamus, production of gonadotropic hormones, thyroid and adrenal function, biological rhythms, etc.	
	7. HOLLWICH EFFECT. Long absence of light (caves, blindness) changes carbohydrate metabolism, urinary volume, blood sugar level, reduction of pituitary and adrenal function.	

TABLE 28—continued

VI: IMPORTANT BIOMETEOROLOGICAL STIMULI DURING SPACE FLIGHTS

1. GRAVITY FIELDS. Acceleration of 2g and more seems to affect electric potentials in plant tissue (Brauner); changes growth in mice, hamsters and birds; 2g acceleration during 11 days increases heart weight, decreases weight of spleen, haemoglobin and blood haematocrit in animals and affects behaviour of man. Effects of very long periods of weightlessness on man have been studied extensively in space flights.
2. MAGNETIC FIELDS. Very strong fields ($>1,500$ Gauss) affect cell rotation, growth of tissue, growth of mice, haematological changes in blood of mice, etc.; influence on orientation of gastropods, planarias and drosophila.
3. CORPUSCULAR SOLAR RADIATION. Electrons, protons, heavy cores of atoms etc., but particularly proton streams during solar flares, could have dangerous biological effects due to deep penetrative capacity.
4. RADON EFFECTS. Inhalation of high concentration of radon aerosols or drinking of water rich in radon could affect functions of cell and endocrines, blood pressure etc.

TABLE 29

The Effect of Meteorological Stimuli on Normal Physiological Processes

PHYSIOLOGICAL PARAMETER	OBSERVED BIOMETRICAL RELATIONSHIPS*
HAEMOGLOBIN	After heat stress, increased plasma volume, decreased Hb; after cold stress, increased Hb. Hb lower in summer than in winter. On sea trip from Liverpool to Peru, Hb decreases.
LEUCOCYTES	Increase after steep barometric fall. Decrease after föhn wind, after previous intracutaneous injection of NaCl. Increased adhesion. Maximum: Oct.–Feb.; max.: April; min.: August.
THROMBOCYTES	In period 1958–1962 in the Netherlands: max. March–April; min. in August.
EOSINOPHILS	Increase from winter to spring, normal in summer. In 1958 in The Netherlands, low from May–Sept. (min. July–Aug.); high from Nov.–April (max. in March).
PROTHROMBIN	Min. in children in Hungary between Sept.–Dec. (Banos). In adults min. in winter and spring (Waddell, Lawson, Lehmann). Fluctuation in prothrombin index with airmass and barometric pressure changes.

* For full references, *see* Tromp (1963).

TABLE 29—continued

The Effect of Meteorological Stimuli on Normal Physiological Processes

PHYSIOLOGICAL PARAMETER	OBSERVED BIOMETEOROLOGICAL RELATIONSHIPS
SERUM PROTEINS	Total serum protein usually decreases from winter to summer from 8·5–7·5 gr/100 cc serum. Albumin level high in summer, low in winter; γ-globulin usually high in winter, low in summer.
FIBRINOLYTIC PROPERTIES OF SERUM	Blood-clotting time very low shortly before passage of cold front. After cold front passage strong fibrinolysis. Fall in fibrinogen after strong cooling.
BLOOD VOLUME	Increases with heat stress, decreases with cold (cold fronts, polar air).
PACKED RED CELL VOLUME	Max.: Feb.–March; Min.: July and Aug. in Japan.
SPECIFIC-GRAVITY RED BLOOD CELLS	Seems to rise in summer.
OXYGEN CAPACITY OF BLOOD	Increases from Jan.–May (in children <6 years); from Feb.–Aug. (in older children).
CARBON DIOXIDE CAPACITY OF BLOOD	Max. absorption of blood around 21 Dec.; min. around 21 June.
SERUM CALCIUM	Min. In Feb.–March (8·5 mg/100 cc); max. in August (11 mg/100 cc).
PHOSPHATE SERUM	Min. in Feb.; max. in summer and autumn (average level 3–5 mg/100 cc).
MAGNESIUM SERUM	In Japan, min. in Feb. (2·12 mg); max. in Dec. (2·85 mg).
COPPER SERUM	In cows in Holland increasing from 76 μg% in Sept. to 104 μg% in May: decreasing again in summer.
SERUM ASCORBIC ACID	Low in winter, high in summer. Rats living under rhythmical light–dark changes have max. values during the light periods and min. values during darkness; ascorbic acid secretion always precedes the secretion of cortico-steroids.
SERUM IODINE	Min. in winter (Dec.–April) (8·35 mg%); max. in summer (July–Aug.) (12·85 mg%).

TABLE 29—continued

The Effect of Meteorological Stimuli on Normal Physiological Processes

PHYSIOLOGICAL PARAMETER (*contd.*)	OBSERVED BIOMETEOROLOGICAL RELATIONSHIPS (*contd.*)
ERYTHROCYTE SEDIMENTATION RATE (ESR)	Relatively low in winter, higher in summer. Short-term fluctuations: low ESR values after influx of cold polar air.
BLOOD COAGULATION	Blood clotting time particularly low shortly before cold front passage. Clotting time increases after cooling.
BLEEDING AFTER TREATMENT WITH ANTICOAGULANTS	Max. in Jan.–Feb.; min. in July.
17-OXTOSTEROID EXCRETION IN URINE	Increase after cold stress, e.g., cold fronts, polar air masses; decrease with increasing temp. During very strong heat stress, also increasing. Corticosteroid secretion of mice reaches max. values 2 hr before the beginning of the dark period, decreasing during the following dark period. Corticosteroid secretion preceded by ascorbic acid secretion.
DIURESIS	Decreasing after heat stress; increasing during fall in temp. (cold front, polar air influxes).
METABOLISM	In children max. during autumn. General metabolism decreasing sharply during winter. In diabetics, during late spring, summer and early autumn, higher metabolism. In winter low metabolism, high blood-sugar content, high insulin consumption.
DIASTOLIC BLOOD PRESSURE	High in winter months (particularly Feb.); low in summer.
THYROID FUNCTION	Increased thyrotrophin production and general increase in activity after cold stress followed by temporary hyperthyroidism. After continued cold stress, effect becomes less.
RELATIONSHIP TO DRUGS	Digitalis more toxic during steeply falling barometric pressure; toxicity of morphine sulphate highest during cold front passages, great atmospheric turbulence etc. (Nedzel, Sargent).
PERMEABILITY TISSUE	After cold fronts, decreasing; after warm fronts or föhn winds, increasing.
CAPILLARY RESISTANCE	Increasing after cold fronts, decreasing after warm fronts.
SKIN CAPILLARY TEST	Capillary structure of the skin changes rapidly in spring; excitability of peripheral nerves in skin increases as shown by dermographic tests.

TABLE 29—continued

The Effect of Meteorological Stimuli on Normal Physiological Processes

PHYSIOLOGICAL PARAMETER (*contd.*)	OBSERVED BIOMETEOROLOGICAL RELATIONSHIPS (*contd.*)
ADRENALINE AND ACETYL-CHOLINE SKIN TESTS	Acetylcholine test shows the influence of cold front passages (8 hr and more) earlier than adrenalin tests.
DIRECT CURRENT RESISTANCE OFFSKIN	Very high during falling barometric pressure and föhn.
MUSCLE METABOLISM	Phosphoric acid and glycogen content of muscles of rabbits high with cold fronts, low with föhn winds and warm fronts.
MUSCLE STRENGTH	Changing during different weather conditions.
GASTRIC ACIDITY	Hyperacidity high in winter, low in summer, anacidity high in summer.
GROWTH AND WEIGHT OF CHILDREN	Slow growth in winter, rapidly increasing in spring; height increase max. in March–May and Nov.–Jan. in Stockholm. Weight increase in Stockholm max. in Sept.–Nov.
BIRTH WEIGHT	Greatest in June–July; smallest Dec.–March.
BIRTH FREQUENCY	Highest number of conceptions in June (legitimate children) or May (illegitimate); stillborn children maximal in Jan. neonatal deaths max. in Feb.
MORTALITY	In Western Europe max. in Dec.–Jan.; min. in July.
ACCLIMATIZATION	More difficult with respect to cold stress in summer than in winter; acclimatization to heat-stress is more difficult in winter.

stress, the smaller the increase after renewed excitation. This principle, related to the phenomenon of 'accommodation' during electrical excitation of nerves, explains how a certain biometeorological effect created by a cold front or the influx of cold air masses may differ if the cold front is preceded by a warm front or if a number of cold fronts pass an area at short intervals. In the latter case quite contrary reactions may be observed.

Apart from previous biological events the **biological rhythm** of an organism also plays an important part. For instance, various studies suggest that during the night parasympathetic stimulation dominates. Therefore, the same weather stress can have an entirely different result, depending on the hour of the day that a certain meteorological phenomenon occurs. In other words the effect of a stimulus depends on the phase of the biological cycle at the time at which the stimulus affects the body.

Several physiologists have pointed out that **acclimatization** to meteorological stresses (particularly thermal stresses) is probably not a reflex, but is the result of slow physiological adjustments. Previous exposure to thermal stress assists adaptation at a later period. The time interval differs with different individuals and depends on the degree of thermal stress and other factors; it is usually a matter of weeks. This explains why higher sweat rates can be elicited in summer than in winter and why an influx of cold polar air after a period of heat adaptation will have a greater effect than after a previous cold period. Repetition of stimuli, as experienced during acclimatization, may be accompanied by gradual changes in responses to those stimuli, a process described as **habituation.** According to Eccles, Konorski and Young, one kind of habituation seems to be caused by permanent (so called 'plastic') changes in nerve function. This workers showed that prolonged inactivity of synapses led to defective synaptic function, whereas repetitive stimulation improved the functional ability. They found also that the size of neurons and dendrites seemed to depend on the amount and variety of the stimulation to which they were subjected.

A fourth factor is the influence of differences in human psychological and physiological pattern, the so-called **meteorotropic typology.**

Various observations suggest that the effects of an adaptation to the meteorological environment are influenced by the body build (physique) and the psychological pattern of the person involved.

Fine and Gaydos have shown that after cold stress, significant differences in men's rectal temperature changes may be observed, provided the personality pattern deviated widely from the norm. Heavily built men felt warmer during cold stress than light-weight small men. Among healthy men with average pulmonary function, differences in oxygen consumption are observed related to differences in personality pattern and body build (Lucas and Pryor).

Youths with a slender body build and a dolichocephalic skull have higher oxygen consumption and higher basal metabolic rates per kilogram of body weight than do those with broader proportions. The same holds for people with shorter extremities when compared with those with long ones. These and other observations explain some of the differences observed in high-altitude acclimatization and other adaptational processes where oxygen consumption plays an important part. Various thermoregulatory studies, such as those using the Hunting reaction of Lewis as a thermoregulatory efficiency test, suggest that differences in pattern of the Hunting reaction are closely related to the psychological pattern of the person tested.

Peterson found a relationship between the incidence of the common cold and body habitus. In the autumn and winter, boys having a small body surface in proportion to their body weight, i.e. pyknics, have the highest rates of the common cold whereas in the spring tall slender leptosomes have a higher incidence. Sheldon's observations indicating a statistically higher incidence of certain diseases, e.g., peptic ulcer and gall-bladder diseases, in persons with certain somatic types, support the assumption that adaptational processes may be seriously affected by the psychological and physiological pattern of the persons involved.

Other factors which affect the outcome of a meteorological stress are the **age** and **sex** of a subject. Young children and people in old age usually lack an efficient thermoregulatory mechanism. It follows that the result of a thermal stress in these groups will differ from the effect in normal healthy adult subjects. Studies by Karvonen in Finland indicate that the **aerobic power** of women, i.e. the maximum amount of oxygen to be absorbed by the lungs which determines the maximum working capacity is less than in men. In Sweden, in men of 20–29 years, it is 3·01 L/hr of oxygen; in women 2·23. This difference is partly due to difference in length, and partly it is the result of the smaller number of erythrocytes per c.c. in females than in males. As a result the oxygen absorption is less. Studies by Hardy, Kawahata, Sargent, and Wyndham *et al* indicate that at a certain temperature women perspire on the whole less than men per unit of body surface. Differences are small on the cheek, breast, forearm and calf. Usually perspiration starts in women at higher skin temperatures than in men. Shortly before and during menstruation the thermoregulatory efficiency of females is considerably reduced. In view of these thermoregulatory problems in relation to sex and age and considering the major role of the hypothalamus in registering meteorological stimuli it is evident that the effect of a meteorological stress could be entirely different under apparently similar laboratory conditions.

THE EFFECT OF WEATHER AND CLIMATE ON HUMAN DISEASES

Due to changes in thermal stimuli, non-thermal radiation stresses (ultraviolet solar radiation), and humidity-temperature effects, man is continuously subjected to a variety of meteorological stresses which could trigger off a number of so called **meteorotropic diseases.** Such diseases may be classified into four major groups: (1) diseases due to disturbances of the natural biological rhythms; these can be both short and long term (seasonal) disturbances; (2) diseases due to hypothalamic disturbances; (3) diseases due to ultraviolet solar radiation and (4) infectious diseases.

The second group can be divided into diseases due to moderate thermal stresses and those due to excessive thermal stresses (both heat and cold). To the moderate stress type belong conditions such as asthma, bronchitis, rhinitis, rheumatic diseases (particularly rheumatoid arthritis), heart diseases (particularly myocardial infarction and angina pectoris), stroke, certain eye diseases (e.g. acute glaucoma, retinal detachments, acute conjunctivitis) and vascular disorders (e.g. Raynaud's disease). Excessive thermal stresses may cause or trigger off diseases like oedema, syncope, sweat gland disorders, chilblains, and frostbite. All the diseases of this group are mainly due to a serious disturbance of the thermo-regulatory mechanism. In Table 30 a summary is given of the most important meteorotropic diseases which could affect particularly those subjects living under extreme meteorological conditions e.g., in many of the underdeveloped countries.

TABLE 30

Seasonal and other Long-Periodical Relationships observed in Various Meteorotropic Diseases

DISEASE	OBSERVED BIOMETEOROLOGICAL RELATIONSHIPS*
Infectious diseases	
LUDWIG'S ANGINA	Increasing after sudden influx of cold air.
CHOLERA	Rare in Europe but max. in Aug. In India and surrounding countries common in hot humid tropics except during Jan.–Feb. if in these areas the temperature drops considerably.
COMMON COLD	Min. in Sept., max. in Feb.–March (in N. Hemisphere). In Holland and England, increase in incidence with falling temperature and rise in humidity.
DIPHTHERIA AND CROUP	In W. Europe, beginning in Aug., max. in Nov.–beginning of Dec., but still high till Feb.

* For full references, *see* Tromp (1963).

TABLE 30—continued

DISEASES	OBSERVED BIOMETEOROLOGICAL RELATIONSHIPS

Infectious diseases—continued

DISEASES	OBSERVED BIOMETEOROLOGICAL RELATIONSHIPS
DYSENTERY	Bacillary dysentery in Germany: max, in summer; in UK: Feb.–April; in the tropics in the rainy season.
INFLUENZA	In colder parts of N. hemisphere max. in Dec.–Feb.; in S. hemisphere June–Aug.; in 1918 in France and England epidemic related to dry, cold periods. In 1956 in the Netherlands: max. in Sept.–Oct.
LOBAR PNEUMONIA	In N. hemisphere max. Dec.–Feb.; in tropical Africa max. during dry season. High mortality if temperature during preceding week has been lower.
CEREBROSPINAL MENINGITIS	In W. Europe beginning in autumn, high in Dec.–April; in the USA max. in Feb.–March; in tropical W. Africa max. in dry season in dry areas.
POLIOMYELITIS	In N. hemisphere min. in March and beginning of April; increasing in April with max. in Aug.–Sept. In S. hemisphere: min. Sept.–Oct.; near the Equator, no seasonal fluctuations.
SCARLET FEVER	In W. Europe beginning in Aug., max. end of Oct.–beginning of Nov.; ending in December. In New York and Chicago: max. in Feb.–March. In northern USA incidence three times that of southern part; in the tropics rare. Scarlatina and other streptococcal infections highest during year with low humidity.
TUBERCULOSIS	Increased sensitivity to tuberculin test in March.–April; low sensitivity during autumn; mortality increases in spring.
TYPHUS	If transmitted by ticks, peak of infection May–July (in N. Europe) coinciding with peak of breeding-season among ticks.
ENTERIC FEVER	In W. Europe, increasing from May till Sept., followed by rapid decrease from middle of Sept. In Java max. in wet season.
SMALL-POX	In N. hemisphere beginning in Aug. max. in March. In India and Ghana, max. in dry periods.
WHOOPING COUGH	No particular period for all countries in N. hemisphere; different for different years, but often in spring and autumn.

T<small>ABLE</small> 30—continued

Seasonal and other Long-Periodical Relationships observed in Various Meteorotropic Diseases

D<small>ISEASES</small>	O<small>BSERVED BIOMETEOROLOGICAL RELATIONSHIPS</small>

Non-infectious diseases

D<small>ISEASES</small>	O<small>BSERVED BIOMETEOROLOGICAL RELATIONSHIPS</small>
A<small>PPENDICITIS</small>	In 1933 Rappert studied 1,000 cases. During 20 days, the daily number of cases doubled of which 16 coincided with weather-front passages. Maurer, using 1,398 cases in München (period 1933–1935), observed in 78·5 % of the cases change in air-mass (normal chance in this period was 42·3 %). Jaki using 2,185 cases in Debrecen (period 1930–1939) did not observe a meteorotropism in that area. During heat-waves in USA Mills (1934) observed an incidence twice the winter values.
A<small>RTERIOSCLEROTIC HEART</small> D<small>ISEASES</small> (myocardial infarction, angina pectoris, coronary thrombosis); A<small>POPLEXY</small>	In W. Europe and northern part of USA: max. in Jan.–Feb., min. in July–Aug. In southern part of USA highest during hot summer, low during winter.
(N<small>ON-BRONCHITIC</small>) A<small>STHMA</small>	Max. Aug.–Nov. in the Netherlands; highly significant correlation with atmospheric cooling; no consistent relationship with allergens.
B<small>RONCHITIS</small>	Max. in winter, particularly in air-polluted areas; low in spring and summer.
D<small>ENTAL CARIES</small>	Meteorotropic relationships are uncertain. Erpf, Lathrop, McBeath and Zucker noticed seasonal variations in dental caries in the USA, max. incidence in late winter and early spring, min, in summer.
D<small>IABETES</small>	Conditions often deteriorating in late autumn and winter; greater need for insulin; greater tendency to acidosis and coma. According to Henkel, in the period 1927–1931 the distribution of 118 cases of diabetic coma was higher in the period Oct.–April (max. in Dec. and March); this coincides with periods of disturbed carbohydrate metabolism. Similar relationships observed by Chrometzka (1940). Hospital admissions at Cincinnati of 1,094 patients (period 1923–1938) indicated that only diabetics with arteriosclerosis showed a highly significant increase in admissions during winter, probably due to increased vascular troubles. In non-arteriosclerotic diabetics no seasonal effect was observed. According to Owens and Mills, diabetes is most common and most severe in middle temperature latitudes and less frequent in tropical regions.

TABLE 30—continued

Diseases	Observed biometeorological relationships

Non-infectious diseases—continued

PRE-ECLAMPSIA AND ECLAMPSIA	Meteorotropic relationships with cold fronts suggested by a number of studies but not statistically significant. Studies of Jacobs using 666 cases in Germany, Austria and Switzerland rather suggest a cold-front relationship. The only statistically significant study was carried out by Berg in Köln (Germany).
FÖHN (WIND) DISEASE	Common in Switzerland, Austrian Tyrol and S. Bavaria due to föhn weather: complaints observed long before obvious changes in weather are noticed. Storm van Leeuwen could demonstrate that the clinical symptoms are not due to microfluctuations in barometric pressure; Reiter observed a significant correlation with days with considerable disturbances of the EM long waves.
GLAUCOMA	High in winter (max. in Nov.); low in summer.
RETINAL DETACHMENTS	Max. in June (the Netherlands) or March–May (Switzerland); min. in winter.
GALLBLADDER STONES	According to Maurer, 83% of the cases in Munich occur during changes in air masses (normal chance 42·3%). According to Hentschel and Von Knorre (1959) particularly common after influx of cold humid air masses.
KIDNEY STONES	According to Maurer, 85% of the cases in Munich occur during changes in air mass (normal chance, 42·3%).
STONES IN URINARY BLADDER	Hauck found a highly significant correlation between fronts, changes in air mass, and renal colics in Leipzig.
GOITRE	In simple goitre meteorotropism related to fluctuations of the iodine content of the thyroid and serum iodine; max. in winter. In Graves disease studies by Breitner, Hutter, Jacobowitz *et al.* indicate a min. in summer, increasing in winter till spring (max. in March).
HAY FEVER	Different seasons depending on flora of the area; close relationships with atmospheric turbulence.
HERNIA	In 1935 in Kyoto (Japan) Mori observed in 860 three-months-old babies with umbilical hernia that the maximum number occurred in those born in January; the incidence was minimal in April.

TABLE 30—continued

Seasonal and other Long-Periodical Relationships observed in Various Meteorotropic Diseases

DISEASES	OBSERVED BIOMETEOROLOGICAL RELATIONSHIPS
Non-infectious diseases—continued	
HERNIA—continued	Meteorotropism not so far established in cases of inguinal hernia; according to Mori, the condition is most frequent in babies born in winter, and less frequent in those born in April.
MALFORMATIONS	More anencephalics born in Oct.–March (max. Dec.) Patent-ductus arteriosus cases show a marked seasonal fluctuation amongst girls only: max. May–Dec., min. Jan.–April (in Birmingham), or max. Oct.–Jan., min. Feb.–Aug. (in Massachusetts).
MENTAL DISEASES	Max. mental stress in Nov.–Jan.: birth frequency of patients with schizophrenia common in Jan.–March.
MULTIPLE SCLEROSIS	Studies by Allison suggest higher incidence in cold temperate climates and low incidence in sub-tropics. Studies by Multu in Turkey indicate highest incidence in coldest regions (particularly north of the 40° N. parallel), very low south of the 37° N. parallel. Studies by Georgi and Hall in Africa suggest extremely low incidence in various parts of Africa, even those with good medical facilities.
RHEUMATIC DISEASES	Complaints usually more during the cold, humid, windy seasons, particularly in the case of arthritis.
RICKETS	Observed mainly between latitudes 40° and 60°; absent in the tropics except in rich families keeping their young children in the darkness; common in heavily polluted industrial areas; incidence decreases with altitude; in high mountain valleys observed only on the shady side; according to Schmorl and others most frequent, in winter, decreasing from March, min. in July–Aug.
SKIN DISEASES	A high incidence of eczema in spring. High incidence in summer of dyshidrosis, miliaria rubra, dermatomycosis etc. From July–Aug. (often increasing till Nov.–Dec.) higher incidence of furuncles, carbuncles and erysipelas. Excessive sunlight causes skin cancer, light dermatoses, photodynamic diseases and herpes zoster in N. hemisphere, common in July, rare in winter; on Java, max. in April and May, min. in Jan.–March (wet monsoon).

TABLE 30—continued

DISEASES	OBSERVED BIOMETEOROLOGICAL RELATIONSHIPS

Non-infectious diseases—continued

DISEASES	OBSERVED BIOMETEOROLOGICAL RELATIONSHIPS
TETANUS	In recent years with new forms of prophylaxis a very rare condition. Formerly pronounced incidence peak Jan.–March, very low in summer, autumn and early winter, central Europe, and in the USA by Tidsall, Brown and Kelly. According to Moro, tetanus common after drastic weather changes (e.g. föhn conditions after cold weather). Also György observed tetanus particularly on warm sunny days in early spring. Lassen observed increased electric sensitivity of the peripheral nerves in spasmolytic children during meteorologically disturbed days, Idiopathic tetany in adults is rare, but according to Frankl-Hochwart, also highest incidence in March, low in summer.
THROMBOSIS AND EMBOLISM	The start of thrombosis is usually not sharply defined, unlike embolism. Therefore, accurate meteorotrophic studies are only possible with embolism. Despite many studies on post-operative lung embolism and weather, very few statistically significant studies have been reported. Tivadar, using 156 cases in Budapest, found a significant correlation with fronts. Raettig and Nehls studied 489 cases in Pommern, Germany. A statistically significant correlation with both cold and warm fronts and lightning storms was found. Reimann and Hunziker could not confirm a significant front relationship in Basel, Switzerland. On the other hand, Maurer, Kayser, Sandritter and Becker confirmed a statistically highly significant correlation between lung embolism and the passage of weather fronts in different parts of Germany especially with sudden changes of air mass.
PEPTIC ULCER	According to Hutter, in Germany serious complaints most common in Dec.–Feb. According to Apperly in Australia (period 1924–1926), incidence decreasing from Tasmania (40° latitude) to Queensland (25° lat.) and W. Australia (with warm tropical climates). According to EINHORN, in USA rise in recurrences in the fourth week of Feb. and first week of March, with max. in May; from the fourth week in May a decline until second week of June; gradual rise from the third week in Aug., until second week of Sept. (max.). Scheidter observed in Bavaria (period 1928–1932) a significant relationship between passages of warm fronts and perforations of peptic ulcers. Gebhardt and Richter (period 1927–1933) showed for peptic ulcer a min. in April, max. in Oct. Sallström in Sweden (period 1937–1942) found a min. in the number of cases in June, max. in Sept. De Franchis in Italy (period 1935–1945) observed the highest incidence during autumn, followed by winter, spring and summer. Boles analysed all the cases of haemorrhage from gastric and duodenal ulcer admitted at the Philadelphia General Hospital (period 1949–1953). The

Table 30—continued

Seasonal and other Long-Periodical Relationships Observed in Various Meteorotropic Diseases

Diseases	Observed biometeorological relationships
Non-infectious diseases—continued	
Peptic ulcer–continued	incidence was highest in Jan., Feb., April, May and June. Schedel in Munich (period 1948–1951) studied the relationship between perforated ulcer and the weather types of Ungeheuer. Most perforations occurred during phase 1, 2 and 6. A maximum occurred during phase 4; the minimum during phase 6. The correlations were statistically significant.
Duodenal ulcer	According to Hutter, in Germany, max. in May (particularly in men below 40), secondary max. in Nov. (only in males). According to Einhorn, similar to gastric ulcer. In Leipzig Gerhardt and Richter observed a min. in April, max. in Jan., with secondary max. in June–July. Sallström in Stockholm observed a min, in June, max. in Aug. Boles in Philadelphia observed an increase in the incidence of haemorrhage from duodenal ulcer during Jan., Feb. and March, and during Oct., Nov., and Dec.

It is beyond the scope of the present paper to present a detailed discussion of each of the meteorotropic diseases listed. Additional mention is reserved for those which have appreciable social and economic consequences related to their morbidity and mortality levels.

Before doing so, however, brief mention should be made of the **influence of weather and climate on the effect of drugs** administered under different meteorological conditions. In orthodox medicine the environmental conditions under which a drug is given in a certain dose is not taken into account; yet it has been known for some time, amongst scientists working in the field of biological rhythms, that a drug administered at different hours of the day can have markedly differing effects. However, traditional clinicians working in the field of biometeorology usually do not consider such fluctuations in effectiveness and toxicity of certain drugs relative to the weather conditions prevailing at the time of administration of the treatment.

In 1927 Macht in Baltimore observed great differences in the toxicity of digitalis under different weather conditions. Thus the drug became more toxic during severe storms which were accompanied by rapidly falling barometric pressures. Toxicity decreased during quiet high pressure conditions. Macht observed that with the same doses of a toxic drug there was a greater mortality amongst animals at 2 000 m than at sea level. His findings were confirmed by Jarisch working at Innsbruck and other physicians working at high altitudes. Hamburger and Ritte observed that on certain days children required a considerable larger dose of atropine than usual. Nedzel (1937) and Sargent (1939) found a higher mortality rate in mice treated with morphine sulphate after the passage of a cold front than on days with stable meteorological conditions when mortality rates were at their lowest.

Since these findings were reported many similar observations have been made by pharmacologists throughout the world and a new branch of science, **pharmacological biometeorology,** is developing (*see* Tromp, 1963). Weather affects the membrane permeability and speed of absorption at the cellular level, and these physiological processes are mainly responsible for the differing pharmacological effects, particularly under extreme meteorological conditions.

Respiratory Diseases

The major meteorotropic diseases in Northern Europe and the USA are the group of respiratory diseases, in particular **asthma** and **bronchitis.** It is estimated that at least 1 or 2 out of every 100 inhabitants is suffering from these diseases. Studies in different parts of the world and particularly long-range studies in the Netherlands since 1955, by Tromp and Bouma (1968) using large groups of children and adults, have shown the following significant relationships between weather and climate and non-infectious (non-bronchitic) forms of asthma.

1. A highly significant statistical relationship was found between asthma frequency and certain phases of the weather.

(*a*) Asthma frequency increases rapidly after a sudden increase in the general turbulence of the atmosphere if combined with the influx of cold air masses. The increase is most striking when it follows a quiet period. In Western Europe this is the result of high atmospheric pressure. The change in the degree of turbulence is characterized by a sudden rise or fall in atmospheric pressure. Periods of steep fall in barometric pressure, due to rapidly approaching low pressure areas, have a particularly marked asthma-increasing effect in Western Europe, especially if these weather changes are accompanied by a sudden influx of cold polar or continental air masses and one or more 'active' cold fronts. The cold fronts are usually fast-moving and give rise to marked changes in atmospheric pressure and temperature, strong precipitation (rain, snow or hail), appreciable disturbances in the electric field of the atmosphere (fluctuations of electric potential gradient between high positive and high negative values), and sudden changes in wind speed and direction. In consequence there is a sharp increase in the cooling effect of the atmosphere.

(*b*) During periods of slight atmospheric turbulence or with the influx of warm tropical air masses and active warm fronts there is a sharp decrease in the frequency of asthma.

(*c*) The rapid succession of periods of cold and warm air masses, or cold and warm fronts, may create sinusoidal waves with periods of increasing and decreasing asthma frequency but each with a relatively small maximum amplitude. If the succession is very rapid, i.e. within a few hours, often no effect is observed.

(*d*) If there is a sudden change in weather associated with the rapid approach of a depression when warm tropical air is being replaced by cold polar air, and there is a decrease in general atmospheric turbulence, the frequency of asthma increases but the amplitude remains small.

(*e*) The same weather phase with an influx of cold air causes a considerably smaller increase in asthma frequency in winter than in summer probably because the body is adapted to cold in winter.

(*f*) A gradual fall in temperature, as is common during a calm cloudless night during an extended period of several days without any appreciable turbulence, usually does not affect the frequency of asthma because the body slowly adapts to the new environmental conditions.

2. Although certain differences may be observed between the incidence of asthma in boys and girls, mainly in the severity of the attacks, the incidence is generally similar in the two sexes.

3. Each month there are periods of high and low incidence of asthma, the periods varying in length from a few days to a week or even 10 days.

4. The same pseudo-seasonal rhythm in asthma frequency is observed in Western Europe every year, both in children and in adults. This result is of particular interest because in the case of the Netherlands the complaints of the children were observed by their sisters and not recorded by the children themselves, whereas the adults filled in their own questionnaires. In winter and early spring (including May and June) the average asthma frequency is low; in summer and particularly in late autumn it is high.

5. The fluctuations in the daily amplitude of asthma frequency usually increase suddenly at the end of June. Fluctuations reach their maximum in September, October and November. The shift in month of maximum values has proved to be related to a shift at the beginning of the winter, with its autumn storms and strong atmospheric cooling.

6. Fog, except perhaps in cities with much air pollution, has no asthma-producing effect in patients suffering from non-infectious asthma; indeed the contrary is the case. Periods of fog (usually characterized by periods of little atmospheric turbulence and usually high barometric pressure) are associated with a very low asthma frequency.

7. In contrast to those suffering from non-infectious asthma, bronchitic patients have most complaints in winter, particularly in January and February; they are at a minimum in summer.

8. Whereas quiet foggy conditions have no asthma 'triggering' effect in non-bronchitic subjects, the bronchitic patient is extremely sensitive to it. Even if he stays indoors he is immediately conscious of the formation of local fog.

9. A cold fog with fine droplets and a low pH is especially detrimental. This has been shown in England as well as in the Netherlands.

10. Patients suffering from either asthma or bronchitis are very sensitive to atmospheric cooling, asthmatic patients appearing to be the more sensitive. Also heat stress in hot areas increases distress for asthmatics.

A careful analysis of various meteorological factors has shown that although allergens could be an important factor in certain cases of asthma the major triggering agent would appear to be thermal stress (Tromp, 1966). This assumption arises from the fact that asthmatics are extremely sensitive to both cooling and heat stress. This sensitivity is due to the poor thermoregulatory efficiency of all asthmatic, bronchitic and many allergic subjects, e.g. those suffering from rhinitis. If thermoregulatory efficiency is improved, as for example, by high altitude treatment in low pressure chambers, the subjective complaints decrease rapidly or disappear completely despite the presence of allergens. Readers interested in the deeper physiological mechanisms involved in the asthma-weather relationship are referred to Tromp and Sargent (1964).

Heart Diseases

Many empirical studies, especially in W. Europe and the USA, have shown significant correlations between meteorological stress, particularly of heat and cold, and increased incidence of heart diseases such as myocardial infarction and angina pectoris and of cerebro-vascular accidents. These findings have been confirmed in climatic chamber studies, in particular by Burch (1959) in New Orleans. Some of the major observed meteorotropic correlations may be summarized as follows:

1. The mortality rate of coronary heart disease in middle and high latitude countries, considerably higher in males than in females, is invariably highest in the months of January and February and lowest in July and August.

2. The mortality rate for apoplexy, which is higher in females than in males, shows the same seasonal pattern as does coronary heart disease.

3. A comparison between average monthly temperatures and mortality rates for both coronary heart disease and apoplexy reveals the following significant correlations:

(*a*) Exceptionally cold winters are characterized by very high mortality; warm winters have a relatively low mortality.

(*b*) Although mortality is at a minimum in summer, there are variations which appear to be related to peak temperatures. The higher the temperature the greater the mortality from apoplexy and coronary heart disease.

4. In those regions which experience very warm conditions, e.g. the southern part of the USA, highest incidence of mortality is in summer; the lowest is in winter. Extreme cold and heat stress increase mortality rates. The correlations between average cooling and mortality are so significant that approximate mortality predictions can be made, solely on the basis of meteorological data.

Considering the great sensitivity of coronary heart patients to environmental thermal stresses it is logical to assume that these patients have a poor hypothalamic thermoregulating capacity. A study by Tromp and Bouma (1965) confirmed this assumption. It was also found that in elderly heart patients treatment by anticoagulants improved their thermoregulatory functions.

The actual mechanisms involved in the relationships between weather and heart disease are probably related to the effect of meteorological stimuli on, for example, the elasticity and peripheral resistance of the blood vessels, the activity of the sympathetic nervous system, the physico-chemical state of the blood (viscosity, blood clotting time and fibrinogen content), and capillary fragility (see Tromp, 1964).

Mental Diseases

Very few systematic studies have been conducted on the effect of weather and climate on mental diseases. The most extensive study to date was that undertaken by Tromp between 1956 and 1960, in collaboration with seven psychiatric institutes in the Netherlands.

1. Schizophrenia

Two hundred patients in the western part of the Netherlands were studied. Restlessness and ill-temper were recorded daily and were based on a number of criteria.

The daily incidence of restlessness in schizophrenic patients usually fluctuated between 7 and 43 per cent, the average monthly percentages between 2 and 16 per cent. In a large constant and rather homogeneous group of schizophrenics (homogeneity arising from such considerations as sex, age and social class) short and long term cyclic or pseudo-cyclic changes in restlessness were observed as follows:

1. During each of the 4 years of observation the highest degree of restlessness was observed in November, December and January. In some psychiatric institutes a secondary peak of restlessness was seen in July and August. The high degree of restlessness in winter showed up both in the high average monthly percentage and in the greater number of days with a high percentage of restlessness.

2. Statistically significant daily fluctuations of restlessness correlated with periodic or non-periodic influxes of warm continental, tropical maritime or warm maritime air masses. The gradual increase in temperature accompanying these air masses resulted in an increase in restlessness. Influxes of cold air masses had the reverse effect. Unpleasant weather conditions such as heavy rain or snow which had an adverse effect on the mood of the nursing staff, did not appear to affect the restlessness of schizophrenic patients.

Studies carried out in Leiden since 1960 suggest that these short and long term fluctuations, mainly triggered off by meteorological factors, seem to be related to a poorly functioning hypothalamic thermoregulatary mechanism in schizophrenics. These show up in the waterbath test of Tromp (1964), in the test of Henschel (1951) and in the diuresis pattern studies of schizophrenics (Tromp, 1963), Whereas in a healthy, well-thermoregulated subject the temperature of the handpalm (after 2 minutes cooling in water of 10°C) reaches the initial value in about 6 minutes, in schizophrenics it takes 20 minutes or more. Initial temperatures are often very low, the difference between initial temperature and the rewarming temperature after 6 minutes is often 4°C or more.

In Henschel's test, the feet of schizophrenics were immersed in water at 45°C. It was found that the time required for the skin temperature of the fingers to rise was appreciably longer in schizophrenics than in normal subjects. In normal, adequately-thermoregulated subjects, the urinary output increases during atmospheric cooling and decreases when air temperatures rise. In schizophrenics the opposite is the case.

Such tests indicate an abnormal functioning of the hypothalamic thermoregulatary centre in schizophrenics compared with normal subjects. This may be an important factor causing cyclic mental changes in such patients.

2. Oligophrenia

The following observations were made:

1. Daily and monthly percentages of restlessness are of the same order of magnitude as in schizophrenics.
2. Pseudo-seasonal fluctuations, though less pronounced than in schizophrenics when averages for each month are compared, show similar early winter peaks when the average for three-month periods is taken.
3. Daily fluctuations of restlessness curves sometimes coincide with those of the schizophrenics, particularly during the winter season; often they are considerably different. Whereas schizophrenics are hardly affected in their restlessness by unpleasant weather conditions, oligophrenics are considerably affected. This purely psychological effect may also give rise to considerable restlessness when there is an influx of cold air.

3. Epilepsy

Two different weather relationships were studied (1) its influence on the restlessness of epileptics and (2) its effect on epileptic fits.

1. *Influence on restlessness*
   (*a*) The pseudo-seasonal fluctuations suggest a maximum in November–December and a minimum in summer; there may be an occasional secondary maximum in May.
   (*b*) There was no convincing relationship between daily restlessness and weather.

2. *Influence on epileptic fits*

In certain instances the seasonal fluctuations in seizures were very pronounced. The lack of consistent seasonal phenomena was not however surprising since most of the patients observed belonged to a group which had lived for many years in a psychiatric institute and were under continuous treatment with drugs.

4. Migraine

Studies on the possible effect of meteorological stimuli on this condition are extremely rare, partly because so many different factors may be involved in the causation of migraine. A type of migraine following a previous concussion of the brain, and particularly if the patient has not taken sufficient rest, appears to correlate significantly with such drastic changes in the weather as a heavy snow fall and thunderstorms. However, considerably more research is needed to explain such relationships.

5. Suicide and Attempted Suicide

Recently, extensive studies have been initiated by Faust and his colleagues at the Psychiatric Clinic of the University of Basel in Switzerland. Several statistically significant meteorotropic relationships have already been established. Relationships between weather and suicide or attempted suicide are difficult to establish because they do not correlate with such meteorological factors as temperature, atmospheric pressure, hours of sunshine or precipitation. More sophisticated meteorological analyses have revealed certain relationships. De Rudder and Tholuck studied 200 cases of suicide which had been recorded at the Institute of Forensic Medicine at the University of Frankfurt in Germany during 1939. No relationships were established with atmospheric pressure, temperature, hours of sunshine or precipitation but there was a highly significant correlation with both cold and warm weather fronts. Studies by Rohden *et al.* in Switzerland have indicated a significant increase in the number of suicide cases and crimes when the hot, dry föhn wind prevails. Reiter was able to demonstrate a statistically significant correlation between periods of high frequency of suicide and days of strong disturbance of long electromagnetic waves in the atmosphere.

Extensive studies were carried out in the Western part of the Netherlands by Tromp and Bouma during the period 1954–1969 using about 10,000 well established cases of suicide and attempted suicide. The ratio between suicide and attempted suicide ranges from 1:6 to 1:10. Because the number of cases per day was very small the method of clusterday analysis was adopted. In other words, only those days with an incidence more than 300 per cent above the daily average for the year were analysed in relation to the meteorological conditions of the days in question. No simple correlation was found to occur with any one particular meteorological factor. Instead a highly significant correlation was observed between cluster-days and days with strong atmospheric turbulence associated with the approach or passage of depressions, low pressure regions and/or weather fronts. Sudden rises or falls of temperature appreciably above or below the average monthly temperature to which the subject is acclimatized and changes in wind speed (i.e., the cooling index), have this triggering effect. It seems likely that a serious disturbance of the hypothalamic-pituitary system, accompanying drastic changes in weather, contributes to the 'triggering' off of suicide and attempted suicide in those subjects predisposed to a serious mental disturbance following certain psychological, sociological and other forms stress.

Rheumatic Diseases

Extensive reviews of the effects of weather and climate on rheumatic diseases in man have been provided by Burt (1936), Mills (1938), de Rudder (1950), Tromp (1963), Lawrence

(1966) and others. All studies appear to suggest considerable differences in the kind and severity of rheumatic diseases and particularly arthritis within different countries, although there is a little variation in the actual global incidence. However, the degree of subjective complaints and the progress of the disease is much affected by weather conditions and in particular by strong atmospheric cooling (temperature-wind effects). These factors are particularly prevalent in countries of middle and high latitudes. At least three meteorotropic factors are involved in the weather sensitivity of arthritic patients, in particular rheumatoid arthritis.

1. A 4-year study by Tromp and Bouma has shown that the excretion of *hexosamines* in urine decreases with increased atmospheric cooling and increases with decreased cooling. Hexosamines, derivatives of mucoproteins, occur in the joints and cartilages. They are the chief agents producing the viscosity of mucin, a glucoprotein secreted by the mucous membranes. Hunter (1951) demonstrated that the viscosity of the synovial fluids responsible for the lubrication of human joints depends to a considerable extent on the mucin content. Accumulation of hexosamine, as occurs during cooling, raises the viscosity of the synovial fluids and increases the intensity of rheumatic complaints. The mucin content depends also on the pH of the synovial fluid. More mucin is dissolved in alkaline fluids. Atmospheric cooling causes an alkalinization of body fluids. Studies in the Netherlands have shown that the hexosamine excretion in the urine of rheumatic patients is about half the amount observed in healthy subjects. The favourable effect of thermal baths is partly due to an increased hexosamine excretion after the baths.

2. *Thermoregulatory efficiency tests* in the Netherlands in large groups in arthritic patients showed that 79 per cent of the subjects had an inefficient thermoregulatory mechanism and 21 per cent an extremely poor thermoregulatory efficiency, both in summer and in winter. Thus atmospheric cooling affects a rheumatic patient much more than a healthy subject.

3. *Adrenal function* and excretion of *corticosteroids* is usually considerably reduced. Since cooling conditions have a considerable affect on steroid levels in the body, this factor may also be involved in the meteorotropic effects of weather and climate on rheumatic diseases.

Infectious diseases

The problem of the possible influence of weather and climate on infectious diseases is very complex. Infectious diseases can be affected by several meteorological factors which can be classified into two main groups. The first lowers the resistance of the human body to infection, the other affects the ease of spread of an infection. Factors in the first group affect the *local* resistance, i.e. the site of entrance of micro-organisms or viruses into the body and the place of settlement and further development. Others affect the *general* resistance of the body, determined by the specific physico-chemical state of the body at a given time. To the second large group of factors belong (1) the influence of weather and climate on the social habits of man, such as crowding in rooms, shutting of windows and doors and changes in clothing and diet during certain seasons; (2) the influence on the development of micro-organisms and viruses, and (3) the effect of meteorological factors on the spread of these infectious agents through the atmosphere.

Local resistance may be affected in different ways. The permeability and capillary resistance of membranes is affected by thermal stresses. Regli and Stämpfli have demonstrated a decrease in capillary resistance (increase in permeability) following the influx of warm air masses and an increase following cold stress. These observations were confirmed by Lotmar,

Häfelin and Arimatsu. The acid coating on the skin is affected by alkaline aerosols in the atmosphere. Marchionini demonstrated that the acid coating produced by the eccrine sweat glands protects the skin against infections. The pH of the skin is usually from 4 to 6; after evaporation of sweat it is from 3 to 5. Spots with low acidity usually develop in skinfolds due to the lack of evaporation of sweat. All meteorological factors affecting the rate of sweating and evaporation may influence the growth of micro-organisms and their penetration into the skin.

The dryness of membranes (especially the nasal mucosa) is affected by temperature, humidity and ionization of the air. Low atmospheric humidity is believed to cause a with-drawal of water vapour from the body cells which in turn may lead to microfissures in the nasal mucosa, particularly on cold frosty days and often in centrally heated rooms. Exces-sively prolonged cold induces constriction of peripheral blood capillaries and reduced blood flow, causing drying and cracking of the nasal mucosa. Humidity affects the secretion of the nasal mucosa. Armstrong was able to show that a greater secretion of the mucosa parallels increased concentration of antibodies in the nose. Cold dry air stimulates secretion by the nasal mucosa. The gradual drying of the membranes tends to concentrate these anti-body secretions. According to Howitt and Bell the content of antibodies in nasal secretions parallels that in the serum. Ozone, if present in high concentrations, appears to lower the resistance of mice to respiratory infections. Humidity of the air also affects the survival of bacteria and viruses. Gram-negative micro-organisms die more quickly at low humidities; gram-positive bacteria and influenza viruses die more rapidly with high humidities and increasing air movement. Thus winter months with very low humidities and little air-move-ment in centrally heated buildings are favourable for the transmission of infectious respira-tory diseases.

General resistance to infections may be affected indirectly by the meteorological environ-ment as with seasonal changes in diet or directly because:

(a) There may be seasonal or short-term variations in the physico-chemical state of the blood. Cold stress reduces the ESR and albumin level; it increases the haemoglobin and gamma-globulin level of the blood. Change in the gamma-globulin level due to weather stress may affect the immunity of the body to infectious diseases.

(b) The effect of the same chemical substance differs if administered under different meteorological conditions. Rabbits kept at temperatures from 26 to 29°C escaped infection after inoculation with a dose of myxoma virus which was fatal when applied at winter temperatures (−3 to +15°C).

(c) Zimmermann reported a correlation between the 17-oxosteroid level and resistance to infectious disease. Weather stress such as cold periods causing a rise in 17-oxo-steroid production by the adrenal cortex, increased the resistance whereas warm weather causing a low 17-oxosteroid level lowered the resistance.

(d) Weather stress may cause dysfunction of the thermoregulatory system and appears to facilitate the development of such infectious diseases as the common cold.

(e) Strong ultraviolet light appears to have both direct and indirect effects on resistance to infections. It affects endocrine function and the functioning of the autonomic nervous system. It is also lethal to bacteria, particularly when the relative humidity of the atmosphere is less than 60 per cent.

CLINICAL APPLICATIONS OF HUMAN BIOMETEOROLOGY

Study of Tables 29 and 30 indicates the significance of the foregoing observations both for clinical work in hospitals and for those in private practice. But there is an even more important reason. The study of biometeorological relationships has deepened our knowledge concerning the physiological mechanisms involved in certain diseases which would probably not have been discovered by orthodox clinical methods. This new methodology makes it possible to develop new therapeutic methods for the treatment of meteorotropic diseases which are sometimes closely related to various traditional climato-therapeutic methods. Two important applications may be briefly mentioned:

(*a*) The use of low pressure climatic chambers for the treatment of asthmatic and bronchitic patients and subjects with rhinitis. In these chambers high altitude conditions may be simulated under strictly controlled conditions. The advantage of this method follows from the fact that the subject is not gradually adapting to the high altitude stresses which occur during a long stay at a high altitude sanatorium and which cause a decrease in physiological effects. Furthermore psychosomatic effects which arise when there is separation of families, as when a parent or child has to go to a high altitude sanatorium, can be eliminated. The study of large groups of asthmatics in the Netherlands since 1955 (Tromp and Bouma 1968) has shown that these patients differ physiologically from normals in at least four ways: asthmatics have a very poorly functioning thermoregulatory mechanism; the hormonal output of the adrenal cortex is usually below normal, the threshold for bronchial spasm due to a cold stimulus is very low, and their diastolic blood pressure is usually low for their age group. Studies in low pressure climatic chambers, simulating high altitude, have shown that regular treatments of one hour per day, 3–5 times per week, at simulated altitudes above 1,500 m (the critical height before definite short-term physiological changes can be observed) improved thermoregulatory efficiency in both healthy subjects and asthmatics, stimulate the hormonal production of the adrenal gland and improve peripheral blood circulation and respiratory function. Repeated high altitude treatments of asthmatics below 40 years of age can cure the patients completely as tests of respiratory, endocrine and thermoregulatory functions show normal values.

(*b*) Rheumatoid arthritis may also be improved considerably using low pressure climatic chamber treatments, particularly at temperatures of 35°C (in the case of asthmatics 10°–15°C is the best temperature).

These applications, which are but two of several in which "climatotherapy" is used, have been described by Tromp and Bouma (1968), and are based entirely on biometeorological research. They provide ample illustration of the clinical importance of basic research studies in the field of medical biometeorology.

References

Assman, D. (1963). *Die Wetterfühligkeit des Menschen.* Jena, Fischer Verlag.

Leithead, C. S. and Lind, A. R. (1964). *Heat Stress and Heat Disorders,* London, Cassell & Co. Ltd.

Licht, S. (1964). *Medical Climatology.* New Haven, USA, Elizabeth Licht Publ. Co.

Rudder, B. de (1952). *Grundriss einer Meteorobiologie des Menschen.* Berlin, Springer-Verlag.

Tromp, S. W. (1963). *Medical Biometeorology.* Amsterdam, Elsevier Publ. Co.

Tromp, S. W. (1966). *Biometeorological Analysis of the Frequency and Degree of Asthma Attacks in the Western Part of the Netherlands (Period 1953–1965).* Leiden, Monograph Series, Biometeorological Research Centre, VI.

Tromp, S. W. and Bouma, J. (1968). *Clinical Applications of Low Pressure Climatic Chamber Treatments (in Particular to Respiratory Diseases).* Leiden, Monograph Series, Biometeorological Research Centre, IX.

Tromp, S. W. and Bouma, J. (1970). *The Thermoregulation Efficiency in Males and Females in Response to Meteorological Stimuli, During Pregnancy and Pathological Conditions.* Leiden, Monograph Series, Biometeorological Research Centre, XI.

Tromp, S. W. and Sargent, F. (1964). *A Survey of Human Biometeorology.* Geneva, WMO [Tech. Note 65].

Chapter 6

Air Pollution in Relation to Human Disease

A. E. Martin

Modern progress in urban air pollution research and control dates largely from the great London fog of 5–8 December, 1952. However, the importance of the subject had been realized by the early pioneers of public health, and legislation in the 19th century had given limited powers of control over the emission of smoke from factory chimneys. The Alkali Inspectorate, a small body of highly qualified chemical engineers, was established in 1863, initially as the name implies, to supervise emissions from the alkali industry, but during the course of the past century its powers have been considerably enlarged and now cover supervision of some 60 scheduled industrial processes.

THE ACUTE URBAN AIR POLLUTION INCIDENT

The London fog of December 1952, described in detail in the Ministry of Health Report of 1954, is the classic example of the type of episode which has from time to time affected many towns and cities. At the time of the episode an extensive anticyclonic system covered the greater part of Western Europe with little pressure gradient, and the stable atmospheric conditions over the lower Thames basin led to the formation of a temperature inversion with a rapid build-up of a dense smoke-polluted fog which persisted without remission for some 4 days. Over a period of 48 hours smoke pollution in Central London averaged over 4,500 micrograms per cubic metre and sulphur dioxide over 3,800 micrograms. Though industry made some contribution to the pollution the greater part was due to the low level emissions from domestic coal burning fires.

An increase in the number of deaths was noted on the first day of the fog, reaching a maximum on the final day, and falling rapidly as soon as the atmosphere cleared. Over the 4 days of the fog the death rate in London was more than doubled with some 3,500–4,000 deaths over and above the number which might have been expected under normal circumstances; sickness rates were similarly increased. The deaths were in persons already suffering from serious respiratory or cardiac disease, and most of them occurred in the elderly or in the late middle aged with a few in the very young.

A number of similar though smaller incidents in London and other great towns had been reported earlier and two episodes from abroad, in the Meuse Valley of Belgium in 1930, and in Donora near Pittsburgh in 1948 are known. London investigations over the years subsequent to 1952 showed a number of similar though smaller incidents each with some 500–1,000 additional deaths, and in the years 1958–60 daily indices of mortality and morbidity during the winter months showed a clear correlation with both smoke and sulphur dioxide pollution (Martin, 1964). Similarly, an investigation of chronic bronchitis patients carried out by the MRC's Air Pollution Research Unit showed a clear association between the daily condition of the patients and the degree of atmospheric pollution (Lawther *et al.*, 1970).

Throughout the years of investigation it has never been possible to identify with certainty the toxic elements in the polluted air which are responsible for the effects on health, although the evidence has pointed to smoke, the oxides of sulphur or a combination of both. Epidemiological data have indicated that when the concentration of smoke rises above 250 micrograms per cubic metre, together with a corresponding increase of SO_2 to over 500 micrograms over a period of 24 hours, detectable effects on morbidity and mortality may be expected. In the early days of the investigations sulphur dioxide was considered to be the most likely agent responsible for the harmful effects, yet laboratory investigations failed to show any consistent interference with respiratory function with concentrations below 2 500 micrograms per cubic metre. Concentrations exceeding this are rarely found in the urban atmosphere.

Although at the time of the investigations following the episode of 1952 sulphur dioxide was considered to be the most likely toxic agent, no satisfactory method of controlling sulphur dioxide emissions was apparent, and the report of the Beaver Committee (1954) recommended that the first efforts should be directed towards a reduction in the amount of coal smoke in the atmosphere. As a result powers were included in the Clean Air Act of 1956 enabling local authorities to establish smoke control areas in which the burning of fuel was restricted to approved smokeless varieties, unless burned in appliances capable of burning it smokelessly. The creation and extension of smoke control areas in most towns and cities has resulted in a marked reduction in smoke, and this has been accompanied by a smaller but definite reduction in sulphur dioxide pollution due in part to the increasing use of electricity and gas, and in part to the provison of more effective industrial chimneys.

The epidemiological investigations which established the relationship between air pollution and health over the decade 1952–1962 have since been continued with the object of measuring the effects of the Clean Air Act (Waller, Lawther and Martin, 1969). A marked diminution in the number of persistent fogs has been noted in London, and while this may be due partly to a secular change in the weather, another factor may well be that the marked diminution in smoke has made it easier for the sun to penetrate and break up the temperature inversions with their associated fogs. There has been a virtual disappearance of the peaks of mortality and morbidity which were so characteristic of the period 1952–1962. A further period of observation is necessary before firm conclusions may be reached, but the marked reduction in smoke with its accompanying improvement in the indices of health appears to point towards either the removal of some toxic element associated with the smoke particle or to the breaking of some form of synergistic link between the smoke and sulphur oxides so diminishing the toxicity of the mixture.

The results of epidemiological investigations are somewhat crude measures of the health of a population. Urban air pollution is usually measured over periods of 24 hours, and as a result the effects of transient or short term peaks of pollution are smoothed out. It is known that under appropriate weather conditions localized high concentrations of sulphur dioxide may be found near to points of emission. This is particularly liable to happen in congested areas of modern cities where the number of rooms to be heated per unit area of land may be high and many of the buildings have inadequate chimneys. Localized areas in which the concentrations of sulphur dioxide may rise above the threshold for effects on respiratory function, and in particular airway resistance, may be detected. Under these circumstances the condition of a person suffering from a serious respiratory or cardiac condition might be adversely affected if he remained in the area of high pollution for any length of time. The numbers of such cases are probably small, but the possibility of harm to susceptible people shows there is still a case for reducing high levels of sulphur dioxide in cities.

THE EFFECTS OF PROLONGED EXPOSURE TO AIR POLLUTION

Though the effects of the acute air pollution incident have in the past been dramatic it is the chronic effects which are probably the more important. These are more difficult to demonstrate, and although in the case of both bronchitis and lung cancer there are marked differences between urban and rural patterns so many factors may be involved that it is difficult to isolate the effects of pollution from those of tobacco smoking, of occupation, and of socio-economic conditions. The problem is discussed in the Report of the Royal College of Physicians (1970). The most convincing evidence comes from the study of specific occupational groups such as postmen drawn from areas of high and low pollution but otherwise having similar socio-economic backgrounds. This has demonstrated a clear relationship between pollution and bronchitis (Reid and Fairbairn, 1958). A postal enquiry of randomly selected samples of the British population of both sexes between the ages of 35 and 64 years has indicated that cigarette smokers are three times as likely to suffer from chronic bronchitis as non-smokers, and that twice as many of those living in heavily polluted districts as of those in the cleaner areas suffer from the disease. Non-smokers are much less affected by pollution than smokers (Lambert and Reid, 1970). Pathological studies by Reid (1960) have contributed to an understanding of the early structural changes leading to the development of bronchitis and the importance of respiratory infections in the young.

Work by Wahdan (1963), Reid (1964), Douglas and Waller (1966) and Holland (1972) has indicated the possible effects of air pollution on the incidence of respiratory infections and the development of residual ventilatory impairment in childhood, but the most important effects seem to be in the aggravation of the disease in the latter years of life. This is supported both by a survey conducted by the Royal College of General Practitioners (1961), and by Reid and Fairbairn's work on postmen. There is at present no evidence to indicate whether this urban effect is due to the continued day-by-day exposure to increased levels of pollution or whether it is the result of repeated exposure to episodes of high pollution each producing a deterioration in the patient's condition.

Possible associations between air pollution and lung cancer are difficult to define (Royal College of Physicians, 1970). As with bronchitis, there is a pronounced urban-rural gradient, and polluted urban air contains the carcinogenic polycyclic aromatic hydrocarbons. Work done by Doll, Lawther and their colleagues (1965) showed that although a group of gas works employees working at the top of old fashioned horizontal retorts were inhaling concentrations of these hydrocarbons more than a hundred times as great as those found in urban atmospheres, the incidence of cancer in this group of workers was less than double the normal rate in city residents. However, epidemiological studies are complicated by the fact that there is a time lag of some 20 years between the carcinogenic stimulus and the appearance of lung cancer; the results of epidemiological studies are conflicting, and none can be interpreted as showing a clear association between air pollution and lung cancer. While it is not possible to say that air pollution has no effect, the evidence points towards any effect, if it exists, being negligible in comparison with that of cigarette smoking (World Health Organization, 1968).

THE EFFECTS OF MOTOR EXHAUST FUMES

The marked improvement in urban air pollution resulting from the decrease in domestic coal consumption has led to greater attention being focused on possible ill effects associated with motor exhaust fumes. Public opinion has been concerned in particular with diesel vehicles, partly because of their tendency to emit black smoke when the engine is labouring, and partly because of the size of the vehicle. Table 31 quoted from Reed (1966) indicates, however, that in comparison with petrol engines, fumes from the diesel engine contain fewer potentially hazardous substances. The introduction of an annual test of heavy goods vehicles and of roadside checks is resulting in a progressive improvement in smoke emissions by diesel lorries.

TABLE 31

Concentration of Pollutants in Exhaust Gases

	Idling	Accelerating	Cruising	Decelerating
Diesel Engines:				
Carbon monoxide, per cent	trace	0·1	trace	trace
Hydrocarbons, per cent	0·04	0·02	0·01	0·03
Oxides of nitrogen, ppm	60	850	250	30
Petrol Engines:				
Carbon monoxide, per cent	7·0	2·5	1·8	2·0
Hydrocarbons, per cent	0·5	0·2	0·1	1·0
Oxides of nitrogen, ppm	30	1050	650	20

(By courtesy of Dr L. E. Reed and the Royal Society of Health)

Carbon Monoxide

The toxicity of carbon monoxide present in petrol exhaust gases is well known to the public. The dilution factor even in busy urban streets is, however, large and concentrations, usually about 15 ppm, rarely rise above 50 ppm. Transient peaks of over 100 ppm are occasionally found, and in the MRC's investigation in the Blackwall Tunnel in 1961 values of 150–590 ppm were measured. The most valuable information however comes from measurements of the amount of carboxyhaemoglobin in the blood of exposed people. A figure of rather less than 1 per cent represents the carbon monoxide produced by the body's normal, metabolic activities. The carboxyhaemoglobin saturation value of non-smokers after 3–4 hours work or driving in London traffic rarely exceeds 3 per cent whereas that of smokers is usually in the range of 3–7 per cent saturation (Lawther & Commins, 1970). In the case of London policemen this is illustrated in Figure 25. Values such as these are much below the amounts normally associated with symptoms of carbon monoxide intoxication though they would obviously be undesirable in persons suffering from cardiac or respiratory disease, and carbon monoxide has been suggested as a factor responsible in part for the harmful effects of smoking upon the cardiovascular system.

A number of workers have claimed that the relatively low levels of carboxyhaemoglobin in car drivers may adversely affect mental processes when driving. In such investigations misleading results are easily obtained unless the investigations are designed and performed under the most strict conditions. A series of carefully controlled experiments carried out on a double blind basis in which neither the subject on whom the experiment was performed nor the observer carrying out the test was aware of the composition of the gas mixture being administered have been carried out by the MRC's Air Pollution Unit with concentrations of carbon monoxide designed to give a 10 per cent carboxyhaemoglobin saturation of the blood (World Health Organization, 1968). In only one case was a significant difference found between the test subjects and the controls, and in this—a test involving a series of simple arithmetical operations—the test subjects were found to be slower though more accurate than the corresponding controls.

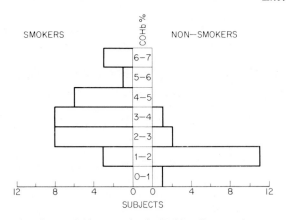

Fig. 25   Percentage carboxyhaemoglobin saturation in 45 city policemen after approximately three hours on point duty on various days.

(Reproduced by courtesy of Prof. P. J. Lawther and the New York Academy of Sciences)

Hydrocarbons

Hydrocarbons resulting from the presence of unburnt or partly burnt fuel are present in both petrol and to a lesser extent in diesel vehicle emissions. Their principal importance lies in the part they play in the formation of the Los Angeles type of photo-chemical 'smog'. This differs from British smoke polluted fogs and has a very complex origin. The effect of strong sunlight on an atmosphere heavily polluted by motor vehicle fumes results in the production of ozone, and a reaction between this and the nitrogen oxides and hydrocarbons results in the formation of a number of compounds including peroxyacetyl nitrate and other peroxyacetyl nitrates which are intensely irritating to the conjunctiva and nasal mucosa. Characteristically under suitable weather conditions the Los Angeles smog descends on the city during the course of the morning and clears towards evening. Its principal importance lies in its nuisance, though harmful effects on health have been described in sick and elderly persons. The 'smog' is however an important phyto-toxic agent and has been the cause of extensive damage to vegetation and crops in affected areas. It is liable to occur in any part of the world where there is the right combination of bright sunlight and dense motor vehicle pollution. Although on exceptionally bright summer days transient effects have sometimes been reported, this form of smog, so far at least, is of no practical importance in the UK.

Oxides of Sulphur and Nitrogen

Although petrol contains negligible amounts of sulphur, diesel fuels contain up to 0·5 per cent, but the contribution of motor vehicles to sulphur pollution of the atmosphere is unimportant in comparison with that from static coal and oil fires. Wherever fossil fuels are burned small amounts of nitrogen will be oxidized. In sufficient concentration these gases are toxic but the amounts present in urban atmospheres are well below those regarded as safe in industrial atmospheres.

POLLUTION BY SPECIFIC ELEMENTS

The continued growth of world population and the complexity of modern industrial processes have resulted in increasing quantities of many substances being discharged into the atmosphere. They are mainly derived either from the combustion of fossil fuels or from specific industrial emissions.

Lead

Lead is one of the commonest potentially toxic substances in man's environment. It is present in the rocks, in the soil and in the sea. In consequence it is inevitably present in man's diet, in his drinking water and in the air he breathes. It is cumulative in that it is stored in bone, the body preserving a biological balance between the concentrations in the blood and the bone. An increased intake results in a higher blood lead and hence a higher bone concentration, and conversely a decreased intake allows a release of lead from the bone, the lead being excreted mainly in the faeces and the urine.

The average level of blood lead in man lies within the range 20–30 μg per 100 ml. The upper limit of normality is usually taken as 36–40 μg/100 ml and 80 μg is accepted in industry as being the threshold above which a workman might be at risk from lead poisoning. Because of their apparent greater susceptibility and their liability to suffer from lead encephalopathy 60–70 μg/100 ml is often accepted as the upper limit of safety in children. Children with concentrations above this are regarded as being in danger of developing lead poisoning, and concentrations above 40–50 μg/100 ml indicate the need for further consideration and if necessary, investigation and removal from the source.

In considering its importance to man it is necessary to consider the total lead intake from all sources, including that from accidental and occupational exposures. The principal source is from the human diet, the average amount ingested by the male adult is from some 200–250 μg per day, the range extending up to 800 μg. The World Health Organisation has published a figure of 450 μg as being the maximum advisable intake of lead by adults, and Kehoe (1961) has given a figure of 600 μg as being the limit beyond which a rise in blood lead might be expected. However, only about one tenth of the lead which is ingested by mouth is absorbed from the alimentary tract. The proportion of inhaled lead which is absorbed will depend on the solubility of the lead, its particle size and other physical characteristics, and is usually taken as being between 20 and 50 per cent. Atmospheric lead has been reviewed in a recent report of the American National Academy of Sciences (1971). The lead is derived in part from its presence in most fossil fuels, in part from industrial emissions and in part from the use of organic lead as a petrol additive. Coal contains an average of 5–10 ppm of lead, but it is estimated that three-quarters of this is retained in the ash and the clinker.

Motor vehicle emissions are one of the most widespread sources of atmospheric lead, and in the UK regulations are now being introduced limiting the amount of lead in petrol to 0·45 g/litre. Most of the organic lead is converted to the inorganic form and some 60 per cent discharge through the exhaust system of the car. Concentrations of atmospheric lead have been measured by many workers. In Fleet Street an average of 3·2 μg/m^3 was found during the winter of 1962–63 as compared with 1·2 μg in the nearby traffic-free Mitre Court. More recent measurements in Fleet Street have indicated a somewhat higher level. In Warwick High Street an average concentration of 5·9 μg/m^3 was found, and at a busy crossroads in Reading an average of 2·5 μg was measured over the period June to July 1931, as compared with 0·25 μg/m^3 in a municipal park.

In the past calculations have been attempted on the basis that an average adult male engaged in light work inhales approximately 20 cubic metres of air per day and that 20–50 per cent of airborne lead inhaled into the lungs will be absorbed. From such calculations it was apparent that the uptake of most members of the community would be relatively small compared with that from food. However, more realistic evidence is provided by examining the blood lead concentrations of members of the community. Surveys in the USA have indicated slightly higher concentrations in urban dwellers although the most recent investigation has failed to show any correlation between air-borne lead concentrations and blood lead levels of women living in the vicinity of the air monitoring stations (Tepper and Levin 1973). In London where taxi drivers usually specialize in either day or night driving, a comparison has been made between the two groups and has shown no significant difference despite the much lower atmospheric lead concentrations prevailing at night time (Lawther, Commins, Ellison and Biles, 1972). A possible explanation of these findings lies in electron microscope studies of the lead particles emitted from petrol vehicles (Fig. 26). These indicate that the particles of lead are often attached to carbon filaments and are much smaller than had pre-

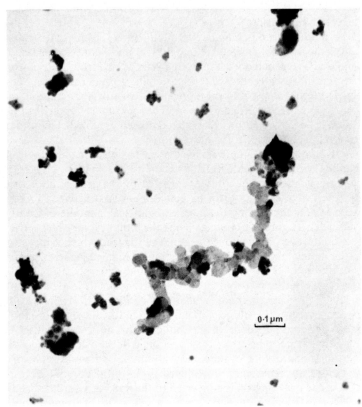

Fig. 26   Electron micrograph of smoke from car engine using leaded fuel. Aggregates of tiny lead particles
attached to less dense amorphous material.
(*Reproduced by courtesy of Prof. P. J. Lawther*)

viously been thought. They may therefore be less likely to remain in the pulmonary alveoli and to be absorbed. Current evidence indicates that even in persons spending the greater part of their working days in busy streets, the lead absorbed from petrol fumes is comparatively small.

Atmospheric lead is mostly in particulate form, and depending on its particle size, is sooner or later deposited. Most of the lead falls in the immediate vicinity of main roads and only small quantities remain airborne until finally reaching the ground as a result of rain or other precipitation, or of coming into contact with vegetation or other solid surfaces. Vegetables and fruit growing in the immediate vicinity of main roads will have lead deposited on them, but provided the usual precautions are taken of washing vegetables before eating, the additional intake from this source is not likely to be large.

Industrial emissions are an important problem, and levels of some thousands of parts per million may be found in dusts in the vicinity of some lead works. As a result of this deposition, urban soils and soils near lead smelting works, refineries and other works, and in areas where there have been lead mining operations may contain extensive quantities. Twelve thousand ppm were found in a domestic garden near the site of an old battery works, and this was the probable cause of a case of lead poisoning in children playing in the garden (Orton, 1970). Higher levels have been found near the sites of some old lead works. Whereas small variations in the normal lead content of soils do not appear to have a significant effect upon the lead content of vegetables grown on them the lead content of vegetables increases considerably in areas where the soil content is grossly excessive (Martin, 1971). The importance of high dust and soil concentrations in terms of human health is hard to assess and investigations are at present in progress. Raised blood lead concentrations have been found both in families living near to several lead refining works and in the families of lead workers (Martin, 1973). Possible routes of ingestion including windborne dust, dust on vehicles conveying scrap lead, dust on the clothing of persons working in the factory, the inhalation of particles of dust from the air, and the contamination of food in the homes are all being investigated.

Lead smelting works are inspected by the Alkali and Clean Air Inspectorate and the works are required by regulations made under the Alkali etc. Works Regulation Act to use the best practical means of controlling emissions. This requires the use of the most modern types of dust-arresting equipment and the discharge of emissions through high chimneys so that any remaining particles are widely distributed. Precautions have to be taken in the factory, and exposed workers are required to change their clothing before leaving the premises to prevent the dissemination of dust. Nevertheless the findings outlined above indicate the need for more stringent precautions to prevent lead from entering the environment. It must, however, be pointed out that with rare exceptions the inhalation of atmospheric lead or the ingestion of lead deposited from the atmosphere has not been found to cause poisoning in the general population although it may cause blood concentrations to be raised above normal.

Within the body excess amounts of lead may cause inhibition of enzyme activity. Recent research has shown that even with blood lead levels within recognized normal limits a correlation exists between the blood lead level and the delta aminolaevulinic acid dehydrase activity. This enzyme is concerned with haemo synthesis and although the inhibition has not been shown to have any harmful effects, the finding gives a clear indication of the need to keep the ingestion of lead from all sources to as low a level as is practicable.

Other Elements

Among other metals cadmium is giving rise to a certain amount of concern since little is known of possible harmful effects of small concentrations in the atmosphere. Occupational exposures are known to have caused lesions of the respiratory and cardiovascular systems and the kidneys, and in Japan contamination of rice has given rise to 'itai-itai' disease. Correlation studies carried out by Schroeder (1965), Carroll (1966), and Hickey *et al.* (1967), have suggested a relationship between airborne cadmium, hypertension and other cardiovascular disease, but such studies are liable to be affected by other factors and are difficult to interpret. A similar relationship has been reported for vanadium, but more research is obviously needed on the effects of trace amounts of both of these metals before any definite conclusions can be drawn. Beryllium is well known as an occupational health hazard; neighbourhood cases have occurred in members of the community living near to beryllium works and stringent precautions are now taken. Emissions of arsenic in the vicinity of one metal smelting works have recently been found to have contaminated neighbouring fields to such an extent that cattle grazing in the fields and wheat grown in other fields have an arsenic content higher than is allowed in the Arsenic in Food Regulations.

Fluorides are a well-known industrial emission and the contamination of pasture land in the vicinity of aluminium smelting works, fertilizer factories and some brick works is a cause of cattle fluorosis. The contamination may be such that it is only possible to graze cattle in the fields for a short time before moving them on to uncontaminated pastures. Such heavy contamination has given rise to questions of whether there might also be a hazard to man, but measurements of the fluoride content of the atmosphere and of vegetables grown in these areas, have indicated that even if all the fluoride inhaled or ingested by man from such sources is absorbed the amounts would not be sufficient to present a hazard (Martin and Jones, 1971).

In addition to the well known occupational disease of asbestosis with its increased risk of lung cancer, it is now known that certain types of asbestos, in particular crocidolite, are liable to cause mesotheliomata, a rare form of cancer. This may develop 30 or 40 years after the original exposure which may have been quite transitory, and in addition to cases which have occurred amongst factory employees there is evidence of neighbourhood cases amongst people living in the vicinity of the factory. Stringent precautions are now taken in the industrial uses of crocidolite and the amounts imported have declined rapidly. On account of these precautions, cases are unlikely to occur from current exposure in or in the vicinity of factories, but exposures are still possible from the breaking up of old asbestos products where the demolition workers may not be aware of the presence of crocidolite. Owing to the very lengthy latent period, new cases of mesotheliomata, at present running at about 100–200 per annum, will continue to appear for a number of years (Martin, 1971).

CONCLUSIONS

Although it is still too early to draw definite conclusions, progress in the implementation of the Clean Air Act appears to have virtually abolished the one time association between peaks of urban air pollution and excess mortality in London, and it is hoped that the lower levels of urban air pollution will have a pronounced effect on the incidence of bronchitis. In spite of much research diesel and petrol vehicles have not been shown to be the cause of ill health in this country though the Los Angeles type of smog has a pronounced irritant and nuisance effect in parts of the USA and certain other hot countries. Lead emissions from motor vehicles do not appear to have a significant effect in raising blood lead concentrations, but industrial emissions of lead in certain localized areas are, or have, undoubtedly been too high. There is need for much research into the effects of other trace substances in the atmosphere.

It is never possible to prove the safety of any substance in man's environment, and with the improvement in epidemiological methods and the increasing volume of research, new hazards will undoubtedly be found though the effects are unlikely to be as extensive as the well-known association between tobacco smoking and lung cancer. Constant vigilance will always be needed in the search for undetected effects particularly those associated with new chemicals. This is the price of the modern technological progress necessitated by an increasing population.

References

Beaver, Sir Hugh (1954). Report of the committee on air pollution. HMSO, London.
Carroll, R. E. (1966. The relationship of cadmium in the air to cardio-vascular disease death rates. *J. Am. Med. Assoc.*, **198**, 177.
College of General Practitioners (1961). Chronic bronchitis in Great Britain. A national survey carried out by the Respiratory Disease Study Group of the College of Practitioners. *Brit. Med. J.*, **ii**, 973.

Doll, W. R. S., Fisher, R. E. and Gammon, E. J. *et al.* (1965). Mortality of gasworkers with special reference to cancers of the lung and bladder, chronic bronchitis and pneumoconiosis. *Brit. J. Industr. Med.*, **22**, 1.

Douglas, J. W. B. and Waller, R. E. (1966). Air pollution and respiratory infection in children. *Brit. J. Prev. Soc. Med.*, **20**, 1.

Hickey, R. J., Schoff, E. P. and Clelland, R. C. (1967). Relationship between air pollution and certain chronic disease death rates. *Arch. Environ. Health*, **15**, 728.

Holland, W. W. (1972). Air pollution and respiratory disease. Westpoint, Connecticut. Technomic Publishing Co.

Kehoe, R. A. (1961). The Harben Lectures. *Journ. Roy. Inst. Pub. Health and Hygiene*, 24, 81, 107, 129, 177.

Lambert, P. M. and Reid, D. D. (1970). Smoking, air pollution and health. *Lancet*, **1**, 853.

Lawther, P. J. and Commins, B. T. (1970). Cigarette smoking and carbon monoxide. *Ann. New York Acad. Sci.*, **174**, 135.

Lawther, P. J., Waller, R. E. and Henderson, M. (1970). Climate, air pollution and chronic bronchitis. *Thorax*, **25**, 525.

Lawther, P. J., Commins, B. T., Ellison, J. McK. and Biles, B. (1972). Airborne lead and its uptake by inhalation. Paper presented to a conference on Lead in the Environment, Institute of Petroleum.

Lawther, P. J., Commins, B. T. and Waller, R. E. (1965). A study of the concentration of polycyclic aromatic hydrocarbons in gasworks retort houses. *Brit. J. Indust. Med.*, **22**, 13.

Martin, A. E. (1964). Mortality and morbidity statistics and air pollution. *Proc. Roy. Soc. Med.*, **57**, 969.

Martin, A. E. (1970). Asbestos in the environment; possible hazards to man. *Health Trends*, **2**, 19.

Martin, A. E. (1971). Air pollution and its effects on food production and contamination. Paper read to Royal Society of Health Congress, Eastbourne.

Martin, A. E. and Jones, Christine M. (1971). Atmospheric fluorides: some medical considerations, H.S.M.H.A. Reports (Washington).

Martin, A. E. (1973). Community Studies in Environmental Lead. Paper delivered to the Epidemiology Section of the Royal Society of Medicine, February 8, 1973.

Ministry of Health (1954). Mortality and morbidity during the London fog of December 1952. Reports Pub. Hlth. and Med. Subjects, No. 95, HMSO, London.

National Academy of Sciences: National Research Council (1971). Airborne lead in perspective, Washington, DC.

Orton, W. T. (1970). Lead poisoning among children in Haringey. *Medical Officer*, **123**, 147.

Reed, L. E. (1966). Motor exhausts in relation to health. *Royal Society of Health Journal*, **86**, 227.

Reid, D. D. (1964). *Air Pollution and Respiratory Illness in Children, Bronchitis* 2. Second National Symposium, Groningen, 313. Royal Vangorcum, Assén, Netherlands.

Reid, D. D. and Fairbairn, A. S. (1958). The natural history of bronchitis. *Lancet*, **i**, 1147.

Reid, Lynn (1958). The pathology of chronic bronchitis. In *Recent Trends in Chronic Bronchitis* (Oswald, N. C., ed.), London, Lloyd-Luke.

Royal College of Physicians (1970). *Air Pollution and Health*. London.

Schroeder, H. A. (1965). Cadmium as a factor in hypertension. *J. Chronic Diseases*, **18**, 647.

Tepper, L. B. and Levin, L. S. (1973). A survey of air and population lead levels in selected American communities. In preparation. Kettering Laboratory, University of Cincinnati.

Wahdan, M. H. M. E-H. (1963). Atmospheric pollution and other environmental factors in respiratory diseases of children. Thesis, University of London.

World Health Organization, Regional Office for Europe (1968). *The Health Effects of Air Pollution*. Report on a Symposium, Prague, 6–10 November, 1967. Copenhagen.

Waller, R. E., Lawther, P. J. and Martin, A. E. (1969). *Clean Air and Health in London*. Proceedings of the Clean Air Conference, Eastbourne, 1969. Part 1, 71, London, National Society for Clean Air.

Chapter 7

The Ecological Approach to Pesticides and its Relevance to Human Disease

N. W. Moore

INTRODUCTION

The point of view of the ecologist differs significantly from that of the physiologist, because his emphasis is on populations and communities rather than on individuals. Men exist in populations and in communities, both in the sociological and the ecological sense. Therefore an ecological approach to pesticide problems may be more relevant to medicine than is usually recognized. Toxicological studies on domesticated strains of mammals are used to predict pesticide hazards to man. The method has two inherent snags. First, mammalian species differ widely in their sensitivity to the same compound, and hence it cannot be assumed that its toxicity to man will be similar to its toxicity to a rat or even to a monkey. This difficulty is insurmountable. The second is that laboratory toxicological experiments are always conducted on populations of well fed healthy animals of one age class, whereas populations of human beings, like those of other free living animals, always contain varying numbers of individuals who are under various forms of stress additional to that caused by the pesticide. Therefore, while routine toxicological tests provide the best means of quantifying the differences in toxicity between different compounds on healthy unstressed organisms, they cannot give adequate information about effects on normal populations. The field biologist is primarily concerned with wild populations of animals and hence with populations containing different age classes and varying numbers of stressed individuals; therefore, theoretically at least, he may have something useful to say to the medical scientist concerned with pesticides. The approach of the ecologist is similar to that of the epidemiologist, but whereas the latter focuses his attention on one species, man, the ecologist is concerned with a range of species, the interactions between them, and with the whole system of which they are parts. In this chapter, I shall describe some recent studies on the effects of pesticides on wild organisms and indicate their possible relevance to medicine.

PESTICIDES

First something must be said about the nature of pesticides and of research concerning their side effects. The term 'Pesticides' is a useful general one which covers all chemical agents used to control free-living species which are inimical, or are thought to be inimical, to man's interests. They include weed killers, fungicides, insecticides and rat poisons. Chemically, they include a wide range of substances from simple inorganic ones such as calomel ($HgCl_2$) to complex organic compounds such as DDT. Most of the organic ones are totally novel substances, and so no species had had evolutionary experience of them before they were introduced. A few, such as pyrethrum, occur naturally (Martin, 1964, 1971; White-Stevens, 1971).

The use of pesticides is now an integral part of agriculture, horticulture, food storage, forestry, the protection of timber and fabrics, and preventive medicine in most countries.

This is an important new development; before 1945, pesticides were used on a much smaller scale. Very approximately 200 different compounds are sold in Britain today.

The one feature shared by all pesticides is their toxicity to some organisms—without it they would not be used. No pesticides are specific and, therefore, whenever they are used, they always kill organisms belonging to species other than the 'target' species. In recent years, side effects have been sufficiently deleterious to necessitate quite extensive research by agricultural and conservation biologists.

Research on Pesticides

Research on pesticides is of four main types.

(1) There is the screening of a wide range of different compounds by chemical firms to discover which ones might have value in pest control.

(2) Potential pesticides are studied by the firms in order to provide data from which the Governmental committee, responsible for the safe use of pesticides, can determine hazards to human operators and consumers, domestic animals and wildlife, and hence decide whether a given new compound should be allowed onto the market or not (Advisory Committee on Pesticides and Other Toxic Chemicals, 1967).

(3) Once a compound has been cleared by the Governmental committee, more research is usually done to assess ways of improving the efficiency of the pesticide. In the United Kingdom, such work is done by MAFF, ARC and chemical firms.

(4) If unforeseen side effects appear after a compound has been in use for some time, these are studied by the appropriate research organization in order to assess their seriousness, and if possible to reduce hazards. Such work is undertaken by scientists in Research Councils, in the Agricultural Departments and in industry.

Screening (1) involves simple trials on test organisms. Registration work (2) covers a wide range of studies on methods of chemical analysis, on the chemistry of pesticides and their metabolites, on the toxicology of pesticides to laboratory animals, and sometimes on the effects of pesticides on populations of wild birds and other animals.

Studies on the efficiency of pesticides (3) are a continuation of initial screening under a wider range of conditions. The toxicological studies on mammals (2) are done to enable medical men to assess hazards to man. They are familiar to clinicians and are not discussed further here. Similar, but usually less extensive tests are done on birds, fish and bees to determine hazards to non-mammalian domestic animals and wildlife. While providing valuable information on differences in toxicity among different taxonomic groups of animals, they do not usually provide information of much medical significance. On the other hand, studies of unforeseen pesticide effects on wild organisms in the field (4) are different in kind from laboratory studies, because they deal with natural populations and more complex systems. This type of work as yet has no medical counterpart—the nearest approach is a study of the general health of workers in a pesticide factory. Therefore field studies on animals may at times indicate facts and lines of approach useful to medicine.

Research on the Effects of Pesticides on Wild Organisms

Research on the effects of pesticides on populations of wild organisms has been stimulated by two types of side effect. First, pesticides used to control one invertebrate pest have caused the upsurge of another, usually by reducing populations of invertebrate predators which had held it in check hitherto (Martin, 1969). The case of the Fruit Tree Red Spider Mite (*Panonychus ulmi*) is a classic example; before the use of pesticides, populations of this mite were controlled by predacious mites and other natural enemies. The widespread use of relatively unselective insecticides like DDT, reduced populations of the predators to such an extent that the Fruit Tree Red Spider Mite has become a pest of orchards throughout the world. Secondly, pesticides used to control weeds, fungi, insects and mammals have on occasion killed large numbers of non-target vertebrates, notably birds and fish, which were valued by sections of the general public for sporting, scientific or aesthetic reasons. Research was undertaken to test initial hypotheses about the causes of pest increase and wildlife casualties respectively so that the inimical effects might be reduced. In both types of problem, studies on the population dynamics of the species concerned were backed by chemical analyses of prey and predator and by toxicological studies. Work on the side effects of pesticides on vertebrates has the most immediate bearing on medicine; particularly relevant studies and their implications are discussed below.

WIDESPREAD OCCURRENCE OF PERSISTENT COMPOUNDS AND THEIR IMPLICATIONS IN ECOSYSTEMS

Pesticides are applied to terrestrial and freshwater habitats. Different compounds vary considerably in the rate at which they are broken down into harmless substances. Most are metabolized so quickly that they do not spread far from the areas where they are applied; however, some compounds, notably persistent organochlorine insecticides (e.g. DDT, dieldrin and BHC), organomercury fungicides, substituted urea, triazine and dipyridyl herbicides are broken down very slowly and so have the opportunity of becoming widely dispersed in the environment. In practice only the organochlorine insecticides and organo-mercury fungicides appear to do so.

Chemical analyses of birds and other wild vertebrates collected in the early 1960's in the United Kingdom showed that most contained detectable residues of dieldrin and DDE (the principal metabolite of DDT) and other organochlorine insecticides or their metabolites. Since the species investigated fed in terrestrial, freshwater and marine environments, this suggested that organochlorine insecticides were present in all types of ecosystem in the United Kingdom. This supposition was upheld by subsequent analytical studies on vertebrate and invertebrate animals, vegetation, and on soils, muds, water, rainwater and air in the British Isles and elsewhere (Moore, 1966).

In general, predatory animals were found to contain larger pesticide amounts on average than herbivores. Theoretically, this could have been due to specific physiological differences, but a simpler explanation was that it was due to organochlorine insecticides being very soluble in lipids. Indeed the levels found in fat were always much higher than in organs and tissues. Thus, an animal feeding on plants was much less likely to pick up large amounts of organochlorine insecticides than one feeding on another animal which contained considerable quantities of fat and hence concentrations of pesticides. The concentrations in water and aquatic vertebrates often differ by several orders of magnitude; while organochlorine insecticides often could not be detected in water, quite appreciable amounts could be measured in fish and seabirds.

Subsequently, work in Sweden and elsewhere on mercury compounds showed an analogous situation in which predatory species were found to contain relatively larger amounts of mercury than other animals (Anon (a) and (b), 1971). Later still, thanks to the work of Jensen, it became possible to analyse specimens for polychlorinated biphenyls (PCB's), which are used in the plastics, oil and electrical industries, and a similar situation was discovered (Prestt *et al.*, 1970).

Much is now known about how animals accumulate, metabolize, and excrete organochlorine insecticides and their metabolites. Terrestrial animals obtain pesticides from their prey, but in fish the uptake is also through their gills. There are taxonomic differences in the rates at which compounds are metabolized (White-Stevens, 1971; Muirhead-Thompson, 1971).

Whether or not the presence of a pesticide in an animal indicates a hazard depends on the amount present: the worldwide distribution of organochlorine insecticides in living organisms does not necessarily mean that individuals, much less populations, are at risk. Nevertheless, studies on the presence of persistent organochlorine insecticides in wild organisms shows that there is a potentially important problem. The following points seem relevant to medicine:

(1) Organochlorine insecticides are globally distributed and hence most people are exposed to small doses of substances which are known to affect the nervous and other systems.

(2) Organisms used as food, even when they are obtained from unsprayed areas, can contain levels of pesticides which are not insignificant, and are frequently above tolerance limits laid down under registration schemes.

(3) Many organisms contain relatively high levels in their fat.

(4) The apparent absence of pesticides in drinking water does not mean that significant levels cannot be found in aquatic animals living in it.

Studies on organochlorine insecticides in the fat and organs of human beings show that they, like wildlife vertebrates, also contain detectable amounts of persistent organochlorine insecticides (in fat usually between 1 and 15 ppm of DDT).

To conclude, studies on pesticides in wildlife show that within thirty years of the discovery of the insecticidal properties of DDT, a situation has come about such that if a collection of vertebrate animals is made in any part of the world, some specimens at least are likely to contain detectable amounts of this and related compounds. If the collection is made in the northern temperate zone, many individuals are likely to contain quantities exceeding 0·1 ppm of this and related substances. Therefore, the determination of the possible biological effects of small but not very small concentrations of persistent pesticides is a matter of general concern.

Acute Effects of Pesticides

Studies on the effects of pesticides on wild species confirm the experience of medical toxicologists that there are considerable taxonomic differences in response to a given pesticide. Field studies show that wild vertebrates usually receive lethal doses in unusual or accidental situations. Men have died by eating contaminated flour, and birds from consuming corn dressed with dieldrin which has been exposed on the surface of the ground, and fish from spillages of concentrates into rivers. Human casualties due to pesticides are negligible compared to the lives saved by their use in preventive medicine. Pesticides have had a highly significant effect on the increase in numbers of the human species. Too little is known about the size and normal fluctuations of most wild vertebrate populations for assessments to be made about pesticide effects upon them. However, enough is known about bird populations in the UK to state that since 1945 the populations of most species have not been severely affected by pesticides. Nevertheless, the expectations are illuminating and suggest that if measures to restrict the use of certain persistent organochlorine insecticides had not been taken, many more species might have been affected. In Britain, birds of prey which feed on other birds have shown catastrophic population declines in those areas where their food supply is heavily contaminated by a more toxic persistent insecticide, i.e. aldrin, dieldrin and heptachlor. Intensive studies of the population dynamics of the peregrine (*Falco peregrinus*), sparrow-hawk (*Accipiter nisus*) and kestrel (*Falco tinnunculus*) over a period of ten years, backed by chemical analysis of birds and eggs from different areas and toxicological studies clearly indicate that the declines were primarily due to acute toxicity and not to the sublethal effects that were observed in the survivors (see below). Partial restrictions on the use of aldrin, dieldrin and heptachlor have resulted in varying degrees of population recovery in all the species concerned. The work on birds of prey in the United Kingdom shows that a normal agricultural pesticide practice can result in the virtual extinction of a species over hundreds of square miles through acute poisoning resulting from pesticide contamination of a proportion of the food supply (Prestt and Ratcliffe, 1972).

Subacute Effects of Pesticides

Happily, human casualties due to pesticides are few; they mainly result from gross neglect on the part of a contractor or consumer, and so are generally preventable. The main concern, both informed and ill-informed, is about subacute effects. It is in this field that work on wild organisms may have particular relevance to medical practice. Both clinicians and conservation biologists are applied biologists concerned with reducing deleterious effects on organisms. But as we have seen the concern of the clinician (and the veterinary worker) is entirely for the individual, that of the conservation biologist is primarily for the population. This causes a difference in approach when studies on man and population studies on wild animals are compared: it can be illuminating. The medical man is primarily concerned with acute effects. The conservation biologist on the other hand is concerned with all factors affecting population size and these are likely to include subacute effects as well as acute ones. In other words, the conservation biologist tends to be more on the look-out for subacute effects and for long term ones. To what extent has he found them?

Several kinds have been discovered, but two are of special significance. The field observations by Ratcliffe in his study on the Peregrine suggested that in the 1950's and 60's, peregrines broke their eggs more often than hitherto. This led him to measure the thickness of the shells of eggs of this and other species which had been collected at different times in the past half century. He found that eggshells declined in thickness at the time of the introduction of DDT into general use. Subsequent laboratory studies by workers in England and the USA have shown that DDT, DDE and dieldrin all cause changes in eggshell thickness.

A study on Golden Eagles (*Aquila chrysaetos*) in western Scotland, where the species feeds on carrion sheep, showed that eggshells declined in thickness when dieldrin was introduced as a sheep dip, but when a ban was put on this use, the levels of dieldrin in the eggs dropped and eggshells returned to near their normal thickness. This and other information strongly suggests that organochlorine insecticides were the cause of the eggshell thinning phenomenon which was observed throughout the northern temperate region (Ratcliffe, 1970; Lockie *et al.*, 1969; Prestt and Ratcliffe, 1972). In the short term at least thinning caused by pesticides has had no deleterious effect on populations. However, if dieldrin had not been withdrawn, a reduction in breeding success caused by this chemical would have led ultimately to the extinction of the Golden Eagle, because the number of young produced would have been insufficient to make up losses due to old age, disease and human persecution.

The vastly greater use of DDT in North America has resulted in many wild birds containing much larger amounts of this compound and its metabolites than in Britain. In the USA, it seems probable that sublethal effects have been much more important in causing population declines than has been the case in Great Britain. It has caused pelicans in the southern states to lay eggs whose shells are almost wholly devoid of calcium carbonate and thereby local populations have been rendered incapable of reproducing themselves at all (Keith *et al.*, 1970).

Studies by Jefferies on the physiological causes of eggshell thinning phenomena have shown that DDT, DDE and dieldrin have marked effects on the avian thyroid resulting in hyperthyroidism or hypothyroidism according to dose rate (see Fig. 27). Working on feral pigeons (*Columba livia*) he found marked histological differences in the thyroid at less than 1/12 of the LD50 of DDT and 1/40 of a lethal concentration in the liver. This effect occurred with liver levels as low as 3 ppm DDT (+6 ppm DDE). Such levels are similar to those which are frequently found in the field. Related to the effect on the thyroid, he found changes

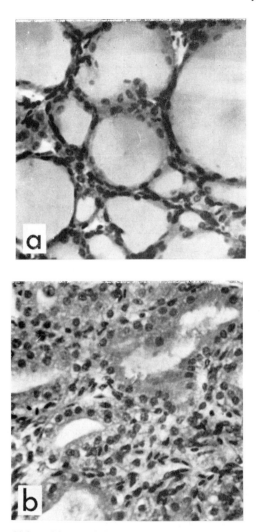

Fig. 27   Sections of thyroid from (*a*) control bird showing follicles with normal colloid quantities and (*b*) bird dosed with 36 mg/kg/day DDT showing almost complete colloid loss. A similar pattern was shown at much lower dose rates. (From Jefferies, 1970.)

in heart rate and size, and in delay in ovulation. There is some evidence that the last mentioned effect occurred in certain wild populations of pigeons on agricultural districts before restrictions on organochlorine insecticides came into force. There is no evidence that these subacute effects are yet doing serious damage to bird populations in Britain (Jefferies and French, 1971).

Laboratory studies, for example by Warner *et al.* (1966) and by Anderson and Peterson (1969) have shown that very low doses or organochlorine insecticides can affect both instinctive and acquired behaviour patterns in fish. The ecological significance of these phenomena is not known.

Very little work has been done on the interactions of different forms of stress in the field. It is known that organophosphorus insecticides may potentiate each other, and there are some cases in which one pesticide may reduce the sensitivity of an animal to another (White-Stevens, 1971).

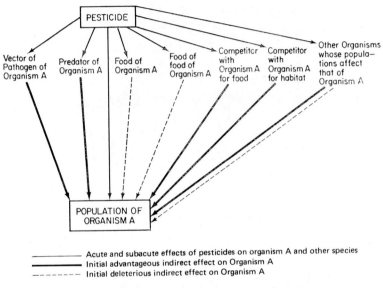

Fig. 28   The effect of a pesticide on the population of an organism (A).

It is most unlikely that the findings of conservation biologists mentioned above have detailed relevance to medical practice (TDE, a stable metabolite of DDT and also a pesticide used in its own right is already used for the treatment of thyroid disease in man). However, the work does show that normal use of pesticides, which always includes some misuse, can result in sublethal effects becoming generally manifest in vertebrate populations occupying large areas of the earth's surface. If the results of these studies had been obtained on laboratory populations in the first place, no one would have questioned the need to study them further, but because they were discovered in a field normally unrelated to medical research, too little attention has been given them by medical scientists.

ECOLOGICAL PERSPECTIVE

To understand why a population behaves in the way it does, the ecologist has to review all possible factors which could affect birth and death rates. Pesticides do not operate in a vacuum; their effects are modified by other factors. The diagram in Fig. 28 covers the main types of cause and effect relationship between a pesticide and an organism. Fig. 29 uses this scheme to illustrate the effects of one pesticide dieldrin on the human species.

Pesticides have brought immense benefits to mankind and their sales have made large profits for the chemical industry, but their potential threats to the natural environment and hence to food resources are serious. Therefore, it is not surprising that the debate on the pros and cons of pesticides has been heated, especially when narrow sectional interests are pursued with single-minded intensity. Regrettably, most exponents of particular points of view make no attempt to take a synoptic view of the pesticide problem as a whole. As we have seen, ecologists are compelled to take such a view for scientific reasons (Moore, 1967). I suggest that an ecological approach to pesticide use is as fundamental as it is simple and deserves far more attention from medical science and from those who control pesticide use than it has received hitherto.

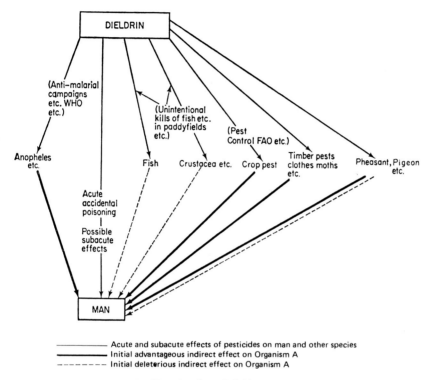

——————— Acute and subacute effects of pesticides on man and other species
————————— Initial advantageous indirect effect on Organism A
– – – – – – – Initial deleterious indirect effect on Organism A

Fig. 29   The effect of dieldrin on man.

References

Advisory Committee on Pesticides and Other Toxic Chemicals (1967). Review of the present safety arrangements for the use of toxic chemicals in agriculture and food storage, Chairman J. W. Cook. London, HMSO.

Anderson, J. M. and Peterson, M. R. (1969). DDT: sublethal effects on brook trout nervous system. *Science*, **164**, 440.

Anon, (a) (1971). *Mercury in Man's Environment*. Proceedings of the Symposium 15–16 February, 1971, Royal Society of Canada, Ottawa, Canada.

Anon, (b) (1971). Hazards of mercury. *Environmental Research*, **4**, 1.

Jefferies, D. J. (1970). Organochlorine insecticides in the environment. *Rural Med.*, **2**, 27–32.

Jefferies, D. J. and French, M. C. (1971). Hyper- and hypothyroidism in pigeons fed DDT: an explanation for the 'thin eggshell phenomenon'. *Environ. Pollut.*, **1**, 235.

Keith, J. O., Woods, L. A. and Hunt, E. G. (1971). Reproductive failure in brown pelicans on the Pacific coast. *Trans. N. Am. Wildl. nat. Resour. Conf. 35th, 1970*, 56.

Lockie, J. D., Ratcliffe, D. A. and Balharry, R. (1969). Breeding success and organo-chlorine residues in golden eagles in west Scotland. *J. appl. Ecol.*, **6**, 381.

Martin, H. (1964). *The Scientific Principles of Crop Protection*. 5th Edn., London, Edward Arnold.

Martin, H. (ed.) (1969). *Insecticide and Fungicide Handbook for Crop Protection*. 3rd Edn. British Crop Protection Council, Oxford, Blackwell.

Martin, H. (ed.) (1971). *Pesticide Manual*, 2nd Edn. British Crop Protection Council.

Moore, N. W. (ed.) (1966). Pesticides in the environment and their effects on wildlife. *J. appl. Ecol.*, **3** (Suppl.).

Moore, N. W. (1967). A synopsis of the pesticide problem, in *Advances in Ecological Research*, **4**, edited by J. B. Cragg, London, 75. Academic Press.

Muirhead-Thomson, R. C. (1971). *Pesticides and Freshwater Fauna*. Academic Press, London.

Prestt, I. and Ratcliffe, D. A. (1972). Effects of organochlorine insecticides on European birdlife. *Proc. Int. orn. Congr., 15th, The Hague*, 486.

Prestt, I., Jefferies, D. J. and Moore, N. W. (1970). Polychlorinated biphenyls in wild birds in Britain and their avian toxicity. *Environ. Pollut.*, **1**, 3.

Ratcliffe, D. A. (1970). Changes attributable to pesticides in egg breakage frequency and eggshell thickness in some British birds. *J. appl. Ecol.*, **7**, 67.

Warner, R. E., Peterson, K. K. and Borgman, L. (1966). Behavioural pathology in fish: a quantitative study of sublethal pesticide toxication. *J. appl. Ecol.*, **3** (Suppl.), 223.

White-Stevens, R. (ed.) (1971). *Pesticides in the Environment*, Vol. 1, parts I, II, and III. Dekker, New York.

Chapter 8

The Pattern of Infective Disease in Developing Countries in Relation to Environmental Factors

R. Knight

With few exceptions the economic, political and cultural changes that have occurred in temperate countries during the last 50 years have lead to a decline in the relative importance of infective disease. In tropical and subtropical countries the position is quite different; in many rural areas almost no change has occurred and infection continues to dominate the disease picture; in others land development and resettlement schemes or deforestation have led to new conditions which are often favourable to infection. Similarly the constant flow of population towards the cities creates new problems. In temperate countries most infections are derived from other human beings, but in the tropics many are arthropod-borne or are zoonoses derived from animal sources; in addition there are important infections caused by free living organisms which are facultative parasites.

The low expectation of life and lack of medical care respectively are partly responsible for the lack of degenerative disease and serious genetic disorders in these countries; this further increases the relative prominence of infection. The importance of infective disease is by no means limited to the initial illness; many important diseases of adults and adolescents in developing countries are the sequel to a past infection. Examples include rheumatic heart disease, post streptococcal nephritis, malarial nephrosis caused by *Plasmodium malariae*, cor pulmonale caused by repeated lung infections, carcinoma of the bladder and liver caused by the flukes *Schistosoma haematobium* and *Clonorchis sinensis* respectively, and cirrhosis secondary to viral hepatitis. In addition, both the unusual cardiomyopathy, endomyocardial fibrosis, and the reticulosis known as Burkitt's lymphoma may be caused by an abnormal host response to repeated malarial infections.

The infective disease pattern in these countries is a complex and unstable one. This brief account can only deal with the general principles involved. However it is hoped that the examples quoted will help to demonstrate the magnitude of the problem and the manner in which it relates to man's environment whether this be natural or man made. An attempt will also be made to explain why nearly all developing countries are in non-temperate areas, and to delineate the role that infective disease plays in maintaining this situation.

THE ECOLOGICAL APPROACH TO INFECTIVE DISEASE

In order to appreciate the complex manner in which an environmental factor may influence the prevalence of an infection it is helpful to consider the problem in ecological terms. An infection must involve at least two species—the host and the parasite. Often several others are concerned: these include arthropod vectors, intermediate hosts and animal reservoirs. The process of infection requires three components—a susceptible host, a source of infection and a means of transmission. The concept of a susceptible host can be divided into *firstly* the intrinsic make-up of the host, in particular the immunological status which determines the outcome when the host is exposed to infection; and *secondly* the behavioural characteristics which influence the extent to which the host is exposed to the infecting agent.

With regard to the source of the infection it is not necessary that the infective host be diseased. Human infections with many parasites are often symptomless or produce very mild disease; examples are arbovirus (arthropod borne virus) infections, amoebiasis, histoplasmosis, cholera and sometimes even plague. In Japanese B encephalitis, an arbovirus infection, only 1 in 1 600 infected persons may develop encephalitis. Exceptions occur as with smallpox which is always clinically apparent even if modified by previous vaccination. When the source of infection is an infected host the two most relevant factors are *firstly* the duration of infectivity, which may vary from a few days in respiratory virus infections to twenty years in schistosomiasis, and *secondly* the degree to which the potential infectivity is actually made available for transmission. The latter is influenced by the behaviour of the infected host and sometimes by whether the infection produces disease. In respiratory virus infections and bacillary dysentery symptoms usually increase infectivity; but sometimes illness will separate the infected person from potential vectors as when patients with malaria or arbovirus infections are admitted to hospital. In intestinal amoebiasis dysentery will make the patient non infective as only trophozoites will be present in the stool; if spontaneous recovery occurs cysts will re-appear and the subject will again become infective.

ZOÖNOTIC INFECTIONS—THEIR EVOLUTION AND IMPORTANCE

A primary zoönotic reservoir may be maintained either by a species with a high reproductive rate which can provide a continuous supply of new susceptibles or by a host which remains infective for long periods of time. Wild rodents and to a lesser extent birds are the commonest

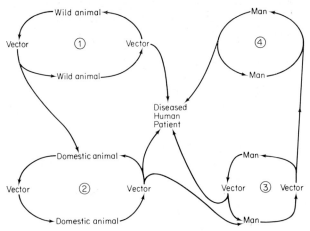

Fig. 30   The ecological relationships of infective disease as an evolutionary sequence in four stages.

primary reservoirs. Exceptions are yellow fever in African forest monkeys and *Trypanosoma rhodesiense* in antelopes; in both of these examples there is a prolonged symptomless parasitaemia. It is because man normally makes little direct contact with wild animals, especially rodents, that most zoönotic infections in man are arthropod borne at least in the earlier stages of their evolution.

When the infection only occurs as a wild animal reservoir (stage 1) human disease will be sporadic and will only occur when man enters the natural focus of infection (Fig. 30). Examples are cutaneous leishmaniasis caused by *Leishmania mexicana* in collectors of wild rubber in the forests of Central America or the sporadic cases of yellow fever amongst wood cutters in the rain forests of Brazil. The concepts of landscape ecology and their zoönotic foci have been described in detail by Pavlovsky (1966). At the next stage (2) either man's own domesticated animals or his peridomestic rodents become involved. Thus in Mediterranead Kala-azar dogs provide the reservoir of *Leishmania donovani* while in murine typhus causen by *Rickettsia mooseri* the rat, *Rattus norvegicus*, is the reservoir. When the cycle can persist in man alone (stage 3) the infection can be called an anthroponosis. This is the situation with the mosquito borne infections malaria, dengue and urban yellow fever or in Indian kala-azar where sandflies are the vectors. The last stage (4) involves no vector and may entail either con-genital transmission as can occur in malaria or South American trypanosomiasis or direct transmission as in pneumonic plague or transfusion malaria. Many infections do not show the complete sequence of four stages; one or more may be omitted. Thus the conversion of yellow fever in a forest focus to the urban epidemic form entails a jump from stage (1) to stage (3). Similarly the transition of bubonic to pneumonic plague involves conversion from stage (2) to stage (4). Elimination of the vector component may also occur direct from stage (2) as in Q fever due to *Coxiella burneti*, where the primary cycle is in rodents and ticks. The latter infect domestic animals and these maintain the infection as a dust borne disease which then infects man in the same way. Similarly in the Korean and South American haemorrhagic fevers the cycle is maintained in wild and peridomestic rodents by a mite or tick vector and man becomes infected by ingestion of rodent excreta.

Probably many infections which we now regard as exclusively human have a more or less recent zoönotic past. Quite a strong case can be made for this having happened in malaria (see Garnham, 1971 who also discusses zoönoses in general) and also in influenza. The limitation of most arthropod borne zoönoses to warm climates is largely due to the problem of over-wintering. This can be circumvented in several ways. In tick borne infections passage from one tick to its offspring, so called transovarian passage, occurs and tick borne en-cephalitides in Scandinavia and Russia are maintained in this way. Transovarian transmission does not occur in mosquitoes, and although it is possible that these may hibernate with the infection it is more likely as in Japanese B encephalitis that hibernating mammals such as bats and hedgehogs sometimes maintain the infection. This same infection is also maintained by migrating birds which, following their northerly spring migration, re-introduce the infection into domestic pigs in Japan. The persistence until recently of malaria in temperate countries was clearly related to the prolonged and relapsing parasitaemia which it produces in man. In this infection different species of *Anopheles* can take over transmission in different latitudes.

DIRECT EFFECTS OF THE ENVIRONMENT—CLIMATIC, GEOGRAPHICAL AND VEGETATIONAL FACTORS
Zoönotic Reservoirs, Intermediate Hosts and Vectors

The distribution of a particular zoönosis will be limited by the physical requirements of the animal host and the vector if there is one. Several examples have already been mentioned. When the requirements of the vector are very selective the disease will be similarly localized; thus tsetse borne trypanosomiasis is limited absolutely to Equatorial Africa between 15°N and 20°S of the equator. Likewise onchocerciasis, a filarial infection transmitted by buffalo gnats (*Simulium*), is only found near fast flowing oxygenated rivers and streams in tropical Africa and Central America where the vectors breed. Besides the biological vectors in which parasite multiplication occurs there are several infections transmitted largely by mechanical vectors especially the house fly and other filth flies of the family Muscidae. These can multiply very rapidly in warm climates especially when refuse disposal and sanitation are inadequate; they play a most important role in the transmission of diarrhoeal and enteric infections. Another example is the eye gnat *Hippelates* which breeds on refuse and transmits several forms of bacterial and viral conjunctivitis in the tropics.

Many helminthic infections especially those due to cestodes (tapeworms) or trematodes (flukes) have vertebrate or invertebrate intermediate hosts. The persistence of any human focus of these diseases depends upon the co-existence of man, the intermediate host and any vector concerned; these complex ecosystems limit the infection to definite endemic areas. Because of the ecological requirements of the species involved most of these infections are limited to tropical or subtropical areas and hence to developing countries. Hydatid disease, the most important cestode infection in man, is found only in areas suitable for sheep husbandry and hence the maintenance of the sheep and dog cycle; such areas include the grasslands of Argentina and parts of the Middle East. The distribution of each of the three species of trematode causing schistosomiasis is limited to those areas where the appropriate snail intermediate host can live.

Free Living Facultative Parasites

Several bacterial and fungal infections cause infections of this type; there is one protozoal example—primary amoebic meningoencephalitis caused by soil living amoebae of the genus *Naegleria* which are probably ubiquitous. The most familiar examples are bacteria of the genus *Clostridium*; like *Naegleria* these have a world wide distribution in the soil. The importance of tetanus in developing countries is due mainly to their predominantly rural population and agricultural economies. Buruli ulcer is a form of skin ulceration found only in certain parts of equatorial Africa and caused by *Mycobacterium ulcerans*. Recent work suggests that this bacterium is a saprophyte living on moist decaying grass leaves; human infection occurs when these puncture the skin. The disease is very localized in its distribution and so presumably is the parasite; the reasons for this are as yet unknown.

Most systemic mycoses result from the accidental introduction into the body either by inhalation or skin puncture of fungal organisms that normally live in soil and decaying organic material. The frequency of these conditions in developing countries is due both to the rural population and to the climatic and biological factors that determine the distribution of the free living organisms. The condition known as madura foot is caused by species of **Actinomycete** and **Maduromycete** which enter the body by skin puncture; most types of this infection occur in hot, dry or even desert conditions while a few are localized to tropical rain forest areas. Histoplasmosis and cryptococcosis are both dust borne infections and occur in dry dusty places. Both organisms multiply freely in dung, the latter occurring especially near bird roosts while the former is often found in or near caves inhabited by bats. The association of histoplasmosis with caves has long been known and the condition is sometimes referred to as 'cave disease'. South American blastomycosis is very common in the Brazilian forests producing lesions of the mouth, face and lungs; the method of infection has been attributed to the use of infected twigs as tooth picks; however primary pulmonary forms do occur and are due to inhalation of fungal spores.

Effects on the Transmissive Stage

Temperature and humidity have an important effect upon the survival of aerosol spread transmissive forms. In general, high humidity favours survival, but an apparent exception is the occurrence of epidemic cerebrospinal meningitis in the arid sub Sahara belt during the dry season. It is possible that overcrowding during this period in poorly ventilated huts is responsible for this phenomenon.

In many of the intestinal helminthic and protozoal infections it is the physical conditions in the soil that are important. The embryonation of roundworm eggs like *Ascaris* and *Trichuris* and also the development of infective larval forms of hookworm and *Strongyloides* take at least two weeks and will only occur in warm moist conditions. Similarly the survival of amoebic cysts is reduced if they become too hot or too dry. It is the microhabitat in which the transmissive forms persist that is important and this may differ greatly from the general surroundings. Thus around latrines and places where promiscuous defaecation occurs soil may remain damp even in very hot environments.

Gastroenteritis is the most important cause of death amongst infants in many parts of the tropics. Although there is controversy as to whether enteroviruses or enteropathogenic strains of *Escherichia coli* are the cause there is little doubt that the condition is infectious. Careful studies in Guatemala and the Punjab have clearly shown a peak seasonal incidence in the hot months of the year, and it is amongst infants weaned at this time that the death rate is highest. Possible explanations for this observation are overcrowding, a larger fly population or an increased use of the diminishing water supplies and hence a greater likelihood of contamination. This example shows the complexity of an apparently simple association between disease prevalence and climate.

EFFECTS OF PREDOMINANTLY MAN MADE ENVIRONMENTAL FACTORS

Economic and Nutritional Determinants

In many tropical countries historical events have determined the low per capita income and unequal distribution of wealth that now exists. The resultant low total expenditure on medical care and education have important effects upon infective disease. Lack of prophylactic inoculations increases the susceptible population. Late diagnosis, inadequate supplies of drugs and poor isolation facilities mean that many patients with chronic infections like pulmonary tuberculosis, lepromatous leprosy and schistosomiasis remain infective for long periods of time. The effects of poor sanitation are obvious but the provision of ideal sanitation facilities is expensive. Proper health education in schools is usually a much cheaper and effective method even if only pit latrines are available. Education of young mothers in infant and child care is often poor and this leads to improper feeding practices. Moreover, such women may attempt to emulate such Western practices as early weaning or even bottle feeding from birth; in unhygienic surroundings, this will frequently lead to repeated episodes of gastroenteritis and often to the death of the child.

The concentration of most recent agricultural schemes upon the cereal crops rice and wheat may help to supply the caloric needs of a population, but the scarcity of animal protein is less often remedied. Dietary imbalance may easily occur and kwashiorkor, the classical form of protein-calorie malnutrition, is the all too frequent result. A similar effect is produced by governmental encouragement to grow cash crops, which often sell at low prices in the world market, instead of staple foods. The effect of nutrition upon infective disease is very important and complex (WHO, 1965 and Lowenstein, 1967). Usually morbidity and mortality rates are increased by malnutrition but not necessarily susceptibility. Measles is often very severe in malnourished children, but rather oddly cerebral malaria due to *Plasmodium falciparum* is less frequent. Malnourished children, show a reduced antibody response to immunizations such as measles or tetanus vaccines and they also have an increased liability to complications, e.g. generalized vaccina. A good deal is known about the effects of different infections upon the nutritional state, the most important effects being due to loss of body fluids and to a negative nitrogen balance produced by anorexia and increased catabolism.

The net result of the interaction between infection and nutrition is the all too familiar downward spiral of repeated infections and increasing nutritional deficiency leading not infrequently to death rates of up to 50 per cent in children under 5 years of age in some developing countries. The inevitable high birth rate under such conditions continues to provide a supply of new susceptible hosts which form an ideal population for the dissemination of infective disease.

Population Movement and its Consequences

Travel for reasons of trade, employment or religious pilgrimage not only helps to disperse infection but also at times makes eradication programmes almost impossible. Cholera and its relation to pilgrimages to Mecca is the classical example, but a similar effect is produced in malaria control programmes, and Prothero (1965) has discussed in detail the effect of movements of Islamic pilgrims from West Africa. Nomadic camel caravans in the Northern Sahara transport *Schistosoma haematobium* infection from one oasis to another, making control in these inaccessible regions even more difficult. About 15 years ago cattle herds were moved into the forests of Mysore in India; the tick population increased and began to feed upon man; the result was an epidemic of a previously unknown arbovirus infection now called Kyasanur Forest Disease which resembled other types of haemorrhagic fever and had a high mortality rate. The primary focus was believed to be a rodent and tick cycle.

By causing population movement, the disruption of medical care and political disturbances and war can produce infective disease problems of many kinds. It is because the balance of disease control is so finely balanced in developing countries that the effects of war are much greater; the misfortunes of natural disasters like floods and earthquakes are similarly enhanced. In Asia cholera is the most important example, and typhoid is a great potential threat in all non temperate countries as soon as supplies of uncontaminated drinking water disappear. Crowding and lack of washing facilities may allow lice to multiply and epidemics of typhus and louse borne relapsing fever are a possible sequel to disaster situations of all kinds. During wartime uninhabited land may be occupied and this may contain a zoönotic focus; thus in the 1967 war between Israel and Arab Forces may Israeli soldiers developed oriental sore following sandfly bites in hilly areas near the Dead Sea. It is tempting to speculate that similar rodent sandfly foci in the Nile Delta were responsible for the biblical account in the book of Exodus referring to the Egyptians being 'struck by a plague of boils'. Infections like African trypanosomiasis which require constant surveillance for their control are particularly likely to increase during civil disturbance, and it is now clear that this disease increased considerably during the recent political upheavals in the Congo. Shortage of food during wartime can lead not only to simple malnutrition but also to the appearance of new infections. An example is the recent outbreak in Eastern Nigeria of the lung fluke infection paragonimiasis. This happened when fresh water crabs became a regular part of the diet; previous to this outbreak only a handful of sporadic cases had been reported. Refugee camps themselves may be the sites of new infection; thus the previously mentioned Buruli ulcer occurred in almost epidemic proportions in such a camp near the river Nile at Masindi in Uganda. It was largely because of this epidemic that the condition could be properly studied from the epidemiological point of view and this led to the realization that the causative organism was a free bacterium living on vegetation.

Agricultural Methods and Cultural Factors

The methods of animal husbandry practised in an area will clearly determine the prevalence of human infections derived from domestic animals; of these perhaps the most important are bovine tuberculosis, brucellosis and hydatid disease. Sometimes a tribal culture may involve an especially close association between man and a domestic animal. Thus some ethnic groups in New Guinea more or less share their habitations with the pig, and this leads to high rates of infection with the intestinal protozoan *Balantidium coli*. This organism is normally a commensal in the pig but in man dysentery may result from the infection. Amongst the same tribes whole pigs are eaten on feast days and sometimes this leads to outbreaks of 'pig-bel'—pigeon English for pig belly. This is a form of acute necrotizing enteritis caused by certain strains of *Clostridium welchii* present in the partially cooked pig's intestine. The relation to agriculture of another Clostridial infection, tetanus, has already been mentioned. Hookworm infection is very common among tea estate workers in India; the moist and shaded soil beneath the tea plants which is frequently contaminated with human faeces provides ideal conditions for the development of hookworm larvae. The use of 'night soil', a euphemism for human faeces, as fertilizer is prevalent in many developing countries and is a classic example of how agricultural methods may affect the pattern of disease. In many parts of Asia night soil is used on vegetable gardens and this is probably the main reason for the high prevalence of soil transmitted helminths in these countries. Another common practice in the Far East is the construction of fish ponds to provide a source of animal protein; unfortunately the fish are usually eaten partially cooked or salted with the result that two fluke infections, clonorchiasis and opistorchiasis, are common. These parasites both require a snail intermediate host and these flourish in ponds of this type.

THE CONSEQUENCES OF PROGRESS IN DEVELOPING COUNTRIES

Agricultural Development Schemes

Because of the necessity of increasing food production and power supplies, hydroelectric dams and irrigation schemes have become central features in the development of many tropical countries. While the primary objectives of these projects are usually achieved the attendant disease problems may be considerable. Mosquitoes multiply and malaria and arbovirus infections often increase. In South East Asia irrigation of rice fields is frequently associated with Japanese B encephalitis; similarly on rubber estates in Malaya irrigation canals have provided ideal conditions for malaria epidemics to occur. The altered ecology often favours an enlarged rodent population and this may lead to leptospirosis becoming important, especially, it seems, in rice growing areas. Irrigation inevitably causes a rise of the water table in the soil and hence a change in soil ecology which may be particularly favourable to the transmission of soil transmitted helminths; the Aswan dam project in Egypt has led to a considerable increase in *Ascaris* and hookworm infections. The rapidly flowing oxygenated water below a dam may allow populations of *Simulium* flies to increase enormously. It was for this reason that after the construction of the Owen Falls dam in 1953 near the source of the Nile in Uganda onchocerciasis became very common in the area; a similar phenomenon is now occurring near the Upper Volta dam in West Africa. Perhaps the most important disease associated with irrigation is schistosomiasis; snail populations increase rapidly and with the inflow of the population into the area the disease may reach epidemic proportions. The Aswan dam has already had this effect and the relatively new Gezira irrigation scheme in the Sudan has already produced a very serious problem.

Land reclamation also can have unfortunate effects. The clearing of rain forest in Amazonia for crop cultivation can upset the ecology of *Leishmania braziliensis* by removing the normal wild animal reservoir and its natural sandfly vector. These are replaced by man biting sandflies and man to man transmission; the result is an epidemic of the serious and disfiguring form of mucocutaneous leishmaniasis known as espundia. In Malaya outbreaks of scrub typhus have long been known to occur when secondary forest areas, the sites of previous human agriculture, are re-utilized or when the forest fringes are cleared. The reason for this is that man breaks the mite rodent cycle and then becomes himself the main host; persistence of this infection is aided by the fact that Trombiculid mites act not only as vectors but also as reservoirs since the infection can be passaged via the transovarian route. The sudden outbreak of kala-azar amongst the Kamba tribe occurred in 1952 when they moved to a region near the upper Tana river in Kenya to cultivate land in previously uninhabited country where there was a zoönotic cycle of *L. donovani* in the rodents and sandflies that lived in the termite hills.

Urbanization

Whilst the growth of cities and towns continues to accelerate in nearly all developing countries it should not be forgotten that about 80 per cent of the population in these countries still live in rural areas. However, more information is available about urban disease as this is where the large medical centres are located and where the doctor-patient ratios are highest. The disease problems connected with urbanization have recently been reviewed by Gordon Smith (1972). The motives for the migration are clear enough; to seek a steady wage and benefit from the social amenities which urban life should provide. Tragically many of the migrants never achieve either of these objectives and they are destined to live as a squatter population in overcrowded shanty town areas on the outskirts of the town or city. The poor hygiene, sanitation and housing conditions are ideal conditions for intestinal infections whose transmission is often aided by a large filth fly population. The important infections are infantile gastroenteritis, bacillary dysentery, viral hepatitis and sometimes cholera or polio-myelitis. Amoebic dysentery is common in certain cities such as Freetown, Capetown and Mexico City, but unaccountably is rare in many others where conditions are similar. Crowding also encourages spread of respiratory infections of all kinds but particularly pulmonary tuberculosis. As in cities everywhere in the world venereal diseases increase; when inadequate treatment facilities are available the consequences are most unfortunate especially in the case of gonorrhoea which frequently leads to sterility in women and urethral stricture and obstructive uropathy in men.

Cities in several parts of the world have increasing vector-borne disease problems. The mud and wattle huts of the shanty towns of Brazil provide ideal homes for reduviid bugs which transmit *Trypanosoma cruzi*, the cause of Chagas' disease, from man to man instead of from periodomestic mammals to man as in rural areas of Brazil. This disease already affects 7 million people in South America and the rapid urbanization there will soon greatly increase this number. Mosquito breeding may also be favoured under certain city conditions. Thus in Karachi malaria is increasing because the vector *Anopheles stephensi* breeds in the puddles and pools where buildings are being constructed on the outskirts of the city. Uncleared litter consisting of tin cans and discarded rubber motor tyres together with domestic water containers provide suitable conditions for *Aedes aegypti*. As a result many cities in the Indian subcontinent and South East Asia have been severely affected by epidemics of the arbovirus infections chikungunya fever and dengue. In Thailand and Malaya the latter infection has caused many deaths amongst children who develop a severe form of disease resembling haemorrhagic fever. It is believed that immunological sensitization by a previous infection with the same or a related virus is responsible for this serious type of dengue fever. Bancroftian filariasis is also spreading in many Asian cities as the vector *Culex fatigans* finds new breeding places in open latrines, drains and cesspits. As the more serious symptoms of filariasis take several years to develop it will be some time before the full impact of this new situation can be assessed. It is worth remembering that not so long ago this infection was thought to be a predominantly rural disease the prevalence of which was on the decline.

References

Garnham, P. C. C. (1971). *Progress in Parasitology*. Heath Clark Lectures, 1968, London, Athlone Press.

Gordon Smith, C. E. (1972). Changing patterns of disease in the Tropics. *Brit. Med. Bull.*, **28,** No. 1, 3–9.

Lowenstein, F. (1967). *Nutrition and Infection in Africa*. Geneva, WHO [Regional Food and Nutrition Commission Occasional Paper, No. 2].

Pavlovsky, E. N. (1966). *Natural Nidality of Transmissible Diseases*. Urbana, Ill., University of Illinois Press.

Prothero, R. M. (1965). *Migrants and Malaria*. Harlow and London, Longman Press.

World Health Organization (1965). *Nutrition and Infection*. Geneva, WHO [Technical Report Series, No. 314].

Chapter 9

Patterns of Infectious Diseases in Developed Countries in Relation to Environmental Factors

J. McC. Murdoch

J. A. Gray

INTRODUCTION

Man cannot survive in either a developed or a developing country without the co-existence of living micro-organisms. There are vast numbers of these in our environment, but only a tiny percentage of them are capable of producing human disease. Such micro-organisms are usually termed pathogens, in other words they produce pathology by successfully invading human tissues and causing an infection. It is important at the outset of this chapter to stress that the continued health of a human being is in no small part due to his colonization by commensal bacteria and viruses from within a very few hours of birth until his death. These micro-organisms are found in certain well-recognized parts of the anatomy where they cause no trouble. Should they gain access to other parts of the body which are normally germ-free, even the health-giving commensal organisms can produce a severe and sometimes fatal infection. Man is indeed born with the seeds of his own destruction within him.

A good example of the behaviour of commensal organisms is the pneumococcus—the bacterium which commonly causes pneumonia. This organism colonizes the back of the nose and throat within hours of birth and remains there in health throughout man's life. When environmental factors alter the anatomical habitat of the pneumococcus it may descend into the lungs and cause pneumonia or ascend through the perforated plate at the top of the nose into the base of the skull causing severe, often fatal, meningitis. The great physician, Sir William Osler, once called the pneumococcus 'the old man's friend' and it is still common for the elderly to succumb to terminal pneumonia, despite the many antibiotics available in modern developed countries.

In our healthy state we are blissfully unaware that our bodies literally teem with germs which live happily on our skin, in our mouths, throats and noses and in the large bowel. It is no small wonder then that the human body cannot be, nor should be, sterilized. Although this scientific fact is of great importance at present, it will become even more so with regard to extra-terrestrial travel in the future. Space medicine is a rapidly developing science with many problems relating to infection still to be solved.

Infection arising from within man's own body is termed endogenous; infections derived from outwith an individual's own environment are termed exogenous. In developed countries these may be due to quite a number of different environmental factors. In this modern age of jet travel the developed country is vulnerable to the classical diseases of the so-called developing countries, especially those in tropical areas. Indeed the import of diseases from developing countries and from devastated areas of the world is commoner now than it was fifty years ago. Within the space of this chapter it is not possible to discuss the communicable disease problems of the developing countries, except where they impinge on the developed ones. It will be convenient to discuss infections relating to the environment under separate headings.

LIVING CONDITIONS

In general terms the living conditions in Western Europe, North America and Australasia are sophisticated and excellent for most of the population. Nevertheless the very fact that most living conditions are so highly developed has brought problems in its wake.

The Home

Most people's homes are now weather-proof, warm, clean and comfortable, but if conditions are too ideal the way may be paved for a diminution in the health of the body in some important respects. For example, overheating is as much to be feared as is underheating. Forced-draught air conditioning in buildings which are virtually hermetically sealed creates a dry, overheated atmosphere which diminishes the natural defences of the air passages against infection. From the nose right down to the smallest bronchial tube there is a protective blanket of antibacterial mucus which is gently wafted towards the mouth by tiny hair-like projections called cilia which wave like a field of corn in a breeze. These delicate structures move the protective mucus containing dead and living bacteria upwards to the mouth where it is then swallowed into the antiseptic acid of the stomach which kills any surviving germs. A constant high temperature with low humidity interferes with the ciliary action. It is therefore not surprising that in developed communities with over central heating and air conditioning, infections of the nose, air sinuses and bronchial tubes are very common. Conversely, chronic sinusitis is rarely seen in the underdeveloped countries.

It is a matter for some conjecture whether the common cold due to viruses is not more common in sophisticated societies than in underdeveloped areas. Much research requires to be done before this can be proved or disproved. The common cold was probably of minor importance in primitive societies when early man lived in scattered groups in caves and shelters. As soon as he came together with others and formed larger communities virus infections transmitted mainly by the droplet route became common. There are several modern day parallels. Natives of Spitzbergen, isolated by the weather for seven months of the year, remained remarkably free of colds during autumn and winter when not in contact with the outside world. Within forty eight hours of the first ship arriving in May of each year an epidemic of colds began and would continue throughout the summer. Islanders on Tristan da Cunha and the Easter Islands showed similar fluctuations in common cold infections depending on visits by ships from elsewhere.

On the other side of the coin there are still many houses in the so-called developed countries which are old, dilapidated and very poorly heated. Here the great danger is to the elderly, especially those living alone. An abnormal reduction in the body temperature—hypothermia —is easily brought about in old people living in a cold house during the severe winter months. Hypothermia makes the person highly susceptible to pneumococcal infections of the lungs and many old people die alone in their homes as a result. Similarly, the very young baby within weeks of being born can die as a result of hypothermic pneumonia if it is neglected by its parents and left alone in a cold room. Even in our modern society with welfare services, this is all too common and is becoming commoner.

The sanitary conditions of homes in the developed countries are generally considered to be adequate. Most houses possess at least one water-closet. Yet if the excreta contain pathogenic germs these can spread by minute particles floating about in the air for several minutes or even hours after flushing the water-closet. In sky-scraper like dwellings the number of water-closets is enormous and the possible degree of particle-contamination of the whole building from multiple flushings around the clock beggars description. Such particles may contain viruses capable of producing various types of infection. The particles may be swallowed, inhaled or may even enter the body through the conjunctival membrane of the eye.

Many viruses live happily in the large bowel: they are known as entero-viruses. The two common groups which cause much mischief in the Western World and in Australasia are the Coxsackie and ECHO viruses. The infections caused by these viruses range from a mild common cold-like illness to a severe meningitis and inflammation of the brain. There seems little doubt that virus meningitis is much commoner today in the developed countries than it was before the Second World War, when bacterial meningitis predominated. Our complex modern dwellings probably produce an environmental milieu highly suitable for the spread of enteroviruses. Such buildings may not only fulfil an architect's happiest dreams but perhaps also those of the enteroviruses.

There is also the problem of sewage disposal from modern dwelling areas especially in the large industrial communities. In some so-called developed countries this remains a very primitive procedure. In maritime nations like Great Britain sewage is still allowed to pour willy-nilly into the surrounding sea. In areas of high tidal change the beaches can be covered with excreta for hours on end. Some bowel pathogens like the germs of dysentery can cause gross contamination of the sewage outfall. It has been estimated that in one area of the city of Edinburgh no less than seven per cent of the population excrete some dysentery germs at any given moment of time. Other cities may equal or even exceed this figure. Most of these people are quite unaware that they are dysentery-carriers and take no precautions to prevent these bacteria from spreading. Instead of a fall in the incidence of dysentery in the British Isles the number of cases reported has been steadily rising since the Second World War despite our so-called modern sanitation in the home and at work. This is partly due to the way in which excreta are disposed of in many areas of the country and partly due to inadequate personal hygiene following defaecation. Many hotels, restaurants and bars have wash-hand basins installed beside the lavatories. In Britain a combination of vandalism and a sorry lack of supervision accounts for the state of disrepair of these basins and the lack of soap and towels. Thus hands are contaminated, and in consequence food becomes contaminated.

Apart from the nuisance value of dysentery to the individual by causing a brisk flux followed by constipation, the rising incidence of dysentery carries with it more sinister implications. Dysentery organisms can carry factors which cause them to become rapidly resistant to a number of antiobiotics. These resistance factors can be transferred from the dysentery organisms to other bacteria, such as *Escherichia coli*, which normally live in the large bowel and which will then also become multiply insensitive to antibiotics. If the *Escherichia coli* then changes its role from a commensal organism of the bowel to a pathogen by, for example, infecting the urinary tract, it will be highly dangerous to the patient because of its resistance to all the commonly used antibacterial drugs. Antibiotic resistance may likewise be transferred to the organisms called salmonellae which commonly cause food poisoning in developed countries. Outbreaks of gastro-enteritis due to any of these bacteria with multiple antibiotic resistance can be disastrous especially if the bowels of very tiny children are infected. There have been many instances in the developed countries of such outbreaks in children. No antibiotic drug can possibly be effective against organisms which have acquired primary resistance to them before infection occurs.

The public are sadly uneducated and ill-informed about personal hygiene, the spread of infection by faecally-contaminated personnel and foodstuffs and our totally inadequate programme of sanitation and sewage disposal. They clamour for antibiotics from their medical attendants, but the thoughtless, repeated and liberal prescription of these drugs for trivial or inappropriate illnesses does much to exacerbate the problem by introducing the sinister mechanism of multiple antibiotic resistance.

Factories, hotels, restaurants, boarding schools and other places where large numbers of people may gather are excellent breeding grounds for the spread of pathogens. This is especially true of virus infections. In the UK the number of working days lost as a result of virus infections of the respiratory tract alone far exceeds the number of days lost through strikes. Despite intensive and brilliant research, no satisfactory vaccine has yet been produced which will give protection against the large number of different types of respiratory viruses which are responsible for the common cold, or coryza syndrome. The respiratory tract viruses are capable of a major degree of adaptation so that the production of effective vaccines is difficult. A vaccine prepared against one type of influenza virus may be completely useless during a subsequent epidemic, by which time the virus may well have changed. In addition to the harm done to the upper airways and lungs by the viruses themselves the tissues they invade are rendered much more liable to infection by bacteria such as the pneumococcus.

Foodstuffs

In the developed countries there has been a marked change of eating habits in the past quarter of a century. Many more people now eat out than ever before and, with the vast increase in overseas travel, tastes have become more varied and exotic. One of the consequences of this has been to intensify foodstuff production, preservation and packaging. This has resulted in some undesirable side-effects in terms of infection.

Chicken and turkey farming together with the intensive rearing of fat stock are now major industries in the developed countries.It has been shown that feeding antibiotics to animals and poultry causes a striking increase in their rates of growth and maturation. Pigs, chickens and turkeys will fatten more quickly if fed small doses of tetracycline. It is now known that tetracycline and other antibiotics alter the bowel flora of animals and poultry which often harbour salmonellae in their intestines. Transference of antibiotic resistance from veterinary strains of salmonellae to human pathogenic and commensal bacteria is an undoubted theoretical risk. Even though this has not yet assumed clinical significance, it may well do so in the future. The whole vexed question of the use of antibiotics in animal husbandry has been the subject of a government enquiry in the United Kingdom and the reader is referred to this for a fuller account of the problem (Report, 1969).

Antibiotic contamination of foodstuffs is hedged around by many different factors—ecological, sociological and economic. For example, fish may be treated with antibiotics before preservation by deep-freezing on board ocean-going trawlers. When ingested these antibiotics may encourage the emergence of resistant bacteria in the human bowel.

Again from the medical point of view even trace amounts of antibiotics in food can have clinical significance for the individual. It is not unknown for a patient to exhibit an allergy to penicillin although he or she has never been treated with this drug for an infection in the past. This can be explained by the production of sensitization to penicillin by the ingestion in early life of milk or milk products which contained penicillin. Although it is illegal in most developed countries to market antibiotic contaminated milk this law is difficult to enforce. Bovine mastitis, a common inflammation of the cow's udder, is usually treated with penicillin. All milk should be discarded whilst the cow is under treatment, but unfortunately this does not always happen. Minute traces of penicillin can then contaminate a bulk supply of milk. These tiny amounts are quite enough to sensitize an individual to penicillin, especially when milk is ingested in quantity in the early years of life. When the patient who has been made allergic to penicillin requires this antibiotic later in life for an infection, there may be a brisk reaction in the form of a skin rash or, much less commonly, severe shock and sometimes death. It has been estimated that over three hundred persons die of penicillin shock every year in the USA alone.

Another important aspect of food processing and its side-effects is the increase in salmonella contamination of dried eggs, desiccated coconut and canned meats. Outbreaks of salmonella infection resulting from this type of processing are usually of nuisance value only, but serious epidemics can and do occur. A very good example was the typhoid outbreak in Aberdeen in 1964 which was traced back to contamination of canned meat imported from Argentina.

Malnutrition in its own right increases the susceptibility to infection often with serious consequences to the individual. Malnutrition in the developed countries is very likely to be due to the excessive intake of foodstuffs and alcohol, leading to obesity and to diabetes mellitus coming on in middle life. Such factors lead to superficial infection of the skin in the form of boils and carbuncles with subsequent blood poisoning and even death. The diabetic is especially liable to infections of all descriptions and, in particular, to tuberculosis of the lungs.

Malnutrition of the elderly in the developed countries is usually due to their economic inability to buy adequate foodstuffs. This can lead to lack of vitamins. Senile scurvy is by no means rare in a developed country because of the poor intake by the elderly of fruit and vegetables containing Vitamin C. Old people suffering from scurvy are predisposed to infected gangrene or pneumonia which can administer the death blow.

The elderly, especially men, now form the main reservoir of tuberculous infection in the developed countries. The old man, with apparent bronchitis, may very well be suffering from tuberculosis. He is capable of causing serious infection to his grandchildren by coughing over them. With the advent of antituberculous drugs the developed countries have tended to think that tuberculosis is no longer a menace to them; however, this is far from being the case. With the increased ageing of the population, it is possible that tuberculosis could become resurgent in these countries. This takes no account of the emergence of strains of tubercle bacilli which have become resistant to one or more of the currently available antituberculous drugs and the importation of tuberculosis from developing countries where drug resistance is even more of a problem than in our own. Fortunately, tuberculous infection derived from cattle has been virtually eradicated in the developed countries, although it remains a menace in the developing ones. If there is to be an interchange of stock to improve the world's meat supplies, adequate precautions will have to be taken to ensure that herds in the developed countries do not become contaminated once again.

Further, brucellosis—undulant fever—derived from infected cow's milk is less uncommon in the general population than might be supposed. Many prolonged illnesses with fever should be investigated with this diagnosis in mind; otherwise it is often missed. The disease could be eradicated from the UK by a full vaccination programme in cattle together with adequate veterinary hygiene. This would not be an unduly costly exercise.

PROTECTIVE PROGRAMMES AND THEIR EFFECTIVENESS

Smallpox

This virulent infection remains endemic in a number of the developing countries. Many years ago smallpox was eradicated from developed countries by vaccination programmes and improved living conditions. There has been a steady decrease in the incidence of world small-pox in the past two decades. Statistics show that in this country, as in other developed countries, the complications of vaccination against smallpox cause more deaths than does smallpox itself when imported from abroad. In the UK compulsory vaccination of children against smallpox has been given up since the Second World War, but only in 1971 was it advised that the routine vaccination of infants should not be carried out at all in the UK. This is a controversial decision as primary vaccination in the adult is much more dangerous than in the child. Should an adult require to be vaccinated because he is going to an area of the world where smallpox is endemic or, if he is passing through it to one of the developed countries, he may be required by the law of that other country to be vaccinated. Severe reactions to compulsory adult primary vaccination are by no means rare and death has occurred as a result. It is very important to keep an open mind on this question. Patients from developing countries may arrive in a developed country by rapid air transport whilst they are incubating smallpox. A vigilant watch must be kept for a possible increase in the importation of this disease from abroad.

Diphtheria

This disease still remains a very serious form of infection in the world as a whole. There have been programmes of protection against it for many years, and as a result diphtheria has almost completely disappeared from developed countries. In the UK, however, the general public have become apathetic regarding diphtheria immunization and there is now a very low herd immunity throughout the country. Indeed in 1970 clinical cases of diphtheria have re-emerged in large urban communities such as Birmingham. The public must be re-educated generation by generation about the danger of diphtheria so that their children will be adequately protected against this serious and often fatal disease.

Whooping Cough

This infection is particularly prevalent in young children and babies and tends to occur in seasonal outbreaks. It is by no means a slight disablement. The illness is severe, prolonged and bedevilled by complications in the unprotected child. Death is not rare and is caused by lung collapse or by brain inflammation. The disease may be prevented or at least made much milder when the child has been protected by whooping cough vaccine. Although there has recently been a suggestion that the vaccine has become less effective because the type of organism has changed, the public have again become apathetic about seeking protection with the vaccine. In this country immunization against whooping cough is usually combined with vaccination against diphtheria and tetanus (lock-jaw) and, just as the herd immunity against diphtheria has decreased, so it has also against whooping cough and tetanus.

Tetanus

This must be the most excruciating of all diseases from which man can suffer. Any doctor who has seen a case of tetanus will have this imprinted on his memory for life. The incidence of tetanus in the developed countries is very low but it remains high in developing areas. The spores of the tetanus germ are hardy, can survive in dust and soil for many weeks, and yet still remain capable of producing disease by passing through small abrasions, cuts or deep wounds. The spores regenerate into bacilli which give off a powerful nerve toxin. Effective protection against tetanus is available by the administration of the combined triple vaccine, which includes diphtheria and whooping cough prophylaxis. Here again there is a distinct danger of a decline in herd immunity in the developed countries, because of public apathy. Re-education programmes to promote the triple vaccine are urgently required.

Poliomyelitis (Infantile paralysis)

The development of a vaccine against this serious and crippling disease by Enders and his co-workers in the United States was a major break-through in the fight against virus infections. The subsequent work of Salk and Sabin produced highly efficient vaccines for protection against poliomyelitis. Regrettably the first wave of enthusiasm for protection by the public has passed away. Once again the degree of herd immunity is dropping in the developed countries. Poliomyelitis is still rife in developing areas of the world, and there is always the danger of the disease becoming resurgent in developed countries.

Measles

An effective vaccine against this disease is now available. No one should believe that measles is of no consequence. Thousands of people in the developing countries die every year as a result of this infection. Mortality in the developed countries is extremely low, but the morbidity and complication rates of measles are still high. This is particularly true of bronchopneumonia and infection of the middle ear with the sequel of deafness or partial deafness for the rest of life.

National measles immunization programmes must aim at complete protection for the whole population; otherwise there is a distinct risk of measles occurring in the adult who may have escaped the disease in childhood. Measles in the adult is a particularly obnoxious illness often complicated by pneumonia, inflammation of the middle ear and sometimes by severe inflammation of the brain. Brain complications can lead to sudden death. If recovery occurs the patient may be left with permanent deafness, blindness, speech defects or paralysis of a limb. It is to be hoped that adequate measles immunization programmes will be introduced during the course of the next decade.

Rubella (German measles)

A great deal of emotional heat has been engendered by this virus disease. If rubella occurs early in pregnancy a deformed child will almost certainly result. Because there is at present no effective treatment for the established infection abortion will normally be recommended and carried out in many of the developed countries if the mother is infected in the first four months of pregnancy. Rubella vaccines are now available which it is hoped will give life-long immunity. They should be given to both boys and girls about the age of puberty if total protection of the herd from German measles is to be achieved in the future.

SOCIAL DISEASES

The Pill

Since the advent of the contraceptive pill and the increase of permissiveness in the developed countries there has been an explosion of venereal disease, especially amongst teenagers. The commonest infection is gonorrhoea. Gonococcal infection in the male is easily recognized by the patient himself. Within a few days of exposure there is an obvious penile discharge associated with pain and discomfort on passing water. In the infected woman, however, obvious symptoms need not occur, and it is the asymptomatic female carrier of the disease who is the main source of spread. The prostitute population is no longer the main reservoir of venereal disease. Today the infected promiscuous teenager is a far greater danger to the community and she is no rarity in developed countries.

There has also been an increase in other venereal infections besides gonorrhoea. Trichomonal vaginitis and so-called non-specific urethritis have become much more common in the past twenty years. These warning signals suggest that the more serious venereal infection, syphilis, is also likely to be on the increase. The late manifestations of this disease cause serious morbidity and early death, especially in the male. Other venereal infections, such as soft sore and lymphogranuloma venereum, which formerly were largely confined to the developing countries are now frequently seen in developed countries as a result of large scale immigration.

The most important and difficult problem in developed countries is to teach the young a sense of social awareness about the seriousness of venereal disease. They must be taught not only to report for treatment, which they undoubtedly do, but to report back after the initial treatment for adequate follow-up and contact tracing purposes. This they are less likely to do as long as they consider gonorrhoea to be no more important than a 'dose of the cold'.

Gonorrhoea is not nearly so easily treated today as it was twenty five years ago. The infecting organism—the gonococcus—is much less sensitive to penicillin than it used to be. Patients may required one very large dose of penicillin or, sometimes even two or three large doses, before they are cured. Inadequate treatment or follow-up often leads to a failure to eradicate the germs. Crippling arthritis, especially of the knee joints, and other unpleasant complications may then occur. In the passage through an infected mother's birth canal her baby's eyes may be so infected by the gonococcus that blindness results.

The standard treatment for gonorrhoea with penicillin may mask the early symptoms of syphilis. The patient must, therefore, be urged to return for a blood check for syphilis six to eight weeks after initial treatment for gonorrhoea. If syphilis is masked and so not recognized and treated shortly after exposure, its delayed complications may not manifest themselves until middle or late life when they are irreversible. Complications of syphilis include insanity, paralysis of the legs and serious disorders of the heart and blood vessels. Syphilis is also readily transmitted from the pregnant mother to her unborn child. Although this congenital infection can easily be prevented by treating the mother early in pregnancy this may not be achieved in the permissive communes of young people where antenatal care is lacking.

The Drug Scene

Controversy rages over the harm or relative lack of harm done by the taking of 'soft' drugs such as marihuana. Whether pot smoking should become legal is not for the authors to state. Nevertheless there is some evidence to support the view that a young person who starts smoking 'hash' may well be tempted to graduate to more potent and addictive drugs such as LSD, cocaine or heroin (see Chapter 17).

From the point of view of infection 'hard' drug taking leads to a rapid decline in personal hygiene with colonization of the skin and hair by abnormal germs, such as fungi and the yeast known as *Candida albicans*. The 'junkie' who repeatedly injects himself runs a distinct risk of these organisms invading his blood stream. Systemic candidiasis is almost invariably fatal unless treated early with the powerful antibiotic Amphotericin B, which is so toxic that it usually damages the kidneys. Another important aspect of the 'junkies' problem is that of sharing syringes. There is a very real danger of the transmission from one addict to another of a noxious virus which causes severe inflammation of the liver. This is the virus of syringe-transmitted hepatitis or Australia antigen positive hepatitis. This disease carries a mortality which can be as high as 30 per cent.

From the environmental point of view drug addicts constitute a grave danger to the community as a whole. Personal contact with addicts who can harbour the hepatitis virus for many months may cause innocent people to contract the disease and indeed to die. In order to make money to buy more drugs, heroin addicts may volunteer to give blood for transfusion purposes in some of the developed countries that pay blood donors. This constitutes a great danger as, if the blood from many different donors is 'pooled', whole batches of transfusion material may be contaminated by the hepatitis virus. Rigorous screening of blood for transfusion is now going on in most of the developed countries to prevent this risk. People who are addicted to 'hard' drugs often develop ulcers on the limbs where they inject themselves; such ulcers are so indolent that amputation of the limb may be required. This means admission to a hospital surgical area where such addicts again constitute a reservoir of cross-infection.

Besides the infectious hazards of the classical 'junkie' drug addict the developed countries now face a massive problem in the form of self-destruction by drugs, notably barbiturates. The suicide problem is generally outwith the scope of the Chapter, but it must be pointed out that even with all the modern resuscitation methods available for suicide patients, infection may administer the *coup de grace*. Suicide patients may stop breathing and require artificial respiration through a tube inserted into the windpipe. The longer the patient remains unconscious and needs artificial respiration the higher is the risk that the bronchial tubes will become infected by abnormal organisms like *Pseudomonas pyocyanea* which are extremely resistant to most antibiotics. Prolonged unconsciousness from drug overdosage often leads to a reduction of body temperature which predisposes the patient to straightforward pneumonia with his or her own pneumococcus. If the patient has attempted suicide in a locked room and has remained undetected for many hours, terminal pneumonia may well finish the event before the drug itself has done so.

MENTAL HEALTH

There has been a vast increase in mental illness in the last thirty years. The institutions which have been built in the past to cope with this problem are no longer adequate. They are usually architecturally out of date, over-crowded with patients and under-staffed with nurses. The standard of personal hygiene in the mentally disturbed patient is often extremely poor, and the risk of endogenous infection in mental institutions is by now well known. Endogenous infection by the tubercle bacillus which used to be a great problem in asylums is now less of a menace although, in keeping with the trend of developed countries, there is the danger of a resurgence of tuberculosis with antibiotic-resistant organisms. Outbreaks of gastro-intestinal infection, especially with dysentery and salmonella organisms, still occur in our mental institutions.

HOSPITAL INFECTIONS

There can be little doubt that the rapid advances made by medical science during the middle part of this century have saved many lives. Nevertheless the price has to be paid in terms of infection for the life-support programmes that we have created. Within our large hospitals there now exists a population of patients who are extremely susceptible to all kinds of germs. Hospital infections are a continuing menace and can destroy months of hard work in an attempt to support life. The sophisticated forms of treatment for malignancy, such as immuno-suppressive drugs and radiotherapy, the use of corticosteroids for common diseases like asthma, rheumatoid arthritis and ulcerative colitis, renal dialysis programmes for kidney failure, the transplantation of organisms and spare-part surgery, all lay the patient wide open to opportunist organisms which tend to lurk in our large hospitals.

Infection spreads by various routes but particularly by intravenous procedures of all kinds. There is a serious danger that the hepatitis virus from transfused blood may enter and live for long periods of time in patients with a lowered resistance to infection. These people then become carriers of the virus and are a continuous hazard to other patients and the hospital staff looking after them. Faecal-oral spread of hospital organisms also occurs, particularly of water-borne pathogens such as *Pseudomonas pyocyanea* which colonizes the sinks, drains and anaesthetic apparatus of hospital wards and surgical theatres with great ease.

Even in developed countries, many hospitals are totally out of date. In some the close proximity of bacteriology laboratories and autopsy rooms to operating theatres is an ever-present source of infection. Isolation rooms for high risk patients are insufficient or even non-existent in some hospitals. Ventilation plants may suck air heavily charged with pathogens into areas which ideally should be free from all micro-organisms—operating theatres and wards caring for burned patients—instead of blowing air in the other direction. Some Scandinavian hospitals now have forced-draught ventilation systems throughout, but these are not without their problems, the least of them being the enormous expense involved in their construction.

Some of the healthy hospital staff carry organisms which can be lethal to the susceptible and debilitated patient. In the 1950's a pandemic of hospital infection by such an organism—the *Staphylococcus aureus*—swept through the developed countries, killing many thousands of patients. This organism colonizes the skin and nostrils of medical and nursing staff, who are apparently perfectly healthy. The mere handling of susceptible patients by such carriers can lead to serious staphylococcal infections and even to death. Fortunately this pandemic has died out, but it might well recur if sufficient vigilance is not maintained.

The most important infective menace in the hospitals of developed countries at present is the bowel organisms. These are termed Gram-negative bacteria because of their staining characteristics in laboratory tests. As has already been stated, they can sometimes show multiple resistance to antimicrobial drugs and they constitute a major problem in causing septicaemia, or blood infections, in debilitated patients. The incidence of Gram-negative septicaemia in the large, modern hospitals of developed countries has increased remarkably since the 1930's. As previously discussed a major reason for this is the production of a highly susceptible hospital population to such infections by the preservation of lives that would formally have been lost even in the 1930's. Susceptible patients should be protected from these infections at all costs. Early diagnosis and effective treatment are essential.

It should be recalled that the second most common cause of shock in hospitalized patients in developed countries is that due to Gram-negative septicaemia. The only other disease which outstrips it as a cause of shock is coronary thrombosis.

AIR POLLUTION

Much could be written about the relation of air pollution to infection in our environment. However, this has already been very well documented and commented upon on the mass media. Most people now accept that tobacco smoke, and especially cigarette smoke, leads to chronic pulmonary infection; diesel fumes and smog favour upper respiratory infections (see Chapter 6).

TRANSPORT

The motor car, besides being potentially lethal in itself and the cause of multiple injuries which can be heavily infected, also favours the air-borne droplet spread of viruses amongst the occupants of the car itself. Indeed it can be said that the motor car is now very much a mixed blessing. The railway train is probably more important in the context of the spread of certain infections than is the aeroplane. This has been well illustrated by the spread all along the trans-Siberian railway of the Asian 'flu virus which has emanated by this route as well as by sea from its origin in S.E. China to cover the whole globe in the past few influenza pandemics. Aircraft do, however, act as important transmitters of diseases such as smallpox which have a longer incubation period. The illness may only declare itself after the immigrant has been in the country for a week or more. By this time he will have made enormous numbers of casual contacts and a great deal of painstaking detective work may be required to trace them all. The recent boom in package-deal air holidays to Mediterranean countries from the UK has led to a regular influx each summer of salmonella infections, especially paratyphoid.

INDUSTRIAL INJURIES

In the developed countries there has been a steady increase in thermal injuries in industry, and it can be predicted that advancing technology may still further increase the number of people who are burned at work. Burns are highly susceptible to infection which can ruin completely many hours of expert plastic surgery in attempts to replace areas of lost skin. Specialized units for the treatment of thermal injuries have been set up in the developed countries, but in most instances these are still not adequate to deal with the problem. Up-to-date specially equipped units should be built in grounds of their own and well away from the environment of the general hospitals with their problems of infection. Full bacteriological support is necessary to minimize the risk of infecting the burnt areas of the body.

CONCLUSION

In developed countries the public have regrettably taken a very negative attitude towards infection in all its forms. Since the discovery of penicillin and the publicity which has accompanied the subsequent vast array of new antimicrobial agents, sophisticated man seems to have been lulled into a sense of false security regarding his own susceptibility to infection. It cannot be overemphasized that only about 15 per cent of the known infections of man are capable of being adequately treated by drugs, and we are as yet only on the threshold of discovering anti-viral agents. The uncanny adaptability of micro-organisms to the development of an artificially hostile environment in their human host must never be forgotten. The ability of bacteria to become rapidly resistant to antibiotics and the way viruses can alter their antigenic structure to outwit previously effective vaccines are well-known examples of the versatility of our pathogenic enemies.

The public must therefore be educated towards the idea of preventing rather than treating infections on a mass population scale, for this is by far the most effective way of reducing infection in the community. The concept of disease prevention does not carry the same glamorous appeal as do heroic curative treatments such as organ transplantation, but in terms of the relief of human suffering and socio-economic improvement, prevention of all infections would be infinitely more rewarding.

References

Anderson, E. S. and Datta, N. (1965). Resistance to penicillins and its transfer in enterobacteriaceae. *Lancet*, **1**, 407.
Andrewes, C. (1965). *The Common Cold*. London, Weidenfeld and Nicolson.
Andrewes, C. (1965). The troubles of a virus. *J. gen. Microbiol.*, **40**, 149.
British Medical Bulletin (1969). Immunization against infectious diseases (Symposium). **25**, 119.
Christie, A. B. (1969). *Infectious Diseases: Epidemiology and Clinical Practice*. Edinburgh, E. & S. Livingstone.
Dixon, C. W. (1962). *Smallpox*. London, J. & A. Churchill.
Du Bos, R. J. (1960). *Mirage of Health*. London, Allen and Unwin.
Proc. roy. Soc. Med. (1964). Section of Comparative Medicine. Hazards of antibiotics in milk and other food products. **47**, 1087.

Postgraduate Medical Journal (1971). Hepatitis (Symposium). **47**, 462.

Report (1964). The Aberdeen typhoid outbreak 1964. Report of the Departmental Committee of Enquiry (reprinted 1970), Edinburgh, HMSO, Cmnd. 2542.

Report (1969). Joint committee on the use of antibiotics in animal husbandry and veterinary medicine, London, HMSO, Cmnd. 4190.

Rich, A. R. (1946). *The Pathogenesis of Tuberculosis*, Springfield, Illinois, Charles C. Thomas.

Stuart-Harris, C. H. (1965). *Influenza and other Virus Infections of the Respiratory Tract*, 2nd Edn. London, Edward Arnold.

Williams, R. E. O., Blowers, R., Garrod, L. P. and Shooter, R. A. (1960). *Hospital Infection*. London, Lloyd-Luke.

Chapter 10

Diseases of Modern Economic Development

D. P. Burkitt

DISEASE AND ENVIRONMENT

Most disease is the result of some factor or factors in the environment. All infections, whether caused by bacteria, viruses, or various parasites, come into this category. Injuries and poisons of all kinds are also environmental hazards, and it has been estimated that over 80 per cent of cancer is due primarily to some environmental factor. Examples of the latter are the relationships between lung cancer and smoking, between skin cancer and sun or X-rays, and between oral cancer in the east and chewing a 'quid' containing tobacco, lime, and other ingredients.

Even malformations present at birth may be the result of some intrauterine environments, as tragically exemplified by thalidomide poisoning or infection with rubella during pregnancy. Very few diseases are, in fact, solely the result of genetic factors.

In view of the environmental dependence of disease, a search for the cause of a particular malady is, in fact, usually an attempt to identify the responsible factors in the environment. Many examples could be cited. Snow, in his classic mapping of cases in a cholera outbreak in London, identified contaminated water as the responsible factor much earlier than the cholera vibrio was discovered. The relationship between fever and the presence of mosquitoes was recorded by David Livingstone long before Rose and his colleagues demonstrated the mode of transmission and life cycle of the malaria parasite. Sleeping sickness was seen to occur in the presence of tsetse flies before Bruce described the transmission of the responsible trypanosome.

In the realm of non-infectious disease, Pott's observation that chimney sweeps were particularly prone to skin cancer preceded knowledge of the carcinogenic action of hydrocarbons and skin cancer was associated with exposure to sun when little was known of the carcinogenic effects of radiation.

Such examples could be multiplied in almost every field of medicine.

THE SIGNIFICANCE OF RELATIONSHIPS

The different effects of any one cause will always tend to be associated with one another. Conversely, if two or more diseases tend to be rare or common together in different communities, common or related causes may be suspected. Diseases due to a common cause will all tend to occur more commonly in individuals most exposed to that causative factor, and as a result these diseases will be found together in individuals more often than otherwise expected. Conversely, the tendency for two or more diseases to occur concurrently in individual patients suggests that they have a common or related cause (Burkitt, 1970).

It is thus evident that studies of the geographical distribution of particular diseases not only help to identify possible causative factors in the environment but also enable the recognition of relationships between one disease and another.

DISEASES WITH A PARTICULAR AND WITH A COMMON GEOGRAPHICAL DISTRIBUTION

Different diseases may have a particular and distinctive geographical distribution or a group of diseases may share a common distribution pattern. The former include conditions such as colloid goitre, related to iodine deficiency, scurvy, due to Vitamin C deficiency, and diseases due to poisons such as arsenic and lead where distribution is related to food contaminated with these substances.

Some of the diseases which have a common distribution resulting from a common causation are:— various skin conditions, (inluding cancer) associated with excessive exposure to sun; bronchitis, lung cancer and nicotine-stained fingers, all caused by cigarette smoking; and liver cirrhosis, delirium tremens and obesity, all resulting from excessive indulgence in alcohol.

Another group tending to have the same geographical distribution comprises many of the most common and serious diseases of the western world (Trowell, 1960; Cleave *et al.*, 1969). It includes:

Coronary heart disease, the commonest cause of death, killing one man in four;
Cancer of the large intestine, the second commonest cause of death from cancer, after tumours of the lung;
Appendicitis—the commonest abdominal emergency.
Diverticular disease of the large bowel—the commonest disease of the intestine;
Gallstones—which are present in some 10 per cent of the adult population.
Over a third of a million gallbladders are removed annually in the USA.
Varicose veins—affecting some 10–15 per cent of the adult population.
Venous thrombosis—one of the major hazards of any serious sickness.
Pulmonary embolism—one of the commonest causes of post-operative death.
Diabetes—the commonest endocrine disorder.
Obesity—the fear of which is now a nationwide obsession; and
Dental caries—so prevalent that it is scarcely regarded as a disease.

These, together with some less well known maladies, can be considered as diseases of modern economic development.

GEOGRAPHY OF DISEASE OF MODERN ECONOMIC DEVELOPMENT

All the diseases listed above are rare or unknown in communities little touched by Western civilization, and Western dietary customs in particular. They are all most prevalent in Europe and North America, and it is particularly significant that in the latter situation they now have a comparable incidence in both the white and coloured communities. In India, Pakistan and the Middle East and, as far as can be ascertained, in rural areas of South America, their incidence is intermediate to that of Africa and America. Japan also has had an intermediate incidence, but in the case of most of these diseases this has been rising, particularly in large cities, since the Second World War, and has risen rapidly in Japanese who have emigrated to California and Hawaii. The situation among the New Zealand Maoris is much closer to that of the population of European descent than it is to the less westernized Polynesian islanders who remain almost free of these diseases.

TIME OF EMERGENCE IN DIFFERENT SITUATIONS

Not only does the pattern of geographical distribution of these diseases point to possible common causes, but the time of their emergence as important clinical entities in different situations also suggests leads as to their cause.

Obesity, diabetes, dental caries and possibly gallstones were well recognized early in the last century, but have become very much commoner during the present century. Coronary heart disease, appendicitis (Short, 1920), diverticular disease (Painter and Burkitt, 1971) and venous thrombosis only emerged as important clinical entities in the first quarter of this century. Experience in developing countries suggests that the same applied to varicose veins (Burkitt 1972), but their incidence was not recorded before they became common.

These conditions were much more common in white than in coloured Americans 30 to 40 years ago, a gap which in most instances has now been bridged. This suggests a causative factor which appeared in the environment of the Caucasians before that of Negroes, but which has increased relatively more quickly in the environiment of the latter than of the former during the past 40 years.

When one turns to the situation in developing countries, of which Africa will serve as an example, these diseases appear to follow the impact of Western civilization, some coming soon after the introduction of sugar, and others only after further dietary changes, particularly those which accompany urbanization.

Diabetes, dental caries and obesity appear relatively early, appendicitis, varicose veins, and femoral thrombosis somewhat later, and intestinal cancer gallstones, coronary heart disease and diverticular disease only long after a high degree of adoption of western habits.

ASSOCIATIONS IN INDIVIDUALS

In the Western World many of these diseases have been observed to occur together in the same individuals more often than would be expected by their incidence in the community as a whole.

As has been explained above, this further suggests that they have a common or related cause.

RELATIONSHIP TO DIETARY CHANGES

The environment responsible for the intestinal diseases enumerated above is most likely to be dietary and therefore changes in food which coincided with or preceded the rise in incidence of these diseases, deserve special attention. The incrimination of the factor or factors responsible for these diseases will suggest possible causes for the associated diseases. One must therefore consider dietary changes:

    (*a*) that occurred in North American and Northern Europe shortly before the turn of the century;

    (*b*) that affected the white before the coloured population in the USA;

    (*c*) that take place in the diet of African villagers when they become educated or urbanized;

    (*d*) that occur in the diet of Japanese on emigration to America.

The main changes that occurred in western diet in the last quarter of the last century and the early years of this century were:

    (*a*) some rise in fat consumption, probably not more than 25 per cent (Antar *et al.*, 1964);

    (*b*) a greater rise in sugar consumption, possibly up to 200 per cent (Antar *et al.*, 1964; Walker, 1971a);

    (*c*) a reduction in cereal fibre, in the region of 5- to 10-fold, following the introduction of roller mills in the 1880s, and also a lower consumption of bread (Trowell 1972a; Robertson, 1972).

Similar changes occur in the diet of Africans following increasing contact with Western civilization. Sugar came in early, but fibre is probably not reduced to a pathological degree until white flour replaces a significant quantity of traditional unrefined cereals and other carbohydrates.

Japanese, emigrating to America, eat less rice and potatoes than when living in Japan, and increase their consumption of meat, eggs and butter, with consequent loss of fibre relative to calorie intake.

CAUSES SUGGESTED BY EPIDEMIOLOGICAL EVIDENCE

It has already been indicated that the similar geographical distribution of the diseases enumerated above and the tendency for them to occur together in individuals suggest common or related causes.

Examination of changes in food habits suggest that sugar excess and fibre depletion might be important causative factors.

Cleave (1956) was the first to point out that sugar excess and fibre depletion tend to be reciprocal to one another. Removal of unabsorbable fibre from natural carbohydrate foods results in concentration of the starch and sugar with consequent over-consumption. When food is diluted with fibre, satiety precedes over-consumption.

This is not the place for a detailed discussion of possible mechanisms whereby carbohydrate excess or fibre lack may cause disease but suggested mechanisms may be summarized as follows:

Deficiency of dietary fibre results in

(a) Delay in transit of intestinal content through the gastrointestinal tract (Walker, 1961, 1971b; Burkitt, 1971a, Burkitt *et al.*, 1972). This results in small formed stools contrasting with the large soft unformed stools characteristic of communities on a high-residue diet.

This constipation, a feature of Western life, results in:

 (i) raised pressures within the lumen of the bowel which are believed to be the cause of diverticular disease (Painter, 1970) and appendicitis (Burkitt, 1971b);

 (ii) altered bacteria in the stools which are believed to form substances that cause an increased incidence of bowel cancer (Aries *et al.*, 1969; Hill *et al.*, 1971);

 (iii) increased pressures within the abdomen during straining at stool, which result in back-pressure in the veins of the legs and may be in part responsible for varicose veins and venous thrombosis (Burkitt, 1972). The latter is the cause of pulmonary embolism.

(b) Diminished excretion in the stools of substances formed from cholesterol, together with increased absorption of cholesterol in the diet through the wall of the intestine. This raises the cholesterol level in the blood, which is believed to predispose to the development of coronary heart disease (Trowell, 1972a, b). It seems likely that in a similar way it influences the formation of gallstones which are largely formed of cholesterol.

(c) Over-consumption of refined carbohydrate, especially in the form of sugar and white flour.

This is largely responsible for the appalling rise in the incidence of dental caries, probably plays a significant role in the causation of obesity and diabetes, and may also contribute to coronary disease and gallstones (Cleave *et al.*, 1969).

CONCLUSION

The geographical distribution and historical development of many of the diseases characteristic of economic development suggests that the most important single factor running through them is the removal of unabsorbable fibre, without which over-consumption is almost impossible.

Because it has no nutritive value, fibre has been the neglected factor, the Cinderella of dietetics. Its retention once again in our flour could make an incalculable improvement in the health of any Western nation, which would be an ultimate fruit of geographical investigation.

References

Antar, M. A., Ohlson, M. A. and Hodges, R. E. (1964). Perspectives in nutrition. Changes in retail market food supplies in the United States in the last seventy years in relation to the incidence of coronary heart-disease with special reference to dietary carbohydrates and essential fatty acids. *Amer. J. Clin. Nutr.*, **14**, 169–178.

Aries, V., Crowther, J. S., Drasar, B. S., Hill, M. J. and Williams, R. E. O. (1969). Bacteria and the aetiology of the large bowel. *Gut*, **10**, 334–335.

Burkitt, D. P. (1970). Relationship as a clue to causation. *Lancet*, **2**, 1,229–1,231.

Burkitt, D. P. (1971a). Epidemiology of cancer of the colon and rectum. *Cancer*, **28**, 3–13.

Burkitt, D. P. (1971b). The aetiology of appendicitis. *Brit. J. Surg.*, **58**, 695–699.

Burkitt, D. P. (1972). Varicose veins, deep vein thrombosis and haemorrhoids: Epidemiology and suggested aetiology. *Brit. med. J.*, **2**, 556.

Burkitt, D. P., Walker, A. R. P. and Painter, N. S. (1972). Effect of dietary fibre on stools and transit-times, and its role in the causation of disease. *Lancet*, **2**, 1408.

Cleave, T. L. (1956). The neglect of natural principles in current medical practice. *J. Roy. Nav. med. Serv.*, **42**, 55–83.

Cleave, T. L., Campbell, G. C. and Painter, N. S. (1969). *Diabetes, Coronary Thrombosis and the Saccharine Disease*. Bristol, John Wright & Sons Ltd.

Hill, M. J., Crowther, J. S., Drasar, B. S., Hawksworth, G., Aries, V. and Williams, R. E. O. (1971). Bacteria and the aetiology of cancer of large bowel. *Lancet*, **1**, 95–100.

Painter, N. S. (1970). Pressures in the colon related to diverticular disease. *Proc. Roy. soc. med.*, Suppl. 63, pp. 144–145.

Painter, N. S. and Burkitt, D. P. (1971). Diverticular disease of the colon: a deficiency disease of the western civilization. *Brit. med. J.*, **2**, 450–454.

Robertson, J. (1972). Changes in the fibre content of the British diet. *Nature*, **238**, 290.

Short, A. R. (1920). The causation of appendicitis. *Brit. J. Surg.*, **8**, 171–186.

Trowell, H. C. (1960). *Non-Infective Diseases in Africa*. London, Arnold, p. 465.

Trowell, H. C. (1972a). Dietary fibre and coronary heart disease. *Eur. J. Clin. Biol. Res.*, **17**, 345.

Trowell, H. C. (1972b). Ischaemic heart disease and dietary fiber. *Amer. J. Clin. Nutn.* **25**, 926.

Walker, A. R. P. (1961). Crude fibre, bowel motility and pattern of diet. *S. Afr. med. J.*, **35**, 114–115.

Walker, A. R. P. (1971a). Sugar intake and coronary heart-disease. *Atherosclerosis*, **14**, 137–152.

Walker, A. R. P. (1971b). Diet, bowel mobility, faeces composition and colonic cancer. *S. Afr. med., J.*, **45**, 377–379.

Chapter 11

Arteriosclerotic Disease in Relation to the Environment

D. D. Reid

Three hundred years ago, a Dr Thomas Brady wrote to the great Sydenham: 'No physician hitherto has attentively considered the force and influence of the atmosphere upon human bodies; nor yet has he sufficiently ascertained the part it plays in prolonging human life.' It is not that the question had not been pondered; Hippocrates had devoted one of his works to the environmental circumstances of 'Airs, Waters and Places' that seemed to affect disease. But, as Greenwood has pointed out, neither the early Greek writers nor the physicians of Rome who followed them applied the numerical or statistical approach to epidemiology that is now the rule. This deficiency 'sterilized the many virtues' of Hippocratic or Galenical thought, and no precise assessment of the effect of environmental circumstances such as climate on heart disease was made until the nineteenth century. By that time, medical data on the weekly toll of disease were being provided by the Registrar-General and when this was related to contemporary information on weather reported by the Astronomer Royal, a new science of medical meteorology was born. In 1840, Dr William Farr noted the sharp rise in mortality that accompanied cold and fog. Early studies by members of the Provincial Medical Association compared changes in weather with those in the incidence of disease and then went on to conduct their own enquiry in 1857–58 on the correlation of meteorological factors with reports on prevalent infectious diseases like dysentry and measles. More precise studies of the effect of weather and other 'atmospheric' or environmental factors on cardiovascular disease did not appear until detailed diagnostic data became available. At the same time, the classification according to the geographic distribution that had proved so helpful in studies of infectious disease began to be applied in epidemiological enquiries on the aetiology of cardiovascular disease.

CLIMATE AND CARDIOVASCULAR DISEASE

Geographical Variation

As a development of geographical epidemiology, Mills (1930) introduced the technique of 'visual correlation' by drawing mortality contours that pin-pointed areas of contrasting mortality experience and relating them to local climatic conditions. As a result he proposed that the geographical distribution of angina pectoris in the USA was related to the stimulating effects of variations in daily temperature and the frequency of cyclonic conditions. A statistically more rigorous approach was taken by Moriyama and Herrington (1938) who used standard correlation techniques to measure the degree of association between meteorological indices, such as mean daily temperature and relative humidity, and the age-standardized death rates from major forms of cardiovascular disease in 44 of the States of the Union during the period 1921–1930 inclusive. They confirmed earlier observations of a concentration of high rates of cardiovascular mortality in the north east coastal areas of the USA. There was, however, no significant and independent correlation between death rates from angina pectoris and variability in temperature and frequency of anticyclones. Indeed, death rates from this cause were particularly high in areas where weather conditions were stable, damp and cold. In general, three climatic circumstances were associated with the groups of cardio-vascular-renal disease covered by this survey (angina pectoris, cerebral haemorrhage and chronic nephritis). These were precipitation rate, which was positively related, and daily temperature range and frequency of anticyclones, which were negatively related to mortality from these diseases. Of this group of conditions, angina pectoris, is the one most affected by meteorological factors, for it is in this condition that cold and damp weather is apparently most lethal. Mortality from cerebral haemorrhage, on the other hand, appears to be associated more with precipitation rate than with any other factor; however, the relationship is far from close. The geography of chronic nephritis in the USA resembles that of cerebral haemor-rhage: both are more frequent causes of death in the hot, damp climate of the Southern States. Since these diseases are especially common amongst black Americans, racial as well as, or rather than, climatic factors may well be involved.

Time Correlations of Climate and Cardiorespiratory Disease

The influence of weather on cardiovascular mortality can also be explored by correlating contemporary changes in the same area in both meteorological conditions and death rates. Early British studies, *e.g.* by the London County Council over the 33 years from 1890 to 1923, have emphasized the marked seasonal swing with winter peaks in mortality from cardio-vascular disease. Prompted by an observation that an excess in sudden death in November was associated with high barometric pressures, Stocks (1925) made a detailed analysis of the data and found, after adjusting for changes in temperature, a small but significant association of that kind. However, a repetition of this approach in Chicago by Bundesen and Falk (1926) showed no such relationship between cardiovascular deaths and barometric pressure. Only low temperature appeared to have a decisive effect. Similar results in repect of temperature were obtained in Britain after the War. Rose (1966) remarked on the similarity between the winter excess in both the onset of myocardial infarction and the risk of death from that disease. He attributed most of the short term fluctuation in mortality from arteriosclerotic heart disease in England and Wales to exposure to low temperatures.

More recently, the technique of time analysis of mortal events in the same area has been rigorously applied to data from Memphis, Tennessee, by Rogot and Blackwelder (1970). They correlated the number of cardiovascular deaths occurring each day in that city over the 3-year period 1959–61 with various indices of local climatic conditions. Multiple-cause coding of all the diseases entered on the death certificate allowed them to classify deaths into those where cardiovascular or lung disease was reported alone and others where both forms appeared together on the same certificate. As a result, they were able to show that daily average temperature was the climatic feature most closely associated with variation in daily rate of death attributed, as the underlying cause, to arteriosclerotic disease. Although respiratory death rates were also related to changes in mean temperature, and both heart and chest disease were often reported on the same death certificate, temperature was clearly involved in the death of individuals whose certificates made no mention of associated respira-tory disease. Vascular lesions of the central nervous system appearing as the underlying cause of death bore no clear relation to temperature; but when counted whenever it was reported, either as underlying or associated cause, the association was similar if less obvious to that already observed in respect of arteriosclerotic heart disease. In these more complicated disorders, general arteriosclerosis and hypertensive disease are responsible for the daily fluctuation in association with low temperature seen in the number of deaths where cere-brovascular lesions are reported. Of the other climatic variables included in the analysis, such as daily precipitation, per cent of possible sunshine and barometric pressure, none could account for any significant variation in the several forms of cardiovascular disease covered by the study. Any correlation initially seen between wind speed or relative humidity and the death rate from cardiovascular disease disappeared when allowance was made for contemporary variations in temperature.

Because of the wide differences that may occur in the lapse of time between onset and death in a mortal cardiovascular illness, Rogot and Blackwelder also related the indices of climatic conditions at various intervals from 1 to 30 days before the date of death with the death rate for that day. Unlike Boyd, who showed that in both East Anglia and London the death rate was most closely associated with average temperature 1–2 weeks before the day of death, they found no particular intervals which gave a closer relation between these variables than the temperature on the same day. As they point out, perhaps especially in the climatic conditions of an inland American state, the temperature from one day to another over short periods is relatively stable. In sudden death, where demise follows onset within 24 hours, the relationship between prevailing temperature and death rate was even more striking than for cardiovascular disease as a whole; although, because of the small numbers involved, the technical significance of the result was lower. It seems possible, therefore, that weather may have an effect on cardiovascular mortality that depends on the clinical type of the disease.

Possible Modes of Action of Climate

Cold weather does not seem to be the only aspect of temperature change that is associated with peaks in the death rate from heart disease. Several authors, e.g. Heyer *et al.* (1952), have indicated that extreme heat may have a similar effect. This sort of observation raises the

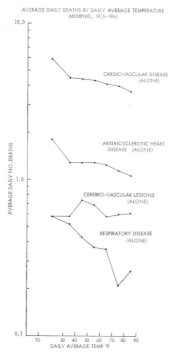

Fig. 31   Mortality from various forms of cardiorespiratory disease in relation to temperature·

question of the true aetiological importance of climate in cardiovascular disease and the way in which it might operate. Mills, for example, pointed to some congruity in the distribution and timing of deaths from cardiovascular and metabolic diseases such as diabetes and exophthalmic goitre, suggesting that the stimulant effect of a variable climate provoked metabolic breakdown and ultimate cardiac failure. The more precise studies of Moriyama and Herrington failed to substantiate this general proposition. Other work emphasized the winter excess in cardiovascular mortality with peaks during days of intense cold, and some of the winter excess is certainly associated with the coexistence of respiratory disease in cardiac invalids. Many experiments in animals have shown how cooling of the body increases susceptibility to respiratory infections that may be fatal. But in Rogot and Blackwelder's study, those peaks in cardiovascular mortality occur, as Fig. 31 shows, among individuals in whom no reference is made on the death certificate to the presence of respiratory disease. Although not so marked as that for deaths with only a respiratory cause reported, the trend is clear in the deaths where arteriosclerotic heart disease alone appears on the certificate. On the other hand, the risk of death from cerebrovascular lesions without mention of other conditions is unrelated to temperature. Other circumstances of life in winter may of course be relevant. Crowding in congested homes or offices can favour respiratory infection; unwonted effort in shovelling snow is a notorious hazard of the middle aged American male. Such incidental aspects of cold weather cannot, however, explain the consistent findings of a close correlation between low temperature and cardiac death on the same day. Coupled with the indications that similar results immediately follow exposure at the opposite extreme of temperature, these findings make it difficult to deny the direct physical effect of unusual ambient temperatures on the outcome of cardiac disease (see Chapter 5).

INTERACTION WITH AIR POLLUTION AND SOCIAL CONDITIONS

Several of the studies already quoted suggest a possible causal interaction in cardiovascular disease between weather and other aspects of urban life. Boyd for example, showed as in Fig. 32, that while the cardiovascular death rate rose when temperature fell in both the London area and in rural East Anglia, the rise was more marked during foggy weather in the congested metropolis where before the enforcement of the Clean Air Act, air pollution reached a high

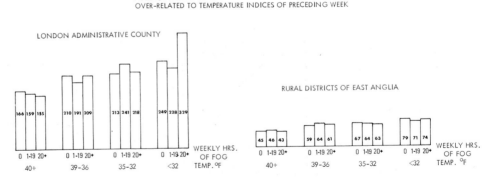

Fig. 32   An urban-rural contrast in the relationship between duration of fog, temperature, and cardiovascular mortality.

level in cold winters. Indeed in the Great London Smog of 1952, deaths from both coronary heart disease and myocardial degeneration, although less affected than bronchitis, doubled in frequency. By correlating the geographic distribution of cardiovascular mortality with the indices of social conditions, Herrington and Moriyama (1938) showed that socio-economic factors such as degree of urbanization or population density were much more closely associated with the risk of cardiovascular death than even the most influential meteorological factor. Nevertheless, factors like extreme cold, although operating even in good socio-economic circumstances, have an enhanced effect in adverse environmental conditions. Increasingly it is being realized that these environmental factors cannot be considered in isolation. Even when their separate influence is small the joint effect of two or more of them can be important in the causation of cardiovascular death.

Water Supply and the Geography of Heart Disease

Of the elements referred to by Hippocrates, the 'Air' or climate is obviously important in cardiovascular disease, but comparison of the mortality experiences of different 'Places' increasingly incriminates some pathogenic feature of 'Water' supply. In 1957, Kobayashi pointed out that the death rates from vascular lesions of the central nervous system in different areas of Japan were highest in districts where the acidity of river water was above average and vice versa. Schroeder (1960) by going into the data more fully, found a negative correlation between cardiovascular mortality and the degree of hardness of the water supply in the same part of Japan. He further found that a similar relationship existed in respect of the USA. These results raised three questions. Does this relationship appear consistently in other countries? Is it a true expression of cause and effect or does it result from a fortuitous common relation to some third factor? And, if water supply is of causal importance, how might differences in the content of drinking water affect the cardiovascular system?

Geographical Consistency in Mortality and Morbidity

Since the initial observations, similar studies have been carried out in several countries. Analysis of vital statistical data in England and Wales, for example, has shown an even

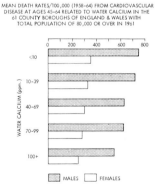

Fig. 33 Cardiovascular disease death rates in areas with water supplies of differing degrees of hardness.

clearer association than that found in the United States between the hardness of drinking water and the risk of death, in both sexes, from cardiovascular disease. (Morris, Crawford & Heady, 1961) Fig. 33 illustrates their relationships. Cerebrovascular disease and myocardial degeneration were particularly closely associated with the nature of local water supply.

Elsewhere, results were less decisive. In Sweden, only among women over the age of 65 was there a significant negative correlation between the hardness of the local water supply and mortality from myocardial degeneration (Biörck *et al.*, 1965). An association found in the Netherlands between the calcium content of drinking water and arteriosclerotic heart disease was restricted to women (Biersteker 1967). In combined data for both Eire and Northern Ireland, (Mulcahy 1968) there was a consistent trend in the same direction although it did not reach conventional levels of statistical significance.

These apparent discrepancies from the general picture seen so clearly in the USA and England and Wales may have several explanations. The most obvious is that in the relatively thinly populated areas of Ireland or Sweden, death rates for specific cardiovascular diseases are based on too small numbers to give statistical stability. Again, the effect may be largely confined to sub-groups of these cardiovascular diseases such as hypertensive heart disease. Moreover, water supply may be important only at certain stages in the evolution of arterial disease. Anderson *et al.* (1969) for example, have remarked on their finding that the greatest difference in cardiovascular mortality between two towns in Ontario in Canada with water supplies of different degrees of hardness lay in the frequency of sudden death. Sudden death was here defined as cardiac death referred to the coroner. Although unable to use the same index, Peterson and his colleagues (1970) found in the state of Washington, USA, parallel results in respect of patients who had either died at home or had been found dead on arrival at hospital. They surmised that, despite this difference in conventions, both studies pointed to a high risk of sudden death in areas where the water supply was relatively soft.

Other information on geographic differences in cardiovascular morbidity leads to equivocal results. Morton (1971) used the Selective Service records for the state of Colorado to show that recruits from areas in that state served by soft water rivers were more frequently found to have either hypertension or hypertensive heart disease than men from the basin areas of hard water rivers. This was in accord with the distribution of cardiovascular mortality in that state. Again, a special pathological survey has been made by Crawford and Crawford (1969) of the hearts of men killed in accidents in London, where the water supply is hard, and Glasgow, where it is soft. Coronary atherosclerosis was, in young men, commoner in Glasgow although there was no such difference among older men among whom healed infarcts were more often noted in Glasgow. Among groups dying suddenly because of heart disease, Glasgow men seemed to succumb to less severe arteriosclerotic lesions. Calcium and magnesium levels in the tissues were higher in the Glasgow men. On the other hand, prevalence surveys of the populations exposed to hard and soft water in both Pakistan (Beg *et al.*, 1970) and South Wales (Edward *et al.*, 1971) revealed no differences in blood pressure levels. In South Wales there was also no difference in the prevalence of ischaemic heart disease between the hard and soft water areas. Both these studies were on a relatively small scale. A different approach was adopted by Comstock (1971) working in areas with markedly different water supplies in Washington County, Maryland. He sampled the domestic water supplies of 198 individuals who had died from arteriosclerotic heart disease while between the ages of 45 and 64 and compared this with the samples for twice as many matched controls. He found no difference in hardness between the samples for these two groups.

CAUSE OR COINCIDENCE?

The evidence for a consistent and possibly causal relationship between cardiovascular disease and some factor in the water supply associated with hardness is thus not yet completely convincing. There are, however, indications that the association should not be too readily dismissed. The problem lies in the fact that a soft water supply can make an area attractive to industry so that the population drawn into man it may be selected in several ways as well as being subjected to the cardiovascular hazards of an industrial environment. Again, as already observed, climatic conditions can affect cardiovascular mortality; and the low rates of death in arid central areas of the USA could be explained by the warmer climate there. In an attempt to see whether other factors such as climatic, socio-economic conditions or air pollution could explain the geographic association between cardiovascular mortality and softness of water supply, Crawford *et al.* (1968) made a detailed statistical analysis of the interrelation of indices of these various factors in the counties of England and Wales. They showed quite clearly that although other social and environmental circumstances were related to cardiovascular mortality, none of these alternative factors could account for the high death rates in soft-water areas. The most striking evidence comes, however, from natural experiments that have occurred in the USA and the UK. Muss (1962) has reported that in New York in 1940, a major change in water supply was made from well water rich in mineral constituents to softer water from rivers. Compared with the preceding five years, this switch seemed to be followed by an increase in the death rate from arteriosclerotic heart disease. In England and Wales, Crawford *et al.* (1971) found eleven county boroughs where the hardness of water had been substantially changed for various reasons during the preceding 30 years. Generally speaking, hardening of the supply had a favourable effect on the trend in the death rate from cardiovascular disease and softening was followed by a reversal. During these changes, the death rate from non-cardiovascular causes remained unaltered. Since no major variation in the physical character of the populations of these areas and no consistent trend in climatic conditions took place during these periods it is difficult to resist the implication that something associated with hardness of drinking water has an influence on the risk of cardiac death.

POSSIBLE PATHOGENETIC MECHANISMS

Like smoke or sulphur dioxide measurement in respect of air pollution, the degree of hardness of water may simply be an index of the level of some closely associated element whose presence or absence is the real cause of cardiac death. Several alternatives are possible:

(1) Some contaminant of water drawn from particular geological formations, perhaps present only as a trace element, could have a toxic effect.

(2) Conversely, the absence of some element essential to efficient cardiac function could be the crucial factor.

(3) Alternatively, the softness of the water, or the acidity associated with it, may increase the risk of constituents of water pipes or cooking vessels being dissolved and ingested in water or prepared food.

The data on the chemical contents of drinking water collected particularly in the USA have been analysed in an attempt to identify those constituents whose presence or absence might be related to a high risk of death from cardiovascular disease. Schroeder (1966) found that arteriosclerotic heart disease death rates were negatively correlated with water levels for vanadium and fluorine but positively related to the concentration of magnesium. Voors (1971) has analysed the relationship between the mineral content of the water supply to 99 large cities in the USA and the age-adjusted mortality rates from arteriosclerotic heart disease for these cities during the period 1959–61. He concluded that deficiency of lithium was related to high arteriosclerotic heart disease in whites but that low levels of vanadium were more important in this respect for blacks. In Colorado, Morton (1971) found that the only constituents of municipal drinking water that were significantly related to local death rates from hypertension were nitrates and potassium which, he stated, are usually the products of animal or plant waste or the residue of chemical fertilizers; and he raised the possibility that, as in workers in the explosives industry, nitrates may cause diastolic hypertension. Masironi (1970) has reported the results of a World Health Organization survey of cardiovascular mortality in relation to α-radioactivity in the water-supplies in the river basins of the USA. Death rates particularly from hypertensive heart disease were consistently lower in areas matched for population density and urbanization, where the water supply was hard and levels of α-radioactivity high. No such trend was obvious for non-cardiovascular disease. At the same time cadmium levels were found to be low in areas where death rates from hypertensive and arteriosclerotic cardiovascular disease were high.

These fragmentary and often contradictory results underline present uncertainties about the role of the constituents of drinking water in the evolution of mortal cardiovascular disease. Nevertheless, the consistency of findings on the association of low levels of hardness and cardiovascular mortality has stimulated various surmises about possible pathogenetic mechanisms. Some element, such as cadmium, may have a toxic action on the kidneys that increases the risk of hypertension. Soft water is plumbo-solvent so that renal damage by lead is another possible cause of hypertension and associated heart disease. Other elements, such as lithium, may have a protective influence against arteriosclerotic changes in the arteries. Again, the fact that water supply is especially relevant to the risk of sudden death raises the possibility of an increased susceptibility of the myocardium to fatal arrhythmia following changes in serum electrolytes. The pathological studies already cited point to low tissue calcium levels as a factor that might affect myocardial irritability. The complexity of the problem of identification of the crucial characteristic of drinking water stems in part from the possible interactions of different ions such as calcium and sodium. Crawford (1971) has speculated on the possibility that sodium retention could develop, with serious vascular and cardiac consequences, where calcium intake is low; and in poorly nourished populations water as a source of calcium could be important. If this were the case, the addition of calcium to the water supply could have a beneficial effect on the expectation of life of cardiovascular invalids.

THE NEXT STEP

This brief review of some of the features of the environment that appear to influence the natural history of cardiovascular diseases has displayed the complexities and inadequacies of current information. Yet it is hard to deny that climate, air pollution and water supply have, particularly when acting jointly, an apparent effect on the onset or outcome of these conditions. The evidence incriminating air pollution as a menace to cardiorespiratory health has been strong enough to lead to the Clean Air Act of which the benefits are becoming daily more evident. Man's increasing ability to control his micro-environment by central heating or air conditioning can protect his cardiovascular system against the influence of extremes of temperature; and the prevention of cardiac death must be taken into account in any assessment of its value to the community. The question of the effect of water supply remains unresolved. Certainly, the vital statistical indices of cardiac disease should be used to monitor the results of changes in the nature of water supply, but the time may be ripe for a carefully controlled field trial of the value of adding calcium to domestic drinking water in the prevention of cardiovascular disease.

References

Anderson, T., Le Riche, W. H. and MacKay, J. S. (1969). Sudden death and ischaemic heart disease. Correlation with hardness of local water supplies. *New Engl. J. Med.*, **280**, 805.

Beg, M. A. *et al.* (1970). Blood pressure findings in relation to hardness and softness of drinking water in a single community in Pakistan. *J. Pakistan Med. Ass.*, **20**, 383.

Biersteker, K. (1967). Medical aspects of softening water. *Tijdochr. soc. Geneesk.*, **45**, 658.

Biörck, G., Boström, H. and Widström, A. (1965). On the relationship between water hardness and death rates in cardiovascular disease. *Acta med. scand.*, **178**, 239.

Boyd, J. T. (1960). Climate, air pollution and mortality. *Brit. J. prev. soc. Med.*, **14**, 123–135.

Bundesen, W. N. and Falk, I. S. (1926). Low temperature, high barometer and sudden deaths. *J.A.M.A.*, **87**, 1987.

Comstock, G. W. (1971). Fatal arteriosclerotic heart disease, water hardness at home, and socio-economic characteristics. *Am. J. Epidem.*, **94**, 1.

Crawford, M. D. (1971). Cardiovascular mortality and drinking water. World Health Organization. Meeting of investigators on trace elements in relation to cardiovascular diseases. Geneva, 8–13 February 1971.

Crawford, M. D. and Crawford, T. (1969). Lead-content of bones in a soft- and a hard-water area. *Lancet*, **1**, 699.

Crawford, M. D., Gardner, M. J. and Morris, J. N. (1968). Mortality and hardness of local water supplies. *Lancet*, **1**, 827.

Crawford, M. D., Gardner, M. J. and Morris, J. N. (1971). Changes in water hardness and local death-rates. *Lancet*, **2**, 327.

Elwood, P. G., Bainton, D., Moore, F., Davies, D. F., Wakley, E. J., Langman, M., and Sweetnam, P. (1971). Cardiovascular surveys in areas with different water supplies, *B.M.J.*, **2**, 362.

Herrington, L. P. and Moriyama, I. M. (1938). The relation of mortality from certain metabolic diseases to climatic and socioeconomic factors. *Amer. J. Hygiene*, **28**, 397.

Heyer, H. E., Teng, H. C. and Barris, W. (1952). The increased frequency of acute myocardial infarction during summer months in a warm climate; a study of 1,386 cases from Dallas, Texas. *Amer. Heart. J.*, **45**, 741.

Kobayashi, J. (1957). Geographical relationship between chemical nature of river water and death rates from apoplexy. *Bar. Ōhara Inst. landiv. Biol.*, **II**, 12.

Masironi, R. (1970). Cardiovascular mortality in relation to radioactivity and hardness of local water supplies in the USA. *Bull. Wld. Hlth. Org.*, **43**, 687.

Mills, C. A. (1930). Geographic or climatic variations in the death-rate from pernicious anaemia, exophthalmic goiter, Addison's disease, and angina pectoris. *Arch. Int. Med.*, **46**, 741–751.

Moriyama, I. M. and Herrington, L. P. (1938). The relation of diseases of the cardiovascular and renal systems to climatic and socio-economic factors. *Amer. Jour. Hyg.*, **28**, 423–436.

Morris, J. N., Crawford, M. D. and Heady, J. A. (1961). Hardness of local water supplies and mortality from cardiovascular diseases. *Lancet*, **1**, 860.

Morton, W. E. (1971). Hypertension and drinking water. A pilot statewide ecological study in Colorado. *J. Chron. Dis.*, **23**, No. 8, 537.

Mulcahy, R. (1968). Mortality and hardness of water supplies. *Lancet*, **1**, 975.

Muss, D. L. (1962). Relationship between water quality and deaths from cardiovascular disease. *J. Amer. Water Works Assoc.*, **54**, 1371.

Peterson, D. R., Thompson, D. D. and Nam, J. M. (1970). Water hardness, arteriosclerotic heart disease and sudden death. *Am. J. Epidem.*, **92**, 90.

Rogot, E. and Blackwelder, W. C. (1970). Associations of Cardiovascular Mortality with Weather in Memphis, Tennessee. Public Health Report, Vol. 85, No. 1, Jan. 1970. US Dept. of Health, Education, and Welfare, Public Health Service.

Rose, G. (1966). Cold weather and ischaemic heart disease. *Brit. J. Prev. Soc. Med.*, **20**, 97–100.

Schroeder, H. A. (1960). Correlation between mortality from cardiovascular disease and treated water supplies. *J.A.M.A.*, **172**, 1902.

Schroeder, H. A. (1966). Municipal drinking water and cardiovascular death-rates. *J. Amer. med. Ass.*, **195**, 81.

Stocks, P. (1925). High barometer and sudden deaths. *B.M.J.*, **2**, 1188.

Voors, A. W., (1971). Minerals in the municipal water and atherosclerotic heart death. *Am. J. Epidem.*, **93**, 259.

Chapter 12

The Environment and Thyroid Disorders

William R. Greig
John A. Thomson
Edward M. McGirr

THE GEOLOGICAL BASIS OF IODINE DEFICIENCY

After the last ice-age vast and varied regions around the world had their iodine-rich surface soil denuded and replaced by soil derived from broken-down crystalline rock. The latter contains only about one tenth of the average iodine content of mature soils (Scrimshaw, 1964). As the expansive glaciers receded, some iodine borne from the sea, for example by evaporation and rainfall, replenished the new iodine-poor soil. In general today, the regions of most severe iodine deficiency are reciprocally correlated with the length of the post-glacial period and the opportunity for iodine replenishment (10,000–20,000 years). The prevalence of severe iodine deficiency is lower at the equator than at the poles and highest in regions of inland mountains far removed from the sea (Figs. 34–40).

More detailed study shows that the iodine content of soil varies with mineral composition of the original rock. Lands exposed to heavy rains and seasonal flooding are continuously losing iodine drained into the local rivers. Cropping depletes the soil of iodine; as much as 3 grams of iodine may be lost from one acre during cropping. In general the iodine content of soil is reflected in the iodine content of drinking water provided that this is derived by surface collection and not from deep natural wells. Synthetic nitrogen fertilizers now contribute some iodine to arable lands in more organized and industrialized countries. It has been suggested too that there is a relationship between the distribution of goitre and the calcium concentration of drinking water, not otherwise explicable by a parallel difference in iodine availability (Murray *et al.*, 1948). Hard water may not only be poor in iodine content but calcium may diminish iodine absorption from the food. This subject remains highly controversial (see Kilpatrick and Wilson, 1964).

In health the iodine requirements are about 150 μg per day with additional needs of about 50 μg per day during adolesence, pregnancy and lactation. Iodine is converted to iodide in the gastro-intestinal tract and the iodide, whether derived from iodine or ingested as iodide, is more or less completely absorbed into the body iodide pool in the upper intestine and thereafter either used by the thyroid to synthesize hormones, or it is lost into the urine. In iodide repletion and balance the daily loss of iodine in the urine is usually greater than 100 μg per day. The remainder is lost into the faeces from the liver as unused hormone. A very small amount of iodide is also lost in sweat.

The plasma inorganic iodide (PII) level can be measured, and is a good index of the state of body iodide balance. In a country such as the UK where iodide is relatively plentiful the plasma inorganic iodide levels vary between 0·04 μg and 0·24 μg per 100 ml. In regions of severe iodine deficiency, levels of PII may be as low as 0·01 μg per 100 ml (Fig. 34).

In countries such as Scotland where endemic iodide deficiency does not exist 80 per cent of the iodide is contained in the food, mainly in milk, meat and sea fish, and 20 per cent is derived from drinking water (the usual content of iodide in water is 10 μg per litre or more). In regions of severe iodine deficiency, of which there are many scattered throughout the world (particularly in inland regions far from the sea and at high altitude—see Figs. 35–41), the total iodide intake may be as low as 20 μg per day. For example, in West Sudan, (Darfur region) no sea fish reach the region, meat is hardly ever eaten, milk is scarce, and the iodine content of the local water is as low as 1 μg per litre, the salt is mined locally and is devoid of

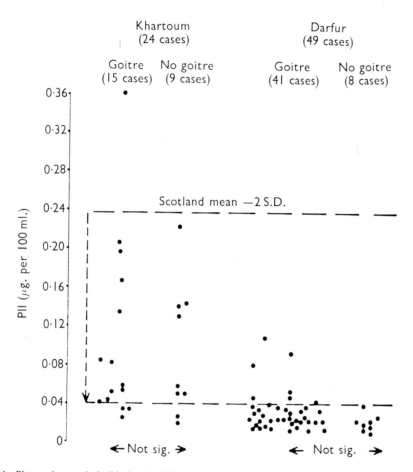

Fig. 34   Plasma inorganic iodide levels (PII) in people from Scotland, Khartoum and Darfur (West Sudan). Not all with low PII levels have goitres. (From Greig *et al.*, 1970.)

iodine. As a result there is a constant and lifelong severe iodine deficiency leading to a high prevalence of goitres, these sometimes being very large.

Where more than 10 per cent of the population is affected this type of goitre is termed endemic. It is simple enlargement of the thyroid in response to iodine deficiency. Simple goitre also occurs sporadically in regions where iodine is adequate. It is questionable whether this type of goitre is due to individual iodine deficiency, and other factors have to be considered. It is useful to discuss endemic and sporadic simple goitre in more detail.

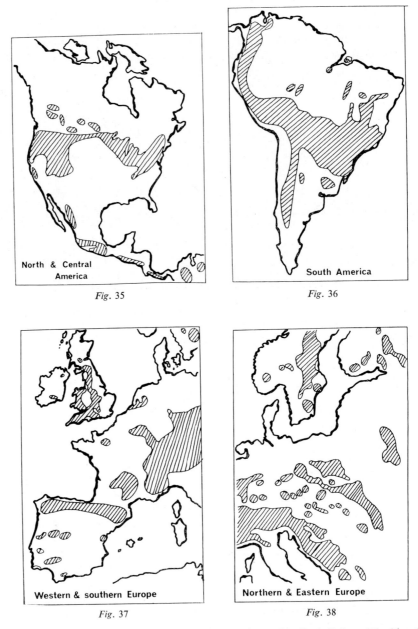

Fig. 35

Fig. 36

Fig. 37

Fig. 38

Figs. 35–38   Shaded areas have community goitre of variable severity. (From Kelly and Snedden, 1960.)

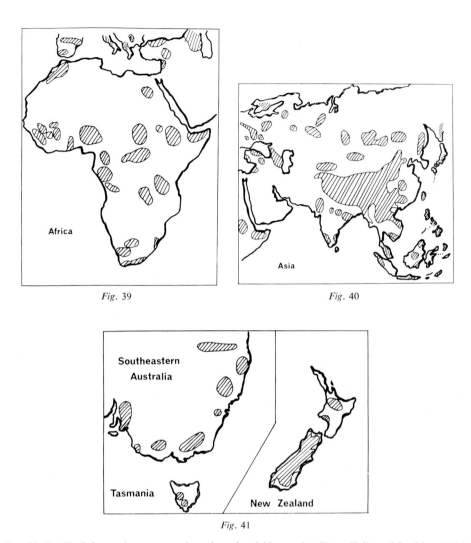

Fig. 39

Fig. 40

Fig. 41

Figs. 39–41   Shaded areas have community goitre of variable severity. (From Kelly and Snedden, 1960.)

SIMPLE GOITRE—ENDEMIC AND SPORADIC

Numerically throughout the world the most common thyroid abnormality is simple goitre. In fact, simple goitre is so common in terms of worldwide prevalence (about 60 million people are affected) that discussion of the epidemiology of thyroid disease in relation to environmental causes has, for practical purposes become almost synonymous with that of the discussion of simple goitre (Kelly and Snedden, 1960). The subject is one which has been under investigation for many years (see McCarrison, 1915).

Simple goitre is best defined as an increase in the size of the thyroid so that it becomes visible and palpable to the trained observer; the definition implies that the patient is euthyroid and does not have neoplastic, inflammatory, autoimmune or malignant thyroid disease, but it does not mean that all simple goitres are necessarily due to one cause, namely iodine deficiency. The goitre is usually symmetrical and diffuse but may be asymmetrical and nodular. In some regions it is very common. In these populations the thyroid may become very large indeed and, as discussed previously, it is usually the consequence of lifelong iodine deficiency. It is curious, however, that in regions of iodine deficiency not all members of the community become goitrous (Fig. 34).

As Kelly and Snedden (1960) have shown the geographical prevalence of simple goitre shows much variation (Figs. 35–41). There are of course, difficulties in comparing the incidence of small goitres because of the different criteria used to define goitre and the inconsistencies of observer variation.

Simple goitre is, however, always several times more prevalent in females and this disparity increases with age. It is certainly more common too during puberty, pregnancy and lactation; these may be times of increased iodide need but suggest too that genetic, hormonal and metabolic influences should not be disregarded.

The cumulative evidence that iodine deficiency is certainly an important cause of endemic goitre may be summarized as follows:

1. Classical epidemiological studies have usually shown a close inverse relationship between the prevalence of goitre and goitre size and the iodine content of the local water, soil, food and salt.
2. In many areas studied throughout the world the administration of iodine prophylactically, either in the form of potassium iodide in drinking water, potassium iodide in the salt, potassium iodate in baking flour or the intramuscular injection of iodized oil, dramatically reduces the incidence of goitre and the size of the goitre.
3. Using radioactive iodine the metabolism of iodine in patients with simple goitre and particularly endemic goitre is consistent with the iodine deficiency theory. For example, in areas of iodine deficiency where the prevalence of goitre is high, the thyroid radioiodine uptake is high, the thyroid radioiodine turnover by the thyroid is high and the absolute iodine content per unit mass of thyroid tissue is low.
4. Goitre can regularly be produced in animals by feeding an iodine deficient diet from which chemical goitrogens have been carefully excluded.

Sporadic Goitre

In the case of sporadic simple goitre the dietary iodine deficiency theory is less clearly applicable. In the UK patients may take an adequate amount of sea fish, meat, milk, cereals, fruit and vegetables—the main sources of iodine in that order—and yet develop simple goitre. Nevertheless, there is controversy as to the importance in the individual of iodine deficiency as a cause of sporadic goitre. For example, in Glasgow, Wayne *et al.* (1964), estimated that the mean dietary intake of iodine of non-goitrous normal women was $273 \pm 26\cdot6\ \mu$g per day whereas in female patients with simple goitre coming to hospital for investigation, the dietary intake appeared to be much lower (130 $\pm$ 19·6 μg per day). They also found that the values for plasma inorganic iodine (PII) and for urinary iodine loss were much lower in goitrous patients than in normal females. All this would suggest that there is some evidence for the iodine deficiency hypothesis as a cause of sporadic goitre in Britain, but retrospective dietary surveys are notoriously difficult and their interpretation is particularly hazardous

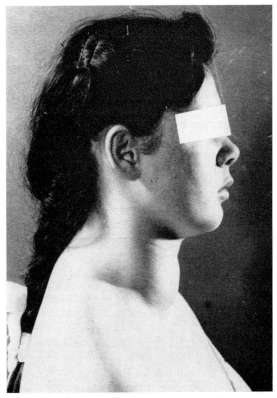

Fig. 42   Type of goitre found sporadically in populations not deficient in iodine (e.g. UK).

when the researcher has prior knowledge of the group of patients in whom he might expect to find a lower dietary intake. Further, it is not clear why, in sporadic areas, only goitrous persons should selectively restrict their iodine intake in contrast to what pertains in endemic areas where all the inhabitants, whether goitrous or not, are involved (see below).

Causes of Simple Goitre other than Iodine Deficiency

Host Factors. Certain unexplained facts emerge in epidemiological studies of the relationship between iodine deficiency and simple goitre. For example, it has been repeatedly shown that in areas of severe iodine deficiency many patients do not develop goitre but have radioisotopic findings consistent with the presence of severe iodine deficiency. This is illustrated by comparing certain indices of iodine deficiency which are determined in goitrous and non-goitrous people from the Sudan where iodine deficiency is very severe. Fig. 34 shows no major differences between goitrous and non-goitrous Sudanese in respect of blood PII levels. We thus believe host factors must be involved even in goitre associated with severe iodine deficiency (Greig et al., 1967, 1970, McGirr and Greig, 1968). The idea that one group of a population is genetically more predisposed to develop goitre in the face of a given degree of iodine deficiency has, of course, been investigated, but neither family studies nor twin studies have shown heredity to be an overwhelmingly important factor. Occasionally some inherited predisposition to goitre formation has been shown. For example, in certain parts of South America where goitre prevalence is very high and both parents have nodular goitre, the statistical chance of their children having a large goitre is increased three times as compared with the children of parents who have no goitre (Stanbury, 1969, 1971). The observation that goitre is more common in females suggests that there are also specific endocrine or metabolic host factors to be considered.

Plant Goitrogens

From time to time interest in plant goitrogens has been renewed. In this connection it has been shown in Finland that at certain seasons grazing pastures may contain a substance, progoitrin, which can be activated to the chemical goitrin, 1–5 vinyl-2-thio-oxizolidone. This substance impairs thyroid hormone production. The conversion of progoitrin to goitrin may take place through enzymatic activity in the plants and appear in water; boiling destroys the enzymes. The release of the goitrogenic substance may also arise due to enzymic activity in the alimentary tract of cows and ultimately appear in milk. Another interesting example of goitrogens in food and milk is found in Tasmania. The suspicion that chemical plant goitrogens may cause goitre arose because of the failure of a regime of iodide supplements to prevent goitre appearing in schoolchildren. It was shown that the goitres occurred more commonly during the spring and early summer and that these may be due to goitrogens in the milk of the cows which grazed in pastures infested with turnip weed. This subject is reviewed by Ramalingaswami (1964).

Some caution is, however, required lest we oversimplify the problem in as much as iodide deficiency which may also be seasonal, may also be involved in these countries. For example, in the Sheffield region it was shown that the mean monthly iodine content of liquid cows milk was high during January, February, March and April but fell to quite low levels throughout the summer months and again rose during the winter (Broadhead et al., 1965). This was because the animals were receiving iodide supplements incidentally in their winter foodstuffs but not from natural grazing during the spring and summer. These workers incidentally also measured the thiocyanate (SCN^-) content of the milk. Thiocyanate, like perchlorate ion ClO_4^-), is known to impair iodine uptake by the thyroid and may thus lead to goitre. Although thiocyanate was present in some samples investigated in Derbyshire the amounts were small, and it was difficult to attribute the goitre in the region to this goitrogen alone. Plant thiocyanates have been shown to cause simple goitre in the Congo, and simple goitre is also caused by a goitrogen in cassava (a plant food) eaten in Nigeria (Delange and Ermans, 1971).

Thus while in certain parts of the world naturally occurring goitrogens are probably more important causes of seasonal goitre than iodine deficiency, on the whole in terms of the total prevalence of endemic goitre their importance is not high.

Other Factors

Neither iodine deficiency nor plant goitrogens could alone or in combination be the sole cause of simple goitre as observed sporadically in countries such as the UK. We have indicated that goitre, especially sporadic goitre, is always more common in females, and shows some familial aggregation. In addition in areas of severe iodine deficiency some people who are severely iodine deficient do not develop goitre. All this leads to the general conclusion that some so far unidentified intrinsic factor or individual host susceptibility must be of importance. While the biochemical pattern of hormone synthesis in simple goitre, whether it be sporadic or endemic, may be slightly deranged, when examined by conventional techniques, no specific defects have been shown in sporadic or endemic goitre. A fair interpretation of present evidence (McGirr, 1960; Stanbury, 1971) would be that the vast majority of patients with simple goitre do not have defects in hormone synthesis. This conclusion may, however, merely underline the relative crudeness of current techniques for showing minor defects in thyroid hormone synthesis which, of course, may be exaggerated in the presence of iodine deficiency or environmental goitrogens. Among environmental goitrogens cobalt ingestion is a rare cause of goitre (BMJ Editorial, 1957).

Testing Trait Hypothesis

It is also worth mentioning a hypothesis for which there has been no good evidence. This hypothesis states that the ability to taste phenylthiourea and related thiocarbonamide antithyroid compounds is dominantly inherited and results in two groups of people, those who have an ability to taste these drugs and those who cannot. In Western Europe about 30 per cent of individuals are non-tasters. It has been claimed that the prevalence of nodular goitre is more common in non-tasters, and it has been proposed that the reason why non-tasters are more liable to goitre is that they are unable to taste certain naturally occurring antithyroid drug compounds, that they have an impaired ability to metabolize them or that their thyroid is more sensitive to antithyroid substances. There is, however, no real evidence for linking the prevalence of simple goitre to the non-tasting trait. In fact, in some areas in South America where the prevalence of goitre is exceedingly high and the goitres are very large the prevalence of non-tasting is extremely small (Stanbury, 1971). (See also hypothyroidism below).

Goitre Associated with too much Iodide

Endemic goitre is found in the north island of Japan, Hokkaido, but paradoxically the affected individuals were found to be ingesting very large amounts of iodine, up to 200 mg (100 times normal requirements) per day. All the cases were euthyroid and PII levels were as high as 44 μg per 100 ml which is 40 times above normal (Stanbury, 1969).

Excessive iodide in some cough mixtures will cause iodide goitre if taken for a long period of time and iodides given to pregnant women can produce iodide goitre in the offspring. The mechanism whereby excessive amounts of iodide block intrathyroidal hormonogenesis which leads to compensatory goitre is poorly understood. It appears that too much iodide interferes with the iodination process and with the release of hormones. It is of importance that hyperplastic thyroids are particularly susceptible, and so regimes of iodide prophylaxis in areas of endemic iodine deficiency should not provide iodide in amounts greater than physiological needs (about 200–300 μg iodide per day).

Endemic Cretinism, Deaf-mutism and Feeble-mindedness

Cretinism is thyroid hormone deficiency starting before or shortly after birth and leading to stunting and mental deficiency; some children may be of short-stature, deaf mutes, or feeble-minded but neither hypothyroid nor goitrous and this distinction is important (Hetzel and Pharoah, 1971).

No country has reliable concordant data on each of these conditions because, especially in poor societies, these individuals are often concealed, die young or are missed in surveys. Nevertheless in regions such as the Alps there was undoubtedly a high prevalence of cretinism and endemic goitre before the regimen of iodide prophylaxis was introduced. It would also appear that when the endemic goitre disappeared so also did the high incidence of cretinism.

Endemic cretinism is regarded by most workers in the field to be an inevitable feature of severe endemic goitre areas, the pathogenic mechanism presumably being that of maternal iodine deficiency leading to foetal iodine deficiency and hypothyroidism. The foetal thyroid may develop abnormally due to the lack of iodine *in utero* so that adequate iodine supplies after birth would be ineffective.

While endemic iodine deficiency appears to be linked to a higher risk of cretinism the relationship is not a straightforward one. For example, the prevalence of cretinism does not co-relate closely with goitre incidence and size within regions of the endemic area. A high degree of consanguinity is a feature of all isolated communities so that genetic influences have to be assessed in association with the iodine deficiency hypothesis.

The apparent link between endemic goitre and dwarfism, mental retardation and deaf-mutism not due to thyroid deficiency, is even more tenuous. There is little doubt that the prevalence of non-cretin but cretin-like defects of growth and neurological development is high in regions of endemic goitre; however, it is not necessarily the same individuals who have goitre. In fact there are quite convincing data to suggest that inbreeding is more important than environment, many of the defects arising from abnormal recessive genes. Because it is difficult or almost impossible to carry out a really satisfactory simultaneous epidemiological study of iodine-deficiency, goitre and genetic studies of goitre, cretinism and non-thryoid dwarfism, deaf-mutism and mental retardation, the true relationships may never be known, nor the role of iodine deficiency *per se*, as distinct from cretinism in the causation of mental retardation and neurological abnormalities in endemic goitre areas, be elucidated.

Summary of the Effect of Environment on the Evolution of Simple Goitre

Simple goitre is easily the most common form of thyroid enlargement and affects about 60 million people in the world. Its incidence varies widely; it is more common in females, and in most areas of the world where the prevalence of large goitre is high there is usually evidence that dietary iodide deficiency is the main reason.

The most common cause of dietary iodine deficiency is deficiency of iodine in the natural soil, vegetation and water in the region. There is no evidence that excessive loss of iodine from the body (with a possible exception during pregnancy) causes iodine deficiency. There is also little evidence, except seasonally in Finland, Tasmania, the Congo and Nigeria that naturally occurring plant goitrogens (chemicals) contribute to the formation of simple goitre. In addition, special defects in hormone synthesis are not numerically important as causes of simple goitre. This applies whether the goitre is sporadic or endemic. There is no good evidence that people who cannot taste phenylthiourea compounds are more prone to nodular goitre. Perhaps the consistent observation that goitre is more common in females has not been given sufficient attention. In this context there are undoubtedly intrinsic endocrine or metabolic goitrogenic influences, particularly during puberty, pregnancy and at the menopause when the prevalence of goitre increases.

It would thus appear that although simple large goitre can undoubtedly be caused by severe iodine deficiency it is not in itself the only cause of this worldwide problem. The causes are likely to be multifactorial and certainly this must be the case in the sporadic simple goitre occurring in the British Isles where neither iodine deficiency nor chemical goitrogens can be definitely implicated.

However, it can be concluded, when severe iodine deficiency is shown to be endemic in a region and large goitre of clinical significance is also endemic, that the supplementation of the dietary iodine can effectively reduce the prevalence of goitre. When possible, dietary iodine can be supplemented by improving communication and food supplies or alternatively adding potassium iodide to salt, water, milk, cows feed, or even to milled flour; alternatively if these procedures are unacceptable a single injection of iodized poppy seed oil can be shown to diminish the prevalence of endemic goitre. One injection can be effective for up to 2–4 years. This approach has been extensively tried in some areas of South America and Borneo.

THE ENVIRONMENT AND OTHER THYROID DISORDERS

Thyroid Dysgenesis—Abnormal Development

Thyroid dysgenesis is the absence of thyroid tissue, athyreosis, or the formation during gestation of only rudimentary thyroid tissue. The latter may be in the position normally occupied by the thyroid gland but sometimes thyroid tissue is found in abnormal positions such as at the base of the tongue or in a line from there down the neck. This kind of thyroid abnormality is responsible for sporadic congenital hypothyroidism or cretinism, or the small amount of thyroid tissue may become inadequate only at a later age producing, for example, juvenile hypothyroidism. As far as we are aware these developmental thyroid defects nearly always arise as isolated examples randomly and by chance and are not influenced by external or maternal environmental factors. However, radiation given during pregnancy either accidentally or incidentally, may lead to abnormal foetal thyroid development, and there is circumstantial evidence that sometimes defective development of a baby's thyroid may be due to genetically abnormal maternal ova, e.g. athyretic cretinism has been observed in sisters and brothers and in twins. There is also evidence that mothers who have in their circulation thyroid autoantibodies, which presumably arise spontaneously, are at risk of producing athyreotic babies. If this is so presumably the maternal antibodies cross the placenta and impair the development of the thyroid tissue in the baby.

Thyroid Dyshormonogenesis—Genetic Enzyme Defects

Impaired intrathyroidal enzyme function may lead to inadequate thyroid hormone production, compensatory thyroid enlargement and cretinism (Fig. 35). These defects are recessively inherited and have been beautifully elucidated by modern radioisotopic and radiobiochemical techniques. Each type has an intrinsic defect in one of the biosynthetic steps of hormonogenesis and the group has been aptly named thyroid dyshormonogenesis (McGirr, 1960). There are, as might be appreciated, several different types classified according to the position at which the hormonogenesis is impaired. It is of interest that each defect is separately inherited, the usual mode being by an autosomal recessive gene so that the parents are heterozygote. It follows that those who are homozygous for the trait are more significantly affected and in addition to the goitre have permanent impairment of thyroid hormone production. The defects, arising as they do in infants and in several members of a family, include absent iodide trapping which is exceedingly rare, and impaired iodide oxidation which is probably the most common; incidentally this type is also often associated with separately inherited nerve deafness and the whole symptom complex is given the name of Pendred's syndrome. There is also a defect which arises because of the failure of coupling of the iodotyrosines to form the iodothyronines (the thyroid hormones). There are also defects which arise because of the production of abnormal thyroglobulin or the release of abnormal thyroglobulin from the thyroid. In the latter condition abnormal iodinated proteins are secreted into the circulation but their hormone content is low. Another type is one in which MIT (monoiodotyrosine) and DIT (diiodotyrosine), not used to make hormones, are lost from the the gland and indeed from the body because of lack of the enzyme, dehalogenase. This enzyme normally removes iodide from MIT and DIT so that it is conserved in the thyroid.

These abnormalities are identified clinically by their early appearance in life with associated hypothyroidism, their familial aggregation and by radioisotopic and biochemical investigations. The severity of any one of the defects would, of course, be increased in the presence of dietary iodine deficiency or if the mother or child were given antithyroid drugs or similar substances. There are, however, family groups with goitre (and hypothyroidism) where even all the elegant modern investigations fail to pinpoint the precise defect in hormonogenesis. For example, families are known in which members have large goitres (sometimes calcified) but no definable biochemical anomaly.

THYROTOXICOSIS

Thyrotoxicosis is due to spontaneous overproduction of thyroid hormones; the disease is seldom fatal now because there are three different but effective treatments. It is in part for this reason especially, that it is difficult to determine the true incidence of the condition in different populations. Any attempts to draw definite conclusions about changes in frequency or geographical distribution of the disease must therefore be interpreted with caution. It would appear that classical thyrotoxicosis (Graves' disease) arises in genetically susceptible individuals; but people with simple goitre are also prone to it especially if they are given too much iodine. The incidence of thyrotoxicosis during the present century appears to have increased.

Pre-existing Goitre and Iodide Intake

It has been shown that simple nodular goitre and thyrotoxicosis were frequently associated in the northern part of the USA but were rare in the south, and more recent studies in Tasmania have also shown an association between the prevalence of nodular goitre and thyrotoxicosis (Connolly, 1970). In that island increased supplies of iodine have resulted in a greater incidence of thyrotoxicosis in people with nodular goitres.

It has been recognized for some time, especially in the European literature, that the administration of iodine to patients with nodular goitres occasionally precipitates an episode of thyrotoxicosis (the so-called Jod-Basedow phenomenon). The phenomenon is especially striking in places where simple goitre is endemic. Tasmania has for some years been recognized as a goitrous part of the world and as an area of relative iodine deficiency. To counteract this, potassium iodate was added to the bread in 1966 in order to ensure an adequate supply of iodine (Connolly, 1971). At approximately the same time—but this was not recognized until later—another source of iodine appeared in Tasmania. This was the use in the dairy industry of iodophors as bactericidal agents both in the washing of the udders and in the milking machines and milk tankers. Coincident with these two sources of iodine being provided, there was a significant increase in iodine intake and in the incidence of thyrotoxicosis in Tasmania.

One problem to be settled is whether this real increase represents an unfolding of a prevalence otherwise low because of the protective effects of iodine deficiency, or a true induction of new thyrotoxicosis. In other words the provision of additional iodine may merely have allowed the incidence of thyrotoxicosis to rise to that expected in a European-derived population. Allowing for the difficulties mentioned in obtaining data, our own estimate is that the new incidence of thyrotoxicosis in Tasmania is now similar to that at present seen in the West of Scotland.

It has also been suggested that relapse of thyrotoxicosis following its apparently successful cure by antithyroid drugs might be correlated with the available supply of iodine. This would be in keeping with the hypothesis that the clinical manifestations of thyrotoxicosis can be subdued by a relative state of iodine deficiency induced by antithyroid drugs and revealed by a great availability of iodine. However, some recent studies, have not confirmed that iodine excess causes recurrent thyrotoxicosis in those so predisposed.

Stress

The incidence of thyrotoxicosis appears to have varied throughout the century: it was common in the period 1920 to 1930, with a fall thereafter even during the Second World War. In Nazi-occupied Europe this is of course a rather surprising finding especially as stress has been emphasized as one of the causes of the disease.

There was, however, one exception to this general rule. This was shown in a study in Denmark in the period 1941 to 1945 during which there was found to be a 9-fold increase in incidence of thyrotoxicosis and nodular goitre and a 4-fold increase in thyrotoxicosis and diffuse goitre. This increase rapidly fell after the war suggesting that an environmental factor such as a temporary rise in iodine intake or stress had been causative.

While there is a great deal of evidence that the Long-acting Thyroid Stimulator (LATS) is implicated in the pathogenesis of thyrotoxicosis it must be admitted that the true primary cause of the condition is not yet known, and as in simple goitre it may depend on the complex interaction of many genetic and environmental agents. Certainly the tendency to develop thyrotoxicosis is often inherited, as twin and family studies show (McDougall and Greig, 1971).

HYPOTHYROIDISM

Adult hypothyroidism (myxoedema) is usually the consequence of spontaneous autoimmune thyroiditis and atrophy of the gland; in other patients the thyroiditis (Hashimoto's thyroiditis) is more chronic and there is a firm goitre. There is evidence that atrophy and goitrous thyroiditis with or without hypothyroidism are different stages of an autoimmune process directed against thyroid tissue. The tendency to develop these conditions is inherited, but the exact mode of inheritance is not known. Certainly both states are more common in females and the risk increases with age. A symptomatic low grade thyroiditis with goitre affects about 2 per cent of women over 50 years of age in Britain and a similar incidence probably exists in most Caucasian populations. There is suggestive evidence that many of these people may eventually develop hypothyroidism as part of a progressive thyroiditis. Not very much is known about the influence of the environment on the initiation or maintenance of thyroiditis due to autoimmunity. It is not a sequel to acute or subacute viral thyroiditis which is occasionally seen complicating mumps (viral parotitis) and influenza. Viral complicating thyroiditis settles quite quickly without permanent residual thyroid damage. Autoimmune thyroiditis is thought to reflect individual immunological instability occurring without extrinsic stimulation. There are no data on the real geographical distribution of thyroid autoimmune diseases but it would certainly be worthwhile obtaining this type of information.

In adults treatment of thyrotoxicosis with radioactive iodine often leads to permanent hypothyroidism (Greig, 1965), and this is also a complication of thyroidectomy. Both treatments are common causes of myxoedema. Paradoxically, excessive iodide intake may lead to hypothyroidism and this is seen in patients taking cough mixtures such as 'Felsol' which may result in the intake of several grams of iodine per day (normal intake is 200 μg per day). Gross iodide excess seems to interrupt normal hormonogenesis and if long continued also leads to goitre (see above). An interesting epidemiological example of this is found in subjects inhabiting the north west shores of Japan whose iodine intake from fish is very large. A number of these people are hypothyroid. No other examples of endemic iodide goitre and hypothyroidism in humans are recorded, but more work needs to be done on the geography of iodide goitre in particular and of hypothyroidism in general. It is of interest, for example, that recently iodide goitre and hypothyroidism was induced in Kentucky race horses in the USA after they were fed kelp which has a very high iodine content.

Dysgenesis and dyshormonogenesis have been discussed above. Unless treatment is commenced very early, both lead to congenital hypothyroidism with stunting and abnormal skeletal growth plus mental retardation. The link between severe iodine deficiency, endemic goitre and cretinism has also been discussed above. The introduction of iodized salt appears to have resulted in a more striking reduction in the incidence of cretinism than of goitre in endemic areas; for example studies in Switzerland and South America demonstrate this fact. Changes in the incidence of disease may not, however, be due to one factor. The possibility of the reduced incidence of cretinism being due to cocomitant opening up of isolated areas allowing wider travel and marriage of individual outside what was formerly a closed community must be considered before iodine deficiency *per se* is implicated as a cause of cretinism.

As will be mentioned below accidental irradiation of childrens' thyroid in Hiroshima and Nagasaki in 1945 and in the Marshall Islands in the Pacific in 1954 resulted in impaired thyroid reserve and hypothyroidism; some of the same two groups of children are now also developing thyroid cancer.

THYROID CANCER

Before considering thyroid cancer as it is seen in the human, it is relevant to consider the considerable body of knowledge about the experimental induction of such tumours in the rat. Goolden (1972) has reviewed evidence that prolonged administration of goitrogens after radioiodine irradiation eventually gives rise to thyroid nodules and frankly malignant thyroid tumours. The postulated mechanism of production of these tumours is through long term secondary hyperplasia coupled to the permanent but latent radiation damage. It is noteworthy that there was found to be a critical dose of radioactivity; doses which exceeded this range resulted in less tumour formation presumably because too much damage kills the cells rather than inducing premalignant changes.

Overall, in the human, cancer of the thyroid is a comparatively rare tumour resulting in approximately 400 deaths per annum in the UK. There is, however, good evidence that the mortality from thyroid cancer varies from one country to another. This may not, however, reflect different prevalences but rather different distribution of the type of cancer because thyroid tumours vary enormously in malignancy. In this respect data based on surgical experience is different from that of clinicians seeing all thyroid cancer cases. For example, the incidence of histological malignancy in solitary thyroid nodules removed at operation may be as high as 15 per cent while in the same region very few people die of thyroid cancer. The problem here is that histological cancer is not necessarily synonymous with invasive cancer.

The classical view has always been that the incidence of thyroid cancer bears a direct relationship to the incidence of endemic or sporadic goitre, and earlier in this century observations on both man and animals were consistent with this viewpoint. These observations would fit in well with the animal experiments mentioned above where goitre and cancer were induced in circumstances leading to an increased TSH production by the pituitary and in the concomitant decrease in thyroid cancer and endemic goitre produced by iodide supplements. With the recent availability of radioimmunoassay techniques for measuring serum TSH, it has been realized however, that in the types of established goitre seen in Europe the serum TSH is not usually elevated, so that TSH stimulation may not necessarily be the common link between simple goitre and cancer. The situation is confused, however, since recent surveys from the USA have produced good evidence against the association of thyroid cancer with simple goitre; similar findings have been reported from Finland. It is of interest that thyroid cancer incidence is as high in Iceland as it is in the UK.

It has also been shown that the fall in goitre incidence coincidental with the introduction of iodized salt in the USA has not resulted in a similar fall in the incidence of thyroid cancer. There is also supportive evidence for this viewpoint from Switzerland. The most striking evidence against this general trend has been data from Colombia in Latin America where there is clearly an increased incidence of thyroid carcinoma in this endemic goitre region (Smithers, 1970).

Since all the causes of endemic goitre are not known it is difficult to decide which of the possible factors including iodine deficiency could be the cause of the thyroid tumours. Perhaps there are unidentified geographical differences in susceptibility to thyroid cancer.

Effect of Irradiation on the Induction of Thyroid Tumours

The study of the effects of radiation has provided the most convincing evidence for one environmental effect on the induction of thyroid cancer (Goolden, 1972). This first came to notice when it was pointed out that of children with thyroid carcinoma a significant number had received X-ray treatment to the chest usually directed to the area of the thymus. This work excited a series of other studies which showed that the administration of therapeutic irradiation in childhood was associated with the statistically increased risk of development of tumours of the thyroid, both simple (adenoma) and malignant. This has been confirmed in other prospective and retrospective series. Indeed in one series it was estimated that 30 per cent of children receiving X-ray treatment to the chest in infancy ultimately developed thyroid nodules. A similar association between external radiotherapy and the subsequent development of thyroid carcinoma has been found in the adult.

Ionizing radiation is of course used to treat thyrotoxicosis. In former years external radiotherapy was employed, and it is of interest that a follow up study of such cases showed no increased risk of thyroid neoplasia. Since the 1940's thyrotoxicosis has been treated by the oral administration of the radioactive isotopes of iodine (^{131}I). Despite the many thousands of patients treated by this method only a very small number have subsequently developed thyroid cancer. This may be due to the fact that ^{131}I so damages the thyroid cell population and uniformly inhibits cell division (Greig, 1965) that no cells survive which are capable of causing cancer. There is however, some evidence of the potentially hazardous effect of ^{131}I therapy in the treatment of childhood thyrotoxicosis in which both benign and malignant thyroid tumours have been produced.

These findings of the thyroid tumour inducing properties of ionizing irradiation given to normal people are reinforced by the follow up of children exposed to immediate and fall-out radiation from the detonation of atomic bombs at Hiroshima and Nagasaki, and among the Marshallese islanders accidentally exposed to fall-out after the Rongelap Atoll atom bomb tests. For example, it has been noted that of the 82 islanders involved 16 benign thyroid nodules and 3 cases of thyroid carcinoma subsequently developed (Goolden, 1972).

References

Broadhead, G. D., Pearson, I. B. and Wilson, G. M. (1965). Seasonal changes in iodine metabolism. *Brit. Med. J.*, **1**, 343.

Connolly, R. I., Vidor, G. I. and Stewart, J. C. (1970). Increase in thyrotoxicosis in endemic goitre area after iodination of bread. *Lancet*, **1**, 500.

Delange, F. and Ermans, A. M. (1971). Further studies on endemic cretinism in Central Africa. *Hormone and Metabolic Research*, **3**, 431.

British Medical Journal—Editorial (1957). Goitre caused by cobalt. *Brit. Med. J.*, **1**, 1293.

Goolden, A. W. G. (1972). The effect of radiation on the thyroid gland with particular reference to the development of tumours. In *Modern trends in endocrinology*, edited by F. G. T. Prunty and H. Gardiner-Hill, London, Butterworths.

Greig, W. R. (1965). Radiation, thyroid cells and ^{131}I therapy—a hypothesis. *J. Clin. Endocr.*, **25**, 1411.

Greig, W. R., Boyle, J. A., Duncan, Anne, Nicol, Janette, Gray, Mary J. B., Buchanan, W. W. and McGirr, E. M. (1967). Genetic and non-genetic factors in simple goitre formation—evidence from a twin study. *Quart. J. Med.*, **36**, 175.

Greig, W. R., Gray, H. W., McGirr, E. M., Kambal, A. and Rahman, I. A. (1970). Investigation of endemic goitre in Sudan. *Brit. J. Surgery*, **57**, 11.

Hetzel, B. S. and Pharoah, P. O. D. (1971). Endemic cretinism. *Monograph Series No. 2*. Institute of Human Biology, Papua, New Guinea.

Kelly, F. C. and Snedden, W. W. (1960). *Endemic Goitre*. WHO Monograph Series No. 4.

Kilpatrick, R. and Wilson, G. M. (1964). Simple non-toxic goitre. In *The Thyroid Gland*, edited by Rosalind Pitt-Rivers and W. R. Trotter. London, Academic Press, **2**, 88.

McCarrison, R. (1915). Collected Papers on Goitre and Cretinism 1913–1914. Calcutta, Thacker, Spink and Co.

McDougall, I. R. and Greig, W. R. (1971). Pathogenesis and treatment of thyrotoxicosis. *Scot. Med. J.*, **16**, 519.

McGirr, E. M. (1960). Sporadic goitrous cretinism. *Brit. Med. Bull.*, **16**, 113.

McGirr, E. M. and Greig, W. R. (1968). Epidemiology of thyroid disease. *Proc. Roy. Soc. Med.*, **61**, 385.

Murray, Margaret M., Ryle, J. A., Simpson, Beatrice W. and Wilson, Dagmar C. (1948). Thyroid enlargement and other changes related to the mineral content of drinking-water. *Medical Research Council Memorandum No. 18*. London, HMSO.

Ramalingaswami, V. (1964). Endemic goitre. In *The Thyroid Gland*, edited by Rosalind Pitt-Rivers and W. R. Trotter. London, Academic Press, **2**, 71.

Scrimshaw, N. S. (1964). The geographic pathology of thyroid disease. In *The Thyroid*, edited by J. B. Hazard and D. E. Smith. Baltimore, Williams and Wilkins.

Smithers, D. (1970). *Tumours of the Thyroid Gland*. London and Edinburgh, Livingstone.

Stanbury, J. B. (1971). Research on endemic goitre in Latin America. *WHO Chronicle*, **24**, 537.

Stanbury, J. B. (1969). *Endemic Goitre*. Pan-American Health Organization, WHO, Washington D.C., USA.

Wayne, E. J., Koutras, D. A. and Alexander, W. D. (1964). Clinical Aspects of Iodine Metabolism. Oxford, Blackwell Scientific Publications.

Chapter 13

Environmental Factors in Relation to Obstetrics

Dugald Baird

The outcome of pregnancy is influenced by the environment in which the mother has grown up, as well as that in which she lives during her pregnancy. In the case of some congenital malformations and biochemical deficiencies, where there is a hereditary predisposition, the incidence may be increased by environmental factors. The ethnic differences in reproductive efficiency may be due to environmental factors as well as heredity.

For some purposes the environment will be taken to include such factors as housing, income, education and obstetric care, all of which are related indirectly and in varying degree to such biological factors as the age of the mother and the number of children.

Formerly when the management of pregnancy and labour was much more conservative than it is today, and breast feeding was the rule rather than the exception it was clear that the pre-requisites for easy and short labour and successful lactation were youth, good physique and health. In primigravidae the peak of reproductive efficiency is maintained for a relatively short time, between the ages of 18 and 22 approximately. If the mother is under 18 and not fully grown she may compete with her foetus for food. If over 22 the reproductive efficiency begins to decline. This applies in widely differing societies.

THE EFFECT OF INDUSTRIALIZATION ON CHILDBEARING

In traditional societies, for example, in West Africa, childbearing begins soon after the menarche because high fertility is essential in a situation where there are very high death rates at all ages in children. Expectant mothers undertake heavy work even during pregnancy, and food may at times be so scarce that even 'physiological' weight gain does not occur. In these circumstances as many as 30 per cent of babies may weigh less than 2,500 g, which although it may reduce the risk of dystocia lessens the chance of survival. Risks are high for both mother and baby especially as medical care is seldom available even in an emergency.

In Britain 40 years ago the situation was even worse for childbearing than in the under-developed countries because of the cold damp climate, lack of sunshine, pollution of the environment and depressing slums in the poorer parts of the large industrial cities. These conditions were much worse in the North of Britain than in the South and were probably worst of all in the industrial belt of Clydeside centred on Glasgow. They not only undermined health but predisposed to emotional disturbances, delinquency and crime.

In the semi-skilled and unskilled classes women were often poorly grown and under-nourished. Childbearing began at an early age, families were often very large and stillbirth and infant mortality rates were high. The wives of professional men were better grown and well nourished but often postponed childbearing for social or economic reasons till they were in their late twenties or early thirties, by which time the risks to the mother and child were much increased, especially since obstetric management was still very conservative.

Britain was the pioneer in the Industrial Revolution which brought her wealth with advantages in many fields, but also deterioration in the environment in the large cities which produced diseases such as rickets, chronic bronchitis and rheumatic fever which made childbearing even more dangerous than it had been before. Most births took place at home with a 'handywoman' in attendance. For the children who survived, life was hard, employed as they were in factories and mines. The Education Act of 1870 was the first effective legislation which helped to stop this exploitation.

In the two World Wars in which Britain was involved (1914–18 and 1939–45) there was great loss of life and property, very serious shortage of food due to the German U boat blockade, dissipation of financial resources and serious disruption of society and of family life. The First World War revealed that the health and physique of many young men recruited to the armed forces were extremely poor. This disturbing state of affairs provided the impetus for a great improvement in social and preventive health measures. After the war there followed social unrest, the General Strike and finally the world trade recession of 1928–32 which increased unemployment until, in Scotland, it was estimated at about 30 per cent of the work force.

Unemployment diminished as preparations for the Second World War got under way. Again there was a German U boat blockade which was relieved to some extent when the USA entered the war after Pearl Harbour. A new feature of this war was a National Food Policy which placed vulnerable groups such as expectant mothers and young children in a priority class for essential foods. Excellent propaganda improved its effectiveness. The result was an overall improvement in nutrition. Those who had been chronically undernourished had a better diet, and those who had been overfed and overweight had their food intake reduced by rationing with consequent improvement in their health.

Birth and Infant Death Rates in Britain

The infant mortality rate in England and Wales, which at the end of the last century had been higher than in Scotland, fell more quickly, so that in 1938 it reached 57 per 1,000 compared to 77 in Scotland. In the years 1934–38 when the birth rate in Croydon was 13·5 per 1,000 and the infant mortality 46, the comparable rates in Glasgow were 19·7 and 99·0 respectively.

The higher rate in Scotland was associated with a lower standard of housing (22 per cent statutory overcrowding compared with 4 per cent in England). Overcrowding is still much worse in Scotland.

A high birth rate is not necessarily associated with a high infant mortality rate, for example in Holland in 1938 although the birth rate was 20·7, the infant mortality rate was only 37 per 1,000. But in Britain in the 1930's the infant mortality rate was highest where the birth rate was highest. Thus in the depressed areas in the North this resulted in an undue proportion of the children born in the 1930's being reared in a poor environment which in due course seriously affected their reproductive efficiency. If, however, the birth rate in these areas had fallen below replacement level as it did in the more affluent South, the sudden large fall in the number of births would have had serious consequences for the age structure of the population. As it was, a Royal Commission on Population was appointed towards the end of the Second World War to examine the problem and recommend policies which would prevent a disastrous fall in the birth rate.

Perinatal Mortality Rates

Stillbirths were first required to be registered in England and Wales in 1928 and in Scotland in 1939. By 1950 the term perinatal mortality (stillbirths plus deaths in the first 6 days of life per 1,000 total births) had come into general use and is now regarded as the most convenient measure of the standard of obstetrics.

The interest in the perinatal mortality rate in the 1950's was stimulated by the fact that deaths in this category had not fallen in the same way as maternal and infant mortality rates.

The decline in the maternal mortality rate began with the introduction of the sulphonamides in 1936, which reduced the number of deaths from puerperal sepsis. Before 1936 childbirth was one of the most dangerous occupations and in fact caused the standard death rate in women between the ages of 20 and 45 to be higher than that in men of the same age, the reversal of the situation in all other age groups. The sulphonamides and later penicillin also prevented many infant deaths from infection and many perinatal deaths from birth trauma by making caesarean section safer in cases of difficult labour. However, poor reproductive efficiency of the mother still remained the most important cause of perinatal mortality.

Causes of Perinatal Mortality

It has been found convenient in studying the influence of the environment on perinatal mortality to divide perinatal deaths into two groups.

(1) An 'environmental' group consisting of 'unexplained' deaths of babies weighing less than 2,500 g., deaths from congenital malformations and those related primarily to diseases in the mother.

(2) An 'obstetrical' cause group consisting of deaths from pre-eclampsia, mechanical complications of labour, 'unexplained' deaths of babies weighing more than 2,500 g., ante-partum haemorrhage and those due to rhesus incompatibility.

Deaths in the environmental cause group are largely unavoidable whereas those in the obstetrical cause group can often be prevented by a high standard of obstetrics. In the environmental group death rates are high at the extremes of maternal age and in the lower social classes. In the obstetrical cause group the rates increase with increasing age of the mother but the social class differences are relatively small. This classification is more helpful than one based on autopsy findings alone for determining why as well as how the baby died (Baird and Thomson, 1969).

Sequential Changes in Perinatal Mortality Rates

In 1928 the perinatal mortality rate in England and Wales was 60 per 1,000, rose to 63 in 1933 and fell slowly till 1940 after which it fell steeply. The wartime fall was greatest in South Wales where the perinatal mortality and unemployment rates had been highest. The fall was least in London and the South East where the perinatal mortality and unemployment rates had been lowest. The wartime fall in the perinatal mortality rate was thought to demonstrate the beneficial effect of improving the diet during pregnancy especially in previously under-nourished women. The fact that 20 years later the children born during these war years were found to be better grown and healthier adults than their parents, tends to confirm this theory. It was disappointing that the perinatal mortality rate remained stationary between 1948 and 1957 during the first 10 years of the National Health Service. The reasons for this will be discussed later.

Regional Differences in Perinatal Mortality Rates

Figure 43 shows that in 1957 the perinatal mortality rate was low in the East, South East and South of England. It was high in the North East and the North West of England, in South Wales and the South West of Scotland. The geographical distribution of perinatal mortality is the same as for diseases known to be closely associated with poverty.

The Perinatal Mortality Survey conducted by the National Birthday Trust in 1958 made it possible to study the geographical changes in more detail. Illsley (1963) divided Britain into three Zones—North, Central and South—roughly on the basis of perinatal mortality rates, and it was for this reason that Wales was included in the North Zone despite its geographical position. He found great differences in socio-economic structure. For example, the North

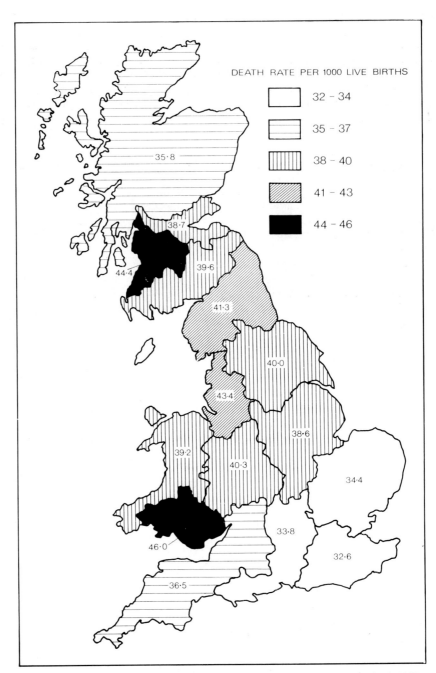

Fig. 43 Perinatal mortality rates. Standard regions of England, Wales and Scotland, 1957.

Zone contained 30 per cent of Britain's professional workers and 43 per cent of its semi-skilled and unskilled workers. The corresponding figures for the South Zone were 43 and 25. The percentage of tall women (defined as 65 in. or more) rose steadily from 26 amongst wives of semi-skilled and unskilled workers to 41 in the professional group. These social class differences in height persist in each region and within each social class. For example, the percentage of tall women in the professional group varies from 37 in the North to 44 in the South. In all social classes, except the professional classes, the mean height of the mother decreases with increasing family size.

In each Zone the professional groups had the lowest perinatal mortality rate, but inside each social class the rate was lowest in the South.

The significance of the mother's height in this context is that many short women have not grown to their genetically determined height, and it seems unlikely that defective growth and development will be confined to the skeleton but will also adversely affect the efficiency of other systems. Table 32 shows that women brought up in a professional home and whose

TABLE 32

Height of wives, educational standard and mortality ratio (all parities) by socio-economic group of father and husband: National Birthday Trust Mortality Survey 1958

Husband's socio-economic group	Father's socio-economic group											
	Professional			Non-manual			Skilled			Unskilled		
	(*a*)	(*b*)	(*c*)	(*a*)	(*b*)	(*c*)	(*a*)	(*b*)	(*c*)	(*a*)	(*b*)	(*c*)
Professional	46	84	(71)	40	71	(72)	37	47	(76)	32	39	(93)
Non-manual	43	73	(76)	36	44	(85)	30	30	(83)	28	18	(89)
Skilled	34	48	(81)	32	32	(89)	28	17	(94)	23	11	(104)
Unskilled	27	42	(117)	28	20	(82)	26	12	(102)	24	10	(124)
All	41	69	(79)	34	43	(85)	29	21	(91)	25	12	(111)

(*a*) % 65 in. or more in height.   (*b*) % Educated beyond the minimum.   (*c*) Mortality ratio.

husband also belongs to a profession are taller than any other group of women, probably because they have been reared in the 'best' environment, have had more education and have the lowest perinatal mortality ratio. Those who move up in the social scale on marriage are taller, have more education and lower perinatal mortality ratios than those who do not rise in the social scale or move down. Today many more move up than down, so that social classes IV and V have decreased in size. This differential social migration may increase the perinatal mortality rate in social classes IV and V or prevent it from falling. The same effect can result from migration between regions. Those who move from depressed to more prosperous regions are taller and have lower perinatal mortality rates than those who do not migrate. They may not influence the rates in the region to which they go but their out-migration will help to maintain the higher rate in the region which they have left.

Fig. 44 shows that in each Zone the death rate from environmental causes was higher in urban than in rural areas, and in both the rates declined steadily from North to South. In the obstetrical cause group death rates from pre-eclampsia fell slightly from North to South. This is in keeping with the finding (MacGillivray, 1960) that the incidence of pre-eclampsia and its severity decrease from North to South in the UK. The death rate from mechanical causes is higher in rural areas in all Zones, especially the Central.

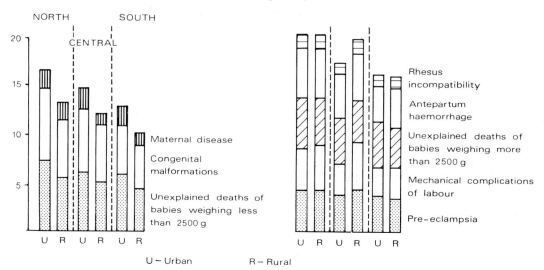

Fig. 44 Perinatal mortality rates in 'environmental' and 'obstetrical' cause groups in three zones in Britain, 1958.

It can be concluded therefore that the higher overall death rate in the North Zone was due more to the lower reproductive efficiency of the women in the North Zone than to differences in the standard of obstetric care. In all Zones the first priority in urban areas was to improve the health and physique of the women, obviously a long-term project, and in rural areas to provide easier access to skilled obstetrics.

STANDARDS OF OBSTETRIC CARE

Figure 45 shows that in Aberdeen the perinatal mortality rate was much lower than in the North Zone and even slightly lower than in the South Zone, evidence of what can be done by a high standard of obstetric and paediatric care, even in women who are much shorter in stature than the average. The rate was lower than that of the Survey as a whole in all pregnancies except the second. Obstetric complications are least common in second pregnancies and a successful outcome depends, more than in any other pregnancy, on the reproductive efficiency of the mother. Therefore there is much less scope in a second pregnancy for reducing the mortality rate by a high standard of obstetric care. This point is well illustrated by a comparison of stillbirth rates in the early 1950's in the city of Aberdeen, where all first and second births take place in hospital, with that of the Netherlands where a majority were confined at home. In primigravidae the stillbirth rate in the Netherlands was 20 per cent higher than in Aberdeen but in second pregnancies the situation was reversed, the rate in the Netherlands being 20 per cent lower than in Aberdeen. The fact that the Aberdeen women were confined in hospital in the care of specialists would obviously be more likely to improve the results in primigravidae. The lower rate in second pregnancies in Dutch women could be explained by their superior health and physique (they were almost 3 inches taller on average than Aberdeen women). In Sweden, where women have an equally high standard of health and physique as in the Netherlands, the perinatal mortality rate is lower, probably due to the very high incidence of hospital confinement (almost 100 per cent).

The probable explanation of the continued fall in the perinatal mortality rate in Aberdeen from 1948 to 1957 when the national rate was stationary is that all first, second and third births (85 per cent of the total) took place in hospital from 1948 onwards. The lack of any improvement in the national perinatal mortality rate between 1948 and 1957 was due to the inevitable delay in providing an efficient modern maternity service throughout the whole country after 5 years of war.

In Aberdeen the downward trend was arrested in 1958 and the rate remained stationary till 1962. This was the result of two factors.

(1) There was now little scope for prevention by good obstetrics except possibly in the 15 per cent of women, mostly of high parity, who still booked for home confinement.

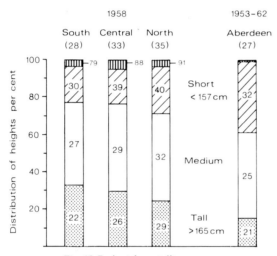

Fig. 45 Perinatal mortality survey.

(2) From 1948 there had been a steady increase in the death rate from malformations of the central nervous system which from 1958 was sufficient to neutralize the fall in the death rate which occurred from other causes.

PERINATAL DEATH RATES FROM MALFORMATIONS OF THE CENTRAL NERVOUS SYSTEM

Perinatal deaths from CNS malformations will be discussed in more detail because there is an association between high death rates in children from this cause and the year of the mother's birth rather than the year of the pregnancy which resulted in the malformation in the child.

Table 33 shows that in Aberdeen the perinatal death rate from CNS malformations was high in primigravidae in 1948–52. This occurred very largely in women between the ages of 18 and 22 from social classes IV and V, and it will be seen that this cohort had greatly increased rates in successive pregnancies during the following 15 years. In 1958–62 a rise also occurred in primigravidae mainly from the upper social classes. They were the wives of non-manual or skilled manual workers who were older and had fewer subsequent children so that the cohort effect seen in the younger women from the semi-skilled and unskilled social groups did not occur.

TABLE 33

Perinatal mortality rate from CNS malformations in 5-year periods by pregnancy number Aberdeen City 1958–70

Pregnancy	1948–52		1953–57		1958–62		1963–67		1968–70		Total			
	No.	Rate	N	R	N	R	N	R	N	R				
1st	28	4·9	13	2·1	20	3·6	11	2·1	9	2·9	81	3·2	25	184
2nd	14	2·8	14	3·0	13	2·7	10	2·2	3	1·2	54	2·5	21	497
3rd	7	2·1	8	2·1	13	4·3	3	1·8	3	2·2	34	2·6	12	797
4th +	14	4·8	8	2·5	15	5·2	15	6·3	4	3·5	56	4·2	13	324
All Preg.	63	3·9	43	2·6	61	3·7	39	2·6	19	2·3	225	3·5	72	802

In Scotland as a whole the death rate from anencephalus rose between 1940 and 1945 reaching 3·1 per 1,000 on two occasions. It rose again from 1951 onwards and remained over 3 per 1,000 from 1957 to 1964. This rise affected all social classes, even social class I.

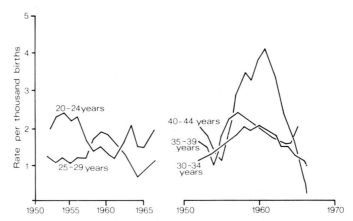

Fig. 46 Stillbirth rate per 1,000 births, anencephalus, Scotland, 1950–68, by age of mother, in Social Classes I and II.

Fig. 46 shows that in social classes I and II in the age group 20–24 it rose from 1953 to 1956 and in the age group 25–29 from 1958 to 1962. In both age groups the women with increased death rates from anencephalus were born between 1931 and 1934. In the age group 35–39 the rate was highest in 1957 in women born in 1920. In the 40–44 age group the rate was highest in 1962 in women born in 1919. In social classes IV and V a similar sequence with increasing age occurred. In the 15–19 age group a sudden rise to over 5 per 1,000 occurred in 1952–53 in women born in 1934 and in the age group 40–44 the rate was over 7 per 1,000 between 1957 and 1959 in women born between 1912 and 1919.

In England and Wales the causes of stillbirth were not registered till 1961, and therefore sequential changes in the stillbirth rate from anencephalus from 1939 onwards could not be studied. Data from Birmingham (Leck and Miller, 1963) suggest that changes in the death rate were very similar to those in Scotland over this period but at a lower level.

The evidence suggests that in Scotland, at least, women born during the First World War and during the worst of the industrial depression in 1928–32 had higher stillbirth rates from anencephalus (and from other malformations of the central nervous system). The timing of the rise in the death rate depends on the year of the mother's birth rather than on the year of her pregnancy. The fact that the reproductive mechanism was shown, 20 years after the mother's birth, to have been seriously impaired suggests that her oocytes may have been damaged either before or soon after she was born. It is known that if the mother contracts rubella during a pregnancy there is a high risk of malformation in the foetus but not involving the CNS. The evidence that other infections, such as influenza, can cause malformations of the central nervous system is conflicting. Leck (1963) found that in outbreaks of influenza there was an increase in the incidence of oesophageal atresia, cleft lip, anal atresia and exomphalos but no increase in malformations of the CNS. In Scotland no association has been found between outbreaks of influenza and the subsequent stillbirth rate from CNS malformations. Nevertheless several reports show that babies with malformations of the CNS are born more often in winter when infections are more common than in the summer. As there is considerable variation in the length of gestation in cases of anencephalus, it might be more accurate to use the date of the last menstrual period rather than the date of birth in studying seasonal incidences.

The high death rate from CNS malformations has been the main reason for the higher perinatal mortality rate in Britain compared with other North European countries. For example in 1961–63 the stillbirth rates in the Netherlands, England and Wales and Scotland were 13·1, 15·3 and 16·2 respectively, but when deaths from CNS malformations are removed the rates for all other causes were 12·8, 13·4 and 13·2 respectively.

LOW BIRTH WEIGHT

Twenty years ago 50 per cent of all perinatal deaths occurred in babies weighing less than 2,500 g. but today this percentage is much higher. For example in 1968–70 in Aberdeen the perinatal mortality rate was 20 per 1,000, and 75 per cent of the deaths were in babies weighing less than 2,500 g. Many of these were the result of such complications of pregnancy as pre-eclampsia, antepartum haemorrhage or twin pregnancy, but many were 'unexplained' obstetrically and were probably predisposed to by social factors. Table 34 shows that in 9,154 first births in Aberdeen in 1958–67 the incidence of babies weighing less than 2,500 g. varied from 4·1 per cent in tall women (>162 cm) in social classes I and II to 10·2 per cent in short women (<155 cm) in social classes IV and V. When the gestation lasted more than 37 weeks there was no social class difference in the incidence of babies weighing less than 2,500 g in women of the same height. When gestation lasted 37 weeks or less the incidence of babies weighing less than 2,500 g. increased with decreasing social status in each height group. Short gestation in the upper social group was more often related to well defined obstetric complications than in the lower social group where the short gestation was more likely to be the result of a poor environment.

TABLE 34

Percentage of babies weighing 2,500 g or less in 9,154 primiparae in Aberdeen 1958–67
by social class and height of mother and length of gestation

Social class	Tall >162 cm		Medium		Short <155 cm		All Heights	
	37 weeks or less	38 weeks or more	37 weeks or less	38 weeks or more	37 weeks or less	38 weeks or more	37 weeks or less	38 weeks or more
I–IIIA	2·2	1·9	3·3	3·8	4·8	5·5	3·1	3·3
IIIB — IIIC	3·1	1·8	3·5	3·0	5·1	4·1	3·9	3·0
IV — V	3·6	1·5	4·6	4·4	5·7	4·5	5·0	4·0
All classes	3·0	1·8	4·0	3·6	5·3	4·6	3·9	3·3
Numbers	2,643		4,521		1,990		9,154	

Table 35 shows that the perinatal mortality rate from 'unexplained' causes in babies weighing less than 2,500 g was high in primiparae in 1948–52 and that over the following 15 years the rate rose in successive pregnancies so that high rates occurred in the 4th+ pregnancy

TABLE 35

Perinatal mortality rate from 'unexplained' deaths in babies weighing <2,500 g.

Aberdeen City 1948–70

Preg. No.	1948–52		1953–57		1958–62		1963–67		1968–70		Total	
	No.	R	No.	R	No.	R	No.	R	No.	R	No.	R
1st	59	10·4	44	7·8	35	6·3	42	8·0	25	8·1	205	5·8
2nd	38	7·6	36	7·7	41	8·4	30	6·8	14	5·6	159	7·0
3rd	17	5·1	19	6·6	17	5·7	26	9·4	10	7·3	89	6·9
4th +	33	11·5	27	8·3	31	10·0	27	10·0	9	8·0	127	9·5
All Preg.	147	9·0	126	7·7	124	7·6	125	8·2	58	7·2	580	8·0

group in 1958–62 and 1963–67. This is very similar to the sequential changes seen in perinatal death rates from CNS malformations in the same population (see Table 33). The short stature of these women is obviously related to deprivation which must have had profound effects on the development of the whole body and thus on reproductive efficiency.

PERINATAL MORTALITY IN BRITAIN IN 1969

The fall in the National Perinatal Mortality rate was renewed in 1958 and is still continuing. This is the result of:

(1) Improvements in the obstetric services, including a better distribution of obstetric specialists throughout the country and better training of both specialist and general practitioner obstetricians.

(2) The fact that women responsible for the high rates from malformations and 'unexplained' prematurity have completed their childbearing and have been replaced by healthier young women who have lower death rates from these causes.

(3) Improvements in methods of contraception and their more widespread use in high risk cases.

Fig. 47 shows that in England and Wales large regional differences in the perinatal mortality rate still exist in the areas where they were highest in 1958. It will be seen that in 1969 the rate was lowest in the Oxford Region (19·3 per 1,000) followed by East Anglia (20·2). The various

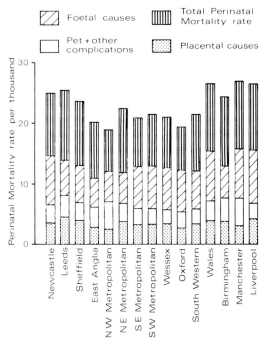

Fig. 47 Perinatal mortality rates in England and Wales, 1964, in standard hospital regions, 1969.

causes of perinatal mortality used in the Registrar General's Report (1964) have been arranged in groups and it will be seen that in the Oxford region the death rate is slightly lower than in East Anglia from complications of pregnancy and labour but slightly higher from foetal causes. The conclusion may be drawn perhaps that in East Anglia the level of health and physique of the women was slightly higher than in the Oxford region but that the standard of obstetric care was not quite so good. The total perinatal mortality rate in the Oxford and East Anglian Regions is approximately 30 per cent lower than in the regions with the highest rates where all causes of death were likely to be multifactorial in aetiology, with a preponderance of deaths from 'foetal causes' which are very difficult to prevent.

The rate in the Oxford region is still slightly higher than the Swedish national rate which is a measure of what still remains to be done in Britain.

SOCIAL AND OBSTETRICAL FACTORS IN MENTAL SUBNORMALITY

In a study of Aberdeen school children the prevalence of ascertained mental subnormality was 12·6 per 1,000, varying from 3·7 in the non-manual occupations (social classes I–IIIa) to 32·6 per 1,000 in unskilled manual workers (social class V). The school population from which the mentally subnormal children came was divided into an upper social group (social classes I–111b) and a lower social group (social classes IIIc–V). The upper social group was approximately 9 per cent larger than the lower.

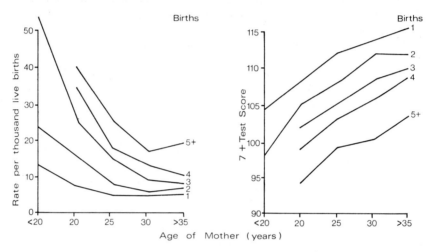

Fig. 48 Right: Mean IQ test scores at age 7+ in Aberdeen school children born 1950–55 by age and parity of mother, Social Class III. Left: Infant mortality 1–11 months by age and parity of mother.

Of the total 92 mentally subnormal children 20 were from the upper social group and 72 from the lower. In the upper social group 17 of the 20 children had signs of damage to the CNS and very low test scores (less than 50 in most cases) and 3 with borderline scores (70–75). The pregnancies and labours were almost all uncomplicated and the baby well grown and not in need of resuscitation. It was concluded that the damage to the CNS was not due to obstetrical factors. In the lower social group 28 of the 72 babies had signs of damage to the CNS and in some of these cases obstetric factors may have been responsible for the damage through an association with low birth weight. In fact 17 of the 72 babies (24 per cent) weighed less than 2,500 g. compared to 1 out of 20 in the upper social group. The remaining 44 cases in the lower social group showed no signs of damage to the CNS and most of them had test scores between 70 and 75, i.e. borderline scores. Siblings were numerous in this group, and their test scores were very little above that of the index case, additional evidence that obstetrical complications were unlikely to be responsible for the mental subnormality in the index case.

Test scores at the age of 7+ were available for the 11,280 children born in Aberdeen between 1950 and 1955, the population from which the mentally subnormal children were drawn. The relationship between these and the age, parity and social class of the mother has been reported by Illsley (1967). The importance of his findings is well illustrated in the following analysis as it affects social class III (manual), the largest and most heterogeneous social class. Fig. 48 shows that at each maternal age the mean scores are highest in first born children. They increase with increasing age of the mother and fall with increasing parity. The highest mean scores occur in children of primigravidae over the age of 30 and the lowest in 5th or later children of mothers under the age of 25. The stillbirth rate is highest in first births and rises in all pregnancies with increasing age of the mother. This suggests that there is unlikely to be any strong significant relationship between difficult labour and mental subnormality.

It will be seen that in the period 1–11 months infant death rates are lowest where the mother is over the age of 30 and highest where she is under the age of 25, especially where there are 3 or more children.

A high post-neonatal death rate and a high incidence of borderline mental subnormality are both characteristic of young mothers of high parity. Since the former is known to be strongly associated with poverty and social deprivation it is also probably true of the latter.

This association of age, family size and test scores occurs in all social classes except I and II. Illsley has pointed out that in social class I parents who have large families have them because they want large families knowing they have the means to support them. In social classes IV and V large families are frequently accidental and due to the parents' inability to control their own environment.

CANCER OF THE CERVIX

The earlier a woman starts childbearing, the more pregnancies she has and the shorter the interval between them, the greater the risk of cancer of the cervix. On the other hand cancer of the breast is more common in nulliparous women and in parous women who were late in having their first child.

Fig. 49 shows that in England in 1969 the mortality ratio in cancer of the cervix was highest in the North West and North East regions. The regional differences resemble closely

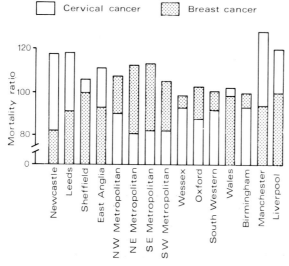

Fig. 49 Mortality ratios in standard hospital regions for cervical and breast cancer, 1969.

these seen in the perinatal mortality rate (Fig. 47). By contrast the mortality ratio in cancer of the breast is lowest in the North and highest in the Metropolitan areas.

In Aberdeen City together with the County of Aberdeen the annual incidence of cervical cancer in married women was 3·4 per 10,000. It varied from 1·5 in social classes I and II to 6·1 in wives of unskilled manual workers and from 1·7 in farmers' wives to 4·5 in the wives of farm workers (Aitken-Swan and Baird, 1969).

Cervical cytology has facilitated the study of the epidemiology of cervical cancer. For example the incidence of positive smears in married women under the age of 30 was 0·03

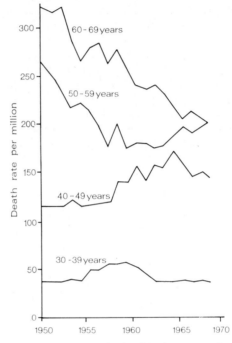

Fig. 50 Cohort death rates per million, England and Wales, cancer of uterine cervix, 1950–70.

per cent in those with one child, 0·69 in those with 4 children and 2·6 per cent in those with 5 or more children. The high rate was not the direct result of the early start to childbearing or the large number of children but rather the result of the type of environment in which young women with this obstetric history usually live. The key factor predisposing to cancer of the cervix seems to be coitus with multiple partners rather than pregnancy. The incidence is less, other things being equal, if a vaginal diaphragm or a condom is used regularly. Since epidemiologically cancer of the cervix resembles a venereal disease one might expect that the incidence would increase in times of great social upheaval such as total war. In England and Wales the death rate from cervical cancer in the cohort of women aged 40–49 rose in 1958 and reached a peak in 1964 after which it fell again (Fig. 50). These women were between 22 and 31 years of age in 1940 and would be at increased risk of cancer from multiple partners during the Second World War. A much smaller rise in the death rate occurred in women aged 30–39 in 1955. These women were aged 15 to 24 and a few of them would also be at risk.

In the age group 50–59 the death rate has risen slightly since 1964, and since these women were aged from 25 to 34 in 1940 they would also be at increased risk to some extent. The steady fall in the death rate from cervical cancer from 1950 to 1968 in the cohort of women aged 60–69 is evidence that death rates have been much higher in the more distant past.

In the City of Aberdeen the number of cases of cervical cancer diagnosed clinically increased in women under the age of 60 in 1958–60, the same years in which the death rate from cervical cancer in England and Wales rose in the cohort 40–49. The number diagnosed clinically also increased in those over the age of 60, but a little later between 1961 and 1966 corresponding in time to the rise in the death rate from cervical cancer in the cohort of women aged 50–59 in England and Wales.

In 1967–69 in Aberdeen the number of clinically diagnosed cases of cervical cancer in women under the age of 60 had fallen to about half the number normally expected because the lesions were now being diagnosed at the pre-clinical stage. Almost 100 per cent of married women in Aberdeen between the ages of 25 and 60 have had at least one cytological examination for cervical cancer. This was a difficult task, but well worth the effort, since the incidence of positive smears increased as the difficulty increased in persuading women to have the test.

It has been possible to calculate that the interval from first coitus to the appearance of cancer cells in the cervical smear is approximately 10 years and that another 10 years will elapse before clinical signs appear. Knowing the number of clinical cases, the incidence of positive smears, the total number of women at risk, it has been possible to calculate that most women with 'cancer' cells in a cervical smear will, if untreated, develop clinical cancer sooner or later. The number of cases diagnosed clinically has decreased at the rate expected on these assumptions.

Conclusions

In the last 150 years industrialization in Britain has brought about great changes in the environment. The increase in the gross national product raised the standard of living for most people but for some vulnerable groups, especially those in the semi-skilled and unskilled occupations in the large cities, living conditions are worse than for those in equivalent jobs in rural areas with consequent deterioration in physical and emotional health. The women are are shorter in stature and have higher perinatal mortality rates especially from congenital malformations of the central nervous system and 'unexplained' low birth weight.

National perinatal and infant mortality rates in Britain compare unfavourably with all other countries in Northern Europe. Sweden has had the lowest death rates from most causes for many years. There are many reasons why this should be so, such as great natural resources, a well educated and socially advanced population, freedom from wars, industrial strife and the slum conditions and pollution from the use of coal which have been characteristic of our large badly planned cities. Norway with a much smaller gross national product has also very low stillbirth and infant mortality rates. In both countries, despite their geographical situation and the long cold and dark winters, the stillbirth rate in the early 1920s was as low as the rate in Britain today.

In Britain we must aim at lowering perinatal mortality rates in all regions to the same level as in the South East of England. The Registrar General's Report for England and Wales for 1969 showed that the stillbirth rate due to 'foetal causes' (i.e. ill-defined causes and malformations) was still high in Wales, Liverpool, Manchester and Newcastle, and the fall in the stillbirth rate was least in these areas because such deaths are difficult to prevent. The situation is still worse in Scotland and Northern Ireland.

The solution to this problem requires more than an improvement in the standard of obstetric services. The undue concentration of new industries in the South of England increases the amount of migration to this favoured area, particularly of those of superior health and physique, with effects on perinatal mortality already described. The loss of the more active and intelligent members of the community from the less favoured areas will reduce the impetus to improve the existing standard of care.

Economic development in the regions which are not so prosperous as the South must be pursued vigorously as well as the development of improved medical services if the differential between regions in perinatal mortality and in health statistics generally is to be obliterated.

The measures necessary to lower perinatal mortality rates would also lower the incidence of cancer of the cervix and of borderline mental subnormality.

References

Aitken-Swan, J. and Baird, D. (1966). Cancer of the uterine cervix in Aberdeenshire. *Brit. J. Cancer*, **20**, 624, 642.

Baird, D. and Thompson, A. M. (1969). *Perinatal Problems: Second Report of the 1958 British Perinatal Mortality Survey*. Edinburgh and London, E. & S. Livingstone Ltd.

Birch, H. G., Richardson, S. A., Baird, D., Horobin, G. and Illsley, R. (1970). *Mental Subnormality in the Community*. Baltimore, The Williams and Wilkins Co.

Department of Health for Scotland (1943). *Infant Mortality in Scotland*. London, HMSO.

Heady, J. A. and Heasman, M. A. (1959). *Social and Biological Factors in Infant Mortality*. London, HMSO.

Illsley, R. (1963), *Perinatal Mortality: First Report of the 1958 British Perinatal Mortality Survey. Survey*. Edinburgh and London, E. & S. Livingstone Ltd.

Illsley, R. (1967). Family growth and its effect on the relationship between obstetric factors and child functioning. In Platt and Parkes (eds.), *Social and Genetic Influences on Life and Death*.

Leck, I. (1963). Incidence of malformations following influenza epidemics. *Brit. J. prev. soc. Med.*, **17**, 70.

Leck, I. and Millar, E. L. M. (1963). Short-term changes in the incidence of malformations. *Brit. J. prev. Soc. Med.*, **17**, 1.

MacGillivray, I. (1961). Eclampsia and pre-eclampsia. 17th Conference Intern. Soc. Geograph. Pathol., London, 1960. *Path. Microbiol.*, **24**, 507–512.

Chapter 14

Environmental Factors in Gynaecological Conditions

M. C. Macnaughton

Environment is defined as the aggregate of all external and internal conditions affecting the existence, growth and welfare of organisms. As far as the effect of environment on gynaecological conditions is concerned, comparatively little work has been done, although there is some information about pregnancy. Women as well as men are intimately affected by psychological, educational, cultural, genetic, hormonal, nutritional, environmental factors and by infectious disease. Gynaecological complaints tend to follow the woman's personality, cultural, hereditary and physical background (Pettit, 1965). Ideally some of the environmental factors such as infections, exposure to drugs and toxic substances and to malnutrition should be preventable. Knowledge of some of the environmental factors affecting women should make it possible to prevent disease due to these factors in an increasingly effective manner.

It is intended, in this chapter, to discuss what is known about some environmental effects which are particularly relevant to gynaecology and the conditions seen in gynaecological departments. Many other conditions, of course, affect both men as well as women but these will not be especially emphasized in this Chapter.

Observations on diseases of women have been made for centuries, and Soranus of Ephesus wrote widely on the diagnosis of gynaecological disease discussing such matters, for example, as the optimum time for conception. Gynaecology has been particularly associated with taboos and some of these are still with us today in certain societies. For instance, the presence of blood after first coitus is still looked for in some cultures. The rise in teenage venereal disease in recent years is a result of changes in cultural mores of society resulting in extramarital sexual relationships becoming more frequent and widespread. The increase in the number of women working in modern times has also meant changes in family life with certain effects on the types of gynaecological condition.

The increase in family planning and the use of newer and more effective methods of contraception have resulted in some new gynaecological problems and the lesser frequency of some older and more familiar ones. For example, while the more frequent use of hormonal types of contraception has meant an increase in some types of menstrual irregularity, the same compounds have also been most helpful in controlling some other types of menstrual abnormality. The reduction in parity, which is a result of better methods of contraception, has caused a lower incidence of gynaecological conditions related to parity, such as prolapse. New methods of screening have also helped to reduce the incidence of severe degrees of cancer of the cervix at the first clinic attendance, and women more often present with the disease at a stage where treatment is more likely to result in permanent cure.

Some general environmental effects on gynaecological conditions will now be reviewed, and then a few particular areas which have been more intensively studied will be discussed in greater detail.

Climate

The effect of climate on the growth and development of female children, with particular reference to the age of onset of menstruation, will be discussed in detail later in this chapter, but this effect appears to have been overestimated. In women, during the reproductive phase of life, a sudden change of climate can result in menstrual irregularity and amenorrhoea but this usually resolves spontaneously. Some women, on the other hand, continue to have the menstrual upset as long as they remain under certain climatic conditions. At the time of the menopause the vasomotor symptoms associated with this period of life may be accentuated, and flushings and headaches, etc., may become more severe.

In older women, past the menopause, there is no particular effect of climate on gynaecological disease except that women with chronic coughs are more likely to develop prolapse at this stage, and the chest condition may be related to the climate in which the patient lives. Since many gynaecological operative procedures are elective, it is sometimes best to postpone operation until spring or summer in temperate climates with a resultant reduction in postoperative complications. This, of course, is a consideration which applies to all older patients who require operation and not only to those having gynaecological procedures.

Altitude

It has been thought for a long time that a combination of cold and high altitude has a harmful effect on the fecundity of apparently acclimatized but not indigenous humans and their domestic animals. This was the prime reason why the Spaniards in 1535 moved the capital city of Peru from Janja, high up in the Andes, to Lima, near the Pacific Coast. In 1639, Father Calancha described how, as a result of a miracle, the first child was born and successfully reared in the settlement of 20,000 Spaniards at Potosi, Bolivia, 53 years after the foundation of that mountain city. It is unlikely that this infertility was due to a prolonged outbreak of celibacy and presumably the conditions of the environment were responsible.

Congenital acclimatization to the Andean high plateaux (2 to 5 km altitude) has produced a climatophysiological variety of human being whose biological characteristics are somewhat different from those of sea level man. Infertility of men and animals brought from the lowlands, as sometimes occurs, means the elimination of the unfit through a process of natural selection. In severe cases of chronic mountain sickness the patients complain of sexual frigidity.

Monge (1943) has investigated the physiology of reproduction at the high plateaux, and it is well known in South America that high altitude exercises a deleterious effect on fertility. Cases have been described where couples were fertile at sea level but infertile at high altitude. It is known that adapted pregnant women often come down to the coast to be delivered because miscarriages and sometimes sudden death of the newborn occur at the heights. The cause of this is not really known but is probably related to the low oxygen tension at these high altitudes. Indeed, one of the cardinal criteria of the acclimatized state is that normal fertility and reproductive powers remain. Thus the physiology of reproduction on the high plateaux is normal in regard to the birth rate of the Andeans which is equal to that of people living at sea level.

Nutrition

The effect of nutrition on women is very variable and, in general, the well nourished woman suffers less from disease than the poorly nourished. Socio-economic and nutritional factors are commonly closely related. Those women of high parity living in a poor socio-economic environment are also likely to have nutritional difficulties which contribute to certain diseases.

Severe malnutrition does seem to result in menstrual disorders and infertility, and these were reported in concentration camps in wartime. However, under the conditions of the camps, the inmates, in addition to suffering severe physical and nutritional deprivation, were also under severe psychological stress, and it is not clear which of these was the main factor in causing the amenorrhoea. These women reverted rapidly to normal when they were released from detention.

It is often said that nutritional difficulties cause abortion and prematurity but the evidence as far as abortion is concerned is poor. There does seem to be a higher incidence of prematurity in women in poor socio-economic circumstances and doubtless the lower nutritional status of the women is a factor. This matter is discussed more fully in the chapter dealing with obstetrics.

What is most important is that good nutritional status should obtain throughout the woman's life. The nutritional status from age 0–20 is probably more important as far as the results of a pregnancy are concerned than what, in fact, the patient eats during her pregnancy. There is evidence, however, that the time at which an individual attains her growth potential is more affected by temporary starvation than is the ultimate stature which is also related to genetic and racial factors.

Since women now live longer than formerly the number of geriatric problems in gynaecology is increasing and her continued satisfactory nutrition is important. This is particularly so in relation to conditions such as utero-vaginal prolapse where the combination of multiparity, poor nutritional status and the menopausal hormone environment with low oestrogen levels all contribute to poor supportive tissues and to an increase in the incidence of prolapse. Diet fads in adolescence may lead to anorexia nervosa with resultant amenorrhoea and this is a well-known sequence of events. There is no doubt that poor nutritional status is an important factor in gynaecological disease as it is in other conditions.

Psychological Environment

Patterns of family background, school and regional differences in behaviour affect every individual. Attitudes towards sex, the onset of menstruation, fertility, and contraception can all affect the growing girl. Anxieties of the mother and difficulties in communication between mother and daughter may lead to dysmenorrhoea and problems with sexual relationships in later life. In some women, 'frigidity' is a result of an aversion to sexual matters which is passed on from a mother to her daughter. The family cultural patterns are intimately involved in the development of a girl's attitudes to sex and reproduction, and gynaecological problems like dysmenorrhoea, dyspareunia and frigidity frequently stem from these patterns. Conflict between a woman's own desires and family pressures may show itself in gynaecological disease. The dislike and fear of gynaecological examination may, in some women, be related to conditioning by adult members of her family.

The psychological problems arising at the menopause are very important. At this time many changes occur due to alterations in the hormone environment. Most women have some instability and nervousness at this time. They attain this period in their lives when their children are growing up and are possibly leaving home. At this time also, their husbands are often in top positions in their work and are frequently at their most productive. As a result women frequently feel unwanted and psychosomatic gynaecological symptoms appear. Proper education and reassurance is necessary for women at this, and if the genesis of the symptoms is explained and organic disease ruled out, the latter being very important, it is possible to help the patient.

For the older patient, an objective to keep her alert and interested is essential and here her home environment is of special significance. Various problems intrude such as housing, and the attitude of her family, the solution of her housing problem, and her day to day care and contacts are also most important.

Socio-economic Factors

These are prominent in the development of many women and have a marked effect on the whole person as well as on her attitude to gynaecological problems.

The attitude to pregnancy varies in this way and fear of pregnancy is more common in the higher than in the lower socio-economic groups. In the higher groups, women are often anxious to complete their families in order to take up their career or to take part in leisure activities whereas in the lower socio-economic groups unwanted pregnancies have been more acceptable. These women tend to have larger families, do not avail themselves of the community facilities that are available and do not attend for diagnostic and screening preventative techniques so that diseases such as carcinoma of the cervix (see p. 189) are in a more advanced state when the patient is first seen. It is this group of women that require most attention in this respect and, for them, facilities such as domiciliary family planning and cervical cytology have to be used. These women come late with most diseases, not only gynaecological.

Occupational Factors

Prolonged exposure to noxious substances may have an adverse affect. For example, radiographers exposed to X-rays are usually women and careful checks are necessary to keep the amount of radiation which these women receive to a minimum. Protective clothing and frequent monitor checks are necessary to prevent excessive radiation.

Exposure of women to lead may lead to foetal death, abortion and sterility. Asbestos may, in some women, lead to ovarian malignancy. However, with proper industrial health monitoring, these conditions should rarely occur although constant watchfulness is necessary to ensure this.

Factors such as heavy manual work, e.g. in factories, may bring on conditions such as prolapse, especially in postmenopausal women. However, the standard of obstetric care of these women is probably of more importance and, as this improves, the number of women damaged at childbirth, will decrease with a resultant drop in later gynaecological complications.

Infection

Gynaecological infection is related to the environment of the patient in many cases. In the female child, faulty hygiene and surroundings may result in a vaginal discharge. Gonococcal vulvo-vaginitis may occur in some children under these circumstances. Pediculus pubis is another condition which is related to hygiene. The sexual environment is frequently associated with gonorrhoea and vaginal infection and, in some women, infertility may result from these infections. In the older women, senile vaginitis may lead to secondary infection with discharge, and a purulent discharge may point to an intrauterine condition such as carcinoma causing a pyometra.

It will be evident that there are some general environmental relationships concerning gynaecological disease. Specific conditions where further information is available will now be discussed.

CARCINOMA OF THE OVARY

The decline in ovarian function at the menopause with elevation of gonadotrophins and cessation of menstrual rhythm could be expected to provide an ideal hormonal environment for ovarian tumour development. Over 50 per cent of cases of human ovarian cancer occur between the ages of 40 and 60 years. It seems, therefore, that a rhythmic hormonal activity may be necessary to prevent the disease. Investigations have also shown that 45 per cent of women with ovarian carcinoma were nulliparous compared with 20 per cent of the general population. Radiation castration in humans does not seem to predispose to the subsequent development of ovarian tumours, but there is some evidence that exposure to asbestos may do so. It is difficult, therefore, to find any other environmental factors associated with ovarian cancer.

CARCINOMA OF THE CERVIX UTERI

Statistics relating to cancer of the cervix uteri and of the corpus are frequently combined under cancer of the uterus making the obtaining of data about each difficult.

Geographical distribution. Cancer of the cervix uteri is more common in British towns than in the country and occurs more frequently in ports than in other towns. It is also common in the wives of servicemen and others who are away from home frequently for days on end, e.g. long distance lorry drivers.

Marriage and intercourse. It has been demonstrated that a close correlation exists between carcinoma of the cervix and sexual experience. The disease is virtually unknown in nuns and is less common in single than in married women.

The important factors seem to be (i) the age at which intercourse starts; a woman beginning before the age of 20 having twice the risk of a woman starting after that age, (ii) the number of partners in intercourse, a woman being married twice or more having twice to four times the risk of a woman married only once. Youth by itself does not predispose to cancer of the cervix. Frequency of intercourse seems to be of less definite importance. Surveys carried out have given conflicting evidence on this point, and it is obviously difficult to establish the basic data relating to intercourse with accuracy.

Promiscuity has been reported to increase the risk of cervical cancer and it has been shown that prostitutes have a high liability to this condition. Syphilis is also commoner in cancer cases than in controls. Squamous cancer of the cervix behaves epidemiologically very like a venereal disease, this being illustrated by the increased incidence with sexual promiscuity.

Size of family. In a recent survey when families were compared, matched for present age of the mother, age at marriage and social class, the actual number of cases of cancer was always found to be lower than the expected number when there were less than five children in the family. In families with more than five children the actual numbers of cases of cancer was always greater than the expected number especially when there were more than seven children. It appears that high parity does not directly increase the risk of cancer but results from a way of life that predisposes to it.

The effect of high parity on the frequency of cancer cells detected in a cervical smear is very similar to that on clinical cancer. In one study the incidence of positive cells in married women under age 30 increased from 0·03 per cent in those with only one child to 0·69 per cent in those with four children and 3·64 per cent in those with six children. The rate of positive cases was also increased with broken marriages and in women who had premarital coitus, or a premarital pregnancy.

Circumcision. Religious sects who practise ritual circumcision, for example, Jews and Muslims in India, show a reduced rate of cancer of the cervix when compared with un-circumcised populations in similar social conditions. Jewish women are well known for their freedom from cervical cancer. Exceptions may occur where social conditions are not strictly comparable. For example, the wives of uncircumcised Parsees have an incidence rate as low as that of the Muslims in Bombay—this has been explained as being the result of the high standard of personal cleanliness of Parsees (Khanolkar, 1950).

There have also been surveys reporting no difference between the incidence of cervical cancer in the wives of circumcised and uncircumcised men. However, some of these have been in cases circumcised for medical reasons. This, of course, is performed at a later age and is less severe than the ritual type—sometimes it has been difficult to decide whether the husband has been circumcised or not.

Social status. There is a steep social gradient in the incidence of cervical cancer, the disease being commoner in lower social classes. This gradient is more evident in younger women and almost disappears after age 65. In one survey the annual incidence per 10,000 married women was 1·5 in social classes 1 and 2 rising steadily with decreasing socio-economic status, till it reached 6·1 in social class 5.

The occupations where the wives have the lowest risk are all of the 'white collar' or professional type; those with a high risk are of a type where there is little incentive or even opportunity for personal cleanliness. As would be expected, prostitutes have a higher risk still and the data available indicates that their risk was six times that of the general population.

Contraceptives. The type of contraceptive used has been reported to be associated with cervical cancer. When patients were classified according to the use of obstructive contraceptives by either sex it was found that there was a lower proportion of cervical cancer patients using this type of contraception than in a parallel series not using such contraception, the difference being just significant (Boyd and Doll, 1964). It does seem, therefore, that the condom or the diaphragm may protect the cervix in some way from cancer.

Bathrooms. A correlation has been found between a high incidence of carcinoma and poor sanitary arrangements. However, the correlation was not particularly close.

The significance of social factors in carcinoma of the cervix. The evidence suggests cleanliness of the husband may play an important part in the genesis of carcinoma of the cervix. At the same time, the age at which a woman begins intercourse and the number of men with whom she cohabits also affect her liability to the disease. It has been suggested that a substantial improvement in personal hygiene in the general population would bring down the incidence of carcinoma of the cervix. This is perhaps something that should be included in any campaign to reduce the frequency of the disease.

AGE AT MENARCHE

The normal age at menarche is a very individual matter and depends on the individual's genes and life history. This is so within limits since it is normal for the human female to start menstruation sometime in the second decade of life. In normal girls signs of sexual maturation usually appear between 8 and 12 years of age and the first menstrual period takes place a little later than this. At the beginning, menstruation may be rather irregular and these initial cycles are frequently anovular. The cycle, however, eventually becomes ovulatory in character and menstruation becomes regular.

Much study has been done on the factors influencing the menarche. From these studies, it has become evident that populations are very different, and the data from one population is not readily applicable to another or even to the same one at a later period in time (Zacharias and Wurtman, 1969).

Comparisons must, therefore, be limited to groups which live under similar conditions and mature at roughly the same time. Many factors are interrelated and it is difficult to separate their effects. A good example of this is the apparent ability of environmental and social factors to override genetic influences shown by the fact that Japanese girls born and brought up in California reached the menarche about one and-a-half years earlier than those born in California but brought up in Japan, and those both born and brought up in Japan (Ito, 1942). A number of factors may influence sexual development and these will now be discussed.

Temperature. There does not seem to be valid evidence for the view that the age of menarche is influenced by climate. It is frequently said that the menarche begins earlier in tropical than in more temperate climates but a survey showed that the menarche occurred at the same age in Edinburgh, Rome and South Russia (Kennedy, 1933). It has, in fact, been suggested by one survey that the menarche may be later in the tropics and that a temperate climate accelerates the onset of the menarche. The consensus of opinion, however, suggests that climate has little effect.

Altitude. It has been reported that height above sea level does have an effect on menarcheal age, and the figure of three months delay for each 100 metres of altitude has been given (Valsik *et al.*, 1963). This, however, is complicated by the fact that economic and nutritional conditions are poorer at high altitudes and there may be differences in calorific requirements. These factors may be more important in determining menarcheal age than the actual height above sea level.

Season. The findings of the few studies made on the relationship between season and the onset of the menarche have not been consistent. It does not seem unreasonable to suppose that this effect of season might be due to the changing length of day. If this were true the changes would be most marked in populations living some distance from the equator. In these countries, as a result, there should be related seasonal variations in the incidence of pregnancy. While there are some times of the year when more deliveries occur than at others various economic factors intrude, and it is not possible to say that season alone is *the* important factor.

Socio-economic factors. It is established from a number of surveys in different parts of the world that major improvements in socio-economic conditions are followed by acceleration of physical growth and sexual development. It has even been suggested that the average age at the menarche might be taken as an index of the general well being of a population (Fluhmann, 1958). The influence of lowering of the age of puberty by good nutrition has been reported from many corners of the world. Temporary nutritional deprivation may also have an effect, and a study of Belgian girls during World War II showed that nutritional deprivation was associated with a delayed menarche in a significant proportion of these girls (Ellis, 1945). There are also a number of reports supporting the hypothesis that during periods of severe food shortage puberty is delayed. It would also appear that while poverty or malnutrition can delay sexual development, improvements in living standards can only accelerate the process up to a point.

Effect of light. It has been found in one survey which compared two populations of blind and non-blind groups that the age at menarche was lower in the blind than in the non-blind group. It was, therefore, suggested that stimulation of the retina by environmental light produced a specific effect on the human ovary (Zacharias and Wurtman, 1964). The data from this survey have not, in fact, been substantiated and differences in environment of the different groups studied seemed likely to account for the differences found.

Family environment and size. In recent surveys in England on how the age at the menarche is affected by family environment no effect of social class or of position in sibship was found. However, the age at the menarche was affected strongly by the size of the family in which a girl grows up, there being a delay in the onset of the menarche in families with many siblings (Roberts *et al.*, 1971). It is possible, however, that this delay in the large families might be explained on nutritional and other factors. It has been shown, for example, that children brought up in a large family group may individually get less to eat than those in small families.

Urban and rural differences. It has been observed, in many parts of the world, that girls brought up in a rural environment, in general, menstruate later than those brought up in urban environments. Whether this is due to nutritional, social or genetic factors has not been established.

Psychological factors. These factors are complicated by other influences such as the emotional stresses of deprivation and again it is difficult to separate the different factors.

Family heredity. Studies have been performed to compare menarcheal ages in mothers and daughters but the data have not really been comparable. Conditions obtaining at the different times of the two generations would invalidate the studies. For what it is worth a few reports do indicate that there is a fairly good correlation between the menarcheal ages of mothers and their daughters (Bolk, 1923).

Change in menarcheal age during the past century. Evidence suggests that in Western Europe and in the USA there has been a trend to earlier sexual development (Tanner, 1962). On the world scene there does not seem to be an overall secular trend. To explain the fluctuations in age of onset of sexual maturity it has been proposed that there is a uniform, relatively early prototype for the menarcheal age which applies to the whole human race, and which becomes delayed as the result of adverse external circumstances. For example, the variations in menarcheal age among European girls between 1800 and 1943 parallel the fluctuations in industrial activity, e.g. the War of 1870 and the following revolutionary movements exercised a retarding influence on sexual development in France (Backman, 1948).

The age at the onset of menstruation, therefore, does not seem to be fixed but varies with population and changes with time. It seems more susceptible to socio-economic factors such as nutrition than to family or climatic considerations.

CONCLUSIONS

Many diseases are affected by environmental factors and gynaecological conditions are no exception. Climate, altitude, nutrition and a number of other factors all influence the menstrual cycle and fertility. Socio-economic status, hygiene, and frequency of intercourse have a particular effect on the incidence of carcinoma of the cervix. Knowledge of these environmental influences is an aid to the further understanding of these conditions and may help toward the prevention of the diseases and associated abnormalities.

References

Backman, G. (1948). Die beschleunigte Entwicklung der Jugend: ver frühte Menarche, verspätete Menopause, verlängerte Lebensdauer. *Acta Anat.*, **4**, 421.

Bolk, L. (1923). Menarche in Dutch women and its precipitated appearance in youngest generation. *Proc. Acad. Sc. Amsterdam Sec. Sc.*, **26**, 650.

Boyd, J. T. and Doll, R. (1964). A study of the aetiology of carcinoma of the cervix uteri. *Br. J. Cancer*, **18**, 419.

Ellis, R. W. B. (1945). Growth and health of Belgian children: during and after German occupation (1940–1944). *Arch. Dis. Childhood*, **20**, 97.

Fluhmann, C. F. (1958). Menstrual problems of adolescence. *Pediat. Clin. North America*, **5**, 51.

Ito, P. K. (1942). Comparative biometric study of physique of Japanese women born and reared under different environments. *Human Biol.*, **14**, 279.

Kennedy, W. (1933). Menarche and menstrual type: notes on 10,000 case records. *J. Obst. & Gynae. Brit. Emp.*, **40**, 792.

Khandkar, V. R. (1958). In *Cancer*, edited by R. W. Raven. London, Butterworths, Vol. 3, p. 272.

Monge, C. (1943). Chronic mountain sickness. *Physiol. Rev.*, **23**, 166.

Pettit, M. De W. (1965). Environment in relation to gynaecologic disease. *Arch. Environ. Health*, **2**, 116.

Roberts, D. F., Rozner, L. M. and Swann, A. V. (1971). Age at menarche, physique and environment in industrial N.E. England. *Acta Paediat. Scand.*, **60**, 158.

Tanner, J. M. (1962). *Growth at adolescence: with a general consideration of the effects of hereditary and environmental factors upon growth and maturation from birth to maturity*, 2nd Edn. Blackwell Scientific Pub. 1962, Oxford.

Valsik, S. A., Stukovsky, R. and Bernatova, L. (1963). Quelques facteurs géographiques et sociaux ayant une influence sur l'age de la puberté. *Biotypol*, **24**, 109.

Zacharias, L. and Wurtman, R. J. (1964). Blindness: its relation to age of menarche. *Science*, **144**, 1154.

Zacharias, L. and Wurtman, R. J. (1969). Age at menarche. *New England J. of Med.*, **280**, 868–875.

Chapter 15

Venereal Diseases in Contemporary Societies

R. R. Willcox

NATURE AND EXTENT OF THE PROBLEM

Syphilis

During the past decade in most countries there has been a noticeable increase in the prevalence of venereal diseases following the sharp decline after the Second World War. Although the reported incidence of primary and secondary syphilis has risen markedly in some countries (Fig. 51) in others it has been more or less contained. In England and Wales, for example, there were 704 cases of syphilitic infections under one year in 1958, 2,118 cases in 1965 and 1,826 cases in 1968. On the other hand, there has been a fall in the numbers of cases of late symptomatic syphilis in countries where statistics are available, and in nearly all countries of early congenital syphilis, but these welcome events represent almost the only brightness in an otherwise sombre scene.

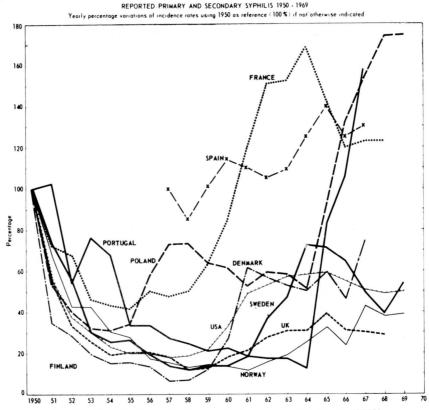

Fig. 51   World incidence of primary and secondary syphilis (Source WHO).

194

Gonorrhoea

Millions of men and women now suffer from gonorrhoea (estimated at 150 millions annually) and the incidence of the condition is mounting steeply in practically all areas. The situation, which is certainly inflationary, has been widely described as 'out of control' and 'an epidemic by any standards'.

The current trends in incidence in a number of countries are shown in Fig. 52. Precise comparisons of one country with another are impossible owing to differences in the provision

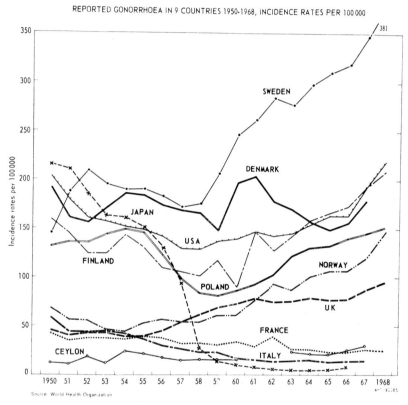

REPORTED GONORRHOEA IN 9 COUNTRIES 1950-1968, INCIDENCE RATES PER 100 000

Source: World Health Organization

Fig. 52 World incidence of gonorrhoea.

of facilities and in reporting. Nevertheless, according to recent figures supplied by the World Health Organization, the current rate in Sweden is 485 per 100,000 while in the USA the annual number of cases is now estimated to amount to 2·0 millions corresponding to an incidence of 1,000 per 100,000 population. Moreover, the majority of these cases escape contract-tracing which is the present mainstay of control, for surveys have shown that seven out of ten infections with both gonorrhoea and primary and secondary syphilis in that country are unreported (Table 36).

Other Diseases

Moreover, this by no means represents the full story, for the problems regarding gonorrhoea and syphilis are but a fraction of the whole for there are many other sexually transmitted

TABLE 36

Reporting on infectious venereal disease in the USA
in 1968

	Primary and secondary syphilis	Gonorrhoea
Estimated number of cases	75,207	1,449,581
Not reported	55,025	1,018,201
Percentage not reported	73·2	70·2

(*Data from McKenzie Pollock, 1970*)

TABLE 37

Sexually transmitted diseases

	Organism	Disease
Spirochaetes	*T. pallidum*	Syphilis
Bacteria	Gonococcus *H. ducreyi* *Donovania granulomatis*	Gonorrhoea Chancroid Granuloma inguinale
Viruses	Chlamydia Other viruses	Non-gonococcal urethritis Lymphogranuloma venereum Herpes simplex Molluscum contagiosum Condylomata acuminata
Protozoa	*T. vaginalis*	Trichomoniasis
Fungi	*C. Albicans* *Epidermophyton inguinale*	Candidiasis Tinea cruris
Parasites	*Acarus scabei* *Phthirus pubis*	Scabies Pediculosis

(Other sexually transmitted organisms whose rôles in relation to disease, if any, are not yet clear include Mycoplasmas, *Haemophilus vaginalis*, Diphtheroids, Mimeae (Moraxellas), *Herellea vaginocola* and Cytomegalic virus.)

diseases (Table 37) caused by spirochaetes, bacteria, viruses, protozoa, fungi and parasites, the numbers of which are not even recorded in most countries. Other organisms are also known to be transmitted sexually, the significance of which is not yet entirely clear.

In England and Wales where, unlike most countries, statistics relating to non-gonococcal urethritis in males have been available since 1951, the numbers with this condition now well

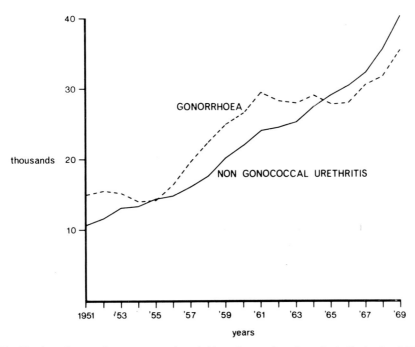

Fig. 53   Number of cases of non-gonococcal urethritis and gonorrhoea in males in England and Wales.

exceed those in that sex with gonorrhoea (Fig. 53). Indeed patients with syphilis and gonorrhoea together comprise only approximately one quarter of new cases in males and one fifth in females attending the venereal disease clinics.

REASONS FOR RESURGENCE OF VENEREAL DISEASE

There are many interwoven factors involved—demographic, socio-economic, and behavioural. These are concerned with population changes, changes in the environment, changes in the host and changes in the responsible organisms.

Demographic Factors

These include more susceptible people owing to increasing populations with more young people, and from a longer life span due to earlier maturity and increased longevity. However, as shown by data from England and Wales (Fig. 54) the increased population accounts for but a small part of the increase in gonorrhoea.

Nevertheless, in the coming years the populations of the world are likely to continue to increase and there will therefore be even further additional venereal disease on this account.

Changes in the Environment

As a result of technical progress there is greatly increased population mobility resulting from, amongst other things, industrialization, urbanization, more female employment,

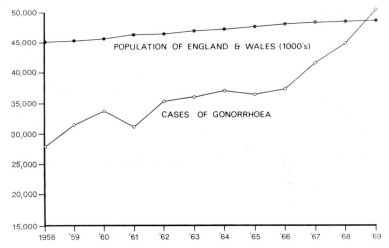

Fig. 54   Increase of population compared with increase in gonorrhoea in England and Wales.

migration of labour both international and within countries (immigrant groups have always shown higher venereal disease rates than the home population) and—above all from increased holiday and business travel at all levels by land, sea and air (Table 38). All of these have combined not only to multiply the opportunity for sexual encounter but have also tended to introduce venereal disease into previously 'closed' circles of sexual expression in the static population.

TABLE 38

Indices of increased travel

Year	World air travel Passengers carried (millions)*	International tourism: Arrivals in 26 countries (millions)†	World shipping tonnage (millions)‡
1958	88	55	118
1968	261	141	194
Percent. increase	197	156	65

* By members of ICAO only.
† Tourism International Bureau, Geneva.
‡ *Lloyd's Register* (1969).

Changes in the Host

Improved social conditions resulting in a better nourished, cleaner, more hygienic host have reduced the number of cases of asexually acquired syphilis and have removed chancroid from a position of importance in developed countries. These gains, however, have been more than offset by the behavioural developments which have occurred.

Changes in attitudes have resulted from and at the same time have themselves induced changes in behaviour with the consequent so-called 'permissive society'. While the guilt has been taken out of sex, in the opinion of some, the increased attention paid to it by the mass media has created a vicious circle by justifying even further attention and consequently increasing promiscuity which is the basic cause of venereal disease.

At the same time (Fig. 55) there has been a diminution of the restraining influences on promiscuity of religion, family and public opinion (the latter being influenced by more divorces, broken homes and high illegitimacy or abortion rates), the fear of venereal disease consequent upon the availability of simple effective treatments which have led to the problem of the chronic 'repeater' (in any venereal disease clinics a small number of patients contribute an unduly high share of infections with gonorrhoea) and of pregnancy. Fear of the latter has been removed by the introduction of the contraceptive pill and the intrauterine device and by the increasing availability of abortion. Unlike the condom the newer methods offer no

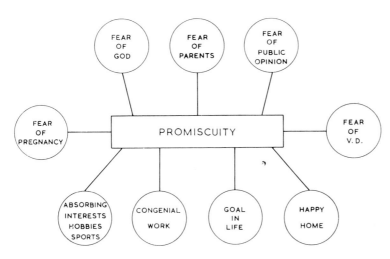

Fig. 55   Factors restraining promiscuity.

protection against the venereal diseases. Also there is now some evidence that they promote promiscuity by increasing the frequency of sexual intercourse and also the number of sexual partners, thus fostering the acquisition of venereal disease.

Today an estimated $17\frac{1}{2}$ million people in Western Societies are 'on the Pill'; there are 1·5 million in Great Britain alone and the numbers continue to rise steeply without any obvious deceleration (Fig. 56). The adverse effects of modern contraceptives in venereal disease control are thus likely to increase even further in the future.

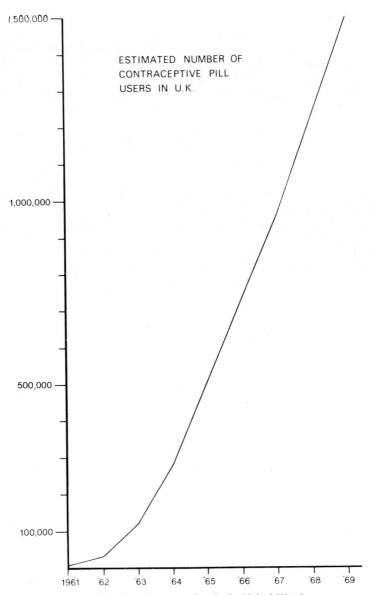

Fig. 56   Use of oral contraceptives in the United Kingdom.

The rise in venereal diseases has affected all age groups particularly those from 20–24 years of age (Table 39). From earlier maturity and other reasons the rates of infection have sometimes risen more steeply in the young (in whom the problem has been observed to be associated with broken homes, juvenile delinquency, illegitimacy, drug taking and—later on in life— with prostitution) than in persons 25 years and over. As patients with venereal disease tend to contract more venereal disease in the future this group merits particular attention.

TABLE 39

Reported VD in the USA

Age group	Primary and secondary syphilis		Gonorrhoea	
	1956	1968	1956	1968
15–19 years	10·7	19·3	415·7	610·6
20–24 years	18·4	38·8	781·8	1,251·1
25 and over	3·6	9·4	106·9	165·6
All age groups	3·9	9·6	135·7	235·1

There have probably also been some changes in sexual practices. Increased numbers of homosexual male patients have been seen in venereal disease clinics particularly those with syphilis who may comprise some 10–90 per cent of males with the infectious form of this disease. Whether this indicates increased homosexuality is not clear, but it probably reflects increased promiscuity in this group. The frequent occurrence of rectal involvement in women with genital gonorrhoea (although such may have always been so) has also been noted, and there have been some recent reports of oral infections of gonorrhoea, both asymptomatic and symptomatic and of the peno-oral transfer of syphilis. However such have been reported in the past and it may be currently fashionable to record them!

Other behavioural changes include altered patterns of prostitution. These first led to the suppression of the brothel with the emergence of the 'call girl', 'car girl', 'lorry girl', or 'hostess', and today the full circle is being completed in some areas by the reintroduction of the brothel in a modernized form (e.g. the Eros Centres of Hamburg, West Germany, and the so-called Ranch Houses in certain parts of the USA).

Changes in the Organism

Lessened virulence. It has been suggested that under chemotherapeutic and antibiotic bombardment both syphilis and gonorrhoea are becoming milder diseases—possibly due to a lowered virulence of the organism resulting from selection by treatment. However, as far as syphilis is concerned, host factors (e.g. better nutrition and improved personal hygiene) may be responsible for fewer exuberant lesions (e.g. condylomata lata) being currently encountered than in former times.

While gonorrhoeal infections have always been frequently asymptomatic in the female, symptomless infection seem to be somewhat more common in males, making it more important than formerly that the male contacts of infected females are also examined as these persons can no longer be expected necessarily to have obvious symptoms. Some patients of both sexes with trivial or asymptomatic genital infections may show septicaemic lesions of skin and joints of a relatively mild nature.

Furthermore, in patients with syphilis, treponemal forms have been found in lymph nodes, in cerebrospinal fluid and in aqueous humour of treated persons; the significance of these observations is not yet clear. It is possible that they may be true *T. pallidum* only in a few instances or they may be a form of *T. pallidum* having a lessened virulence.

Centuries hence, if the reduction in virulence of the causative organisms was carried to its ultimate end-point, both the gonococcus and *T. pallidum* would evolve into non-pathogenic commensals representing no bigger problems than some of the other organisms which are sexually transmitted without producing overt disease! However, this does not help the immediate situation in which gonorrhoea is becoming more difficult to control in both sexes.

Increased resistance. A bacterial change not in dispute is the developing resistance of the gonococcus to antibiotics, particularly to penicillin. This phenomenon is not apparent in relation to *T. pallidum*, but inability to culture this organism renders its precise testing impossible.

TABLE 40

Sensitivity of gonococcus to penicillin
(Percentage MIC in mcg/ml)

Area	Authors	MIC		
		0·05 or less	0·06–0·12	<0·12
EUROPE				
UK (London)	Lynn *et al.* (1970)	65·0	18·5	16·5
	Leigh *et al.* (1969)	63·0	18·0	19·0
Norway	Gundersen *et al.* (1969)	65·5	17·8	16·7
NORTH AMERICA				
Canada	Amies (1969)	34·9	19·0	46·1
USA	Fischnaller *et al.* (1968)	20·9	22·4	56·7
AUSTRALIA	Hatos (1970)	22·0	6·8	71·2
AFRICA				
Uganda	Arya and Phillips (1970)	19·1	7·5	73·4
FAR EAST				
Hawaii	Keys *et al.* (1969)	14·5	11·6	73·9
Japan				

Even making allowances for differences in methods of assessment and in reporting, and by selecting data only referring to routine and not to problem strains, the distribution of the more resistant strains of gonococci is found to be patchy as between continents; for example, the situation in Africa and the Far East, is much worse than in Northern Europe (Table 40; Fig. 57).

Moreover, differences are noted in different parts of the same country, the gonococcus being apparently more resistant in San Francisco than in Philadelphia.

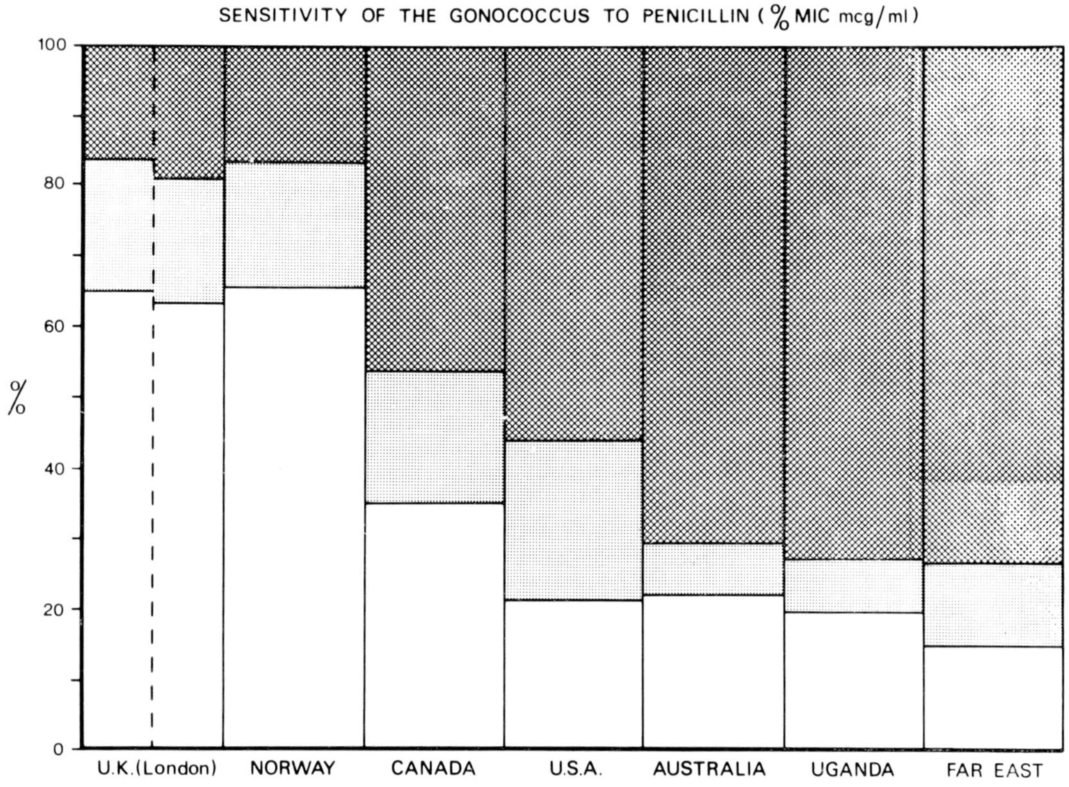

SENSITIVITY OF THE GONOCOCCUS TO PENICILLIN (% MIC mcg/ml)

Fig. 57 Sensitivities of gonococcus to penicillin in different world areas.

In the Far East, particularly in areas near to recent and past war zones, it is evident—as happened nearly thirty years ago with the sulphonamides—that resistance builds up most quickly when two fairly closed promiscuous groups are having repeated sexual exchanges over long periods of time.

The process of ever-increasing resistance has been shown to be capable of being checked or even reversed (e.g. in Denmark and Greenland) when a new and much more effective treatment regimen has been introduced. Nevertheless, the continuing trend is generally upwards towards a greater and greater degree of resistance.

This has necessitated bigger and bigger doses of penicillin if the disease is still to be cured by a single injection. Today 'one shot' treatments with procaine penicillin (e.g. 4·8 mega units) are having to be used. These represent almost the full capacity of the human, especially female, buttock, and even this dose has been shown to fail in about thirty per cent of cases in some parts of the world, as, for example, in the earlier war zones of the Far East.

This disquieting situation is being combated by the use of multiple injections of short-acting penicillin, of other more expensive antibiotics of which there are a considerable number effective in gonorrhoea, or by still using penicillin with the simultaneous use of probenecid which, by delaying the excretion of the antibiotic through the kidneys, enables higher and more prolonged serum levels to be maintained.

At present new effective antibiotics are being discovered which just about keep pace with the deteriorating situation; however, it is by no means certain that such will always be the case. Continued research on new drugs is therefore required. In the meantime the useful life of the relatively cheap non-toxic penicillin is being prolonged by the use of probenecid.

HOPES FOR THE FUTURE

These can only lie in the intensified application of existing techniques, of case-finding, health education, improved diagnostic tests and case-finding methods and in the development of immunizing procedures.

Increased Application of Existing Techniques

There is plenty of scope for increased application of existing techniques, particularly contact-tracing. Britain is in a privileged position in this connection because of the network of venereal disease clinics. In most other countries the majority of venereal diseases are treated by private practitioners, and as a result they usually escape contact-tracing activities particularly regarding sources of infection.

Need for Health Education

There is scope, too, for intensified health education. Whenever surveys have been made they have revealed a widespread ignorance of venereal diseases particularly amongst the young. This ignorance also extends to the medical practitioner and medical students who as a result of competing interests in a crowded curriculum are frequently inadequately taught about their control.

The venereal diseases have been classified as behavioural diseases and therefore, at least in theory, are capable of being influenced if human behaviour can be influenced.

TABLE 41

Countries showing increased per capita cigarette consumption 1956–1966

Austria
Australia
Belgium
Canada
Denmark
Eire
France
Germany (W)
Greece
Iceland
Italy
Japan
New Zealand
Norway
Spain
Sweden
Switzerland
Turkey
United Kingdom
USA

(Reduced consumption was reported only by Argentina,
Brazil, Finland, Netherlands, Mexico and South Africa.)

Although it has been shown that, as a result of health education in schools, young persons are quicker to seek advice when they believe they may have contracted a venereal disease, it is generally admitted that it is difficult to change established norms of behaviour. A full explanation and understanding of the dangers involved is not enough as illustrated in Table 41 from the experience of cigarette smoking. All the countries listed had an increased per capita consumption between 1956 and 1966 when the implications in relation to lung cancer had been very widely discussed and understood. Also, extremely high rates of venereal disease have been reported both from Europe and Africa among the best educated sections of the community (e.g. university students).

Nevertheless, the fact remains that advertisers usually have no trouble in persuading the public that a product is good if it fulfils a need. Perhaps if adequate money were forthcoming and the campaign properly sponsored, using all of the expertise of Madison Avenue or its equivalent, and provided that emphasis was laid on the desirability of maintaining positive venereal health by means of the restriction of sexual contacts to one at a time, something worthwhile might be achieved.

Research Developments

Towards improved diagnostic tests. One of the reasons for the runaway rise in the incidence of gonorrhoea is that, unlike syphilis, there is no reliable serum test which can be applied to screen the population to reveal the reservoir of the disease, particularly in the asymptomatic female. There would be a considerable epidemiological advantage in such a test which could obviate the necessity of a genital examination in the first instance.

The old-fashioned gonococcal complement fixation test was of little value, but improved results were later obtained when multiple strains of gonococci were used as antigen. Following the discovery at the Venereal Disease Research Laboratory, Atlanta, Georgia, USA that colonial typing could be related to virulence, antigens could be made from selected virulent strains. A microprecipitin test was first evolved, a variant of which was reported upon favourably in India. More recently similar or better results have been reported by research workers from the USA, Canada and London by a number of test procedures (see Table 42).

Thus it is now evident that some detectable antibodies are formed in the essentially local disease of gonorrhoea, even in the male and that the availability of an effective serum test is foreseeable. The present difficulty is that with all procedures so far tried a significant proportion for false-positive results are obtained and positivity may persist in those with past infection.

Towards immunizing procedures. Some progress has been made in rabbits towards a vaccine against syphilis using penicillin-killed or irradiated organisms. However, many intravenous injections over long periods of time have so far proved to be necessary and the procedure has no immediate application to man.

It is possible that one day some sort of live vaccine may be introduced against syphilis but, quite apart from the moral and social problems as to who should receive it, there would also be many medical problems. For example, even if the permanent acquisition of a live treponeme was acceptable (with the danger that it might assume some pathogenicity through the years—e.g. under steroid cover) it is unlikely that no antibody response whatever would be produced in the host. Such antibodies would confuse markedly the detection of the disease by a routine screening test and, if treponemal antibodies were involved, would render it impossible (by treponemal immobilization and other existing tests) to distinguish false-positive from true-positive results. Many might decide, therefore, that syphilis was more easy to control using intensified methods of contact-tracing and epidemiological treatment which so far have shown better prospects in the control of that disease than of gonorrhoea.

TABLE 42

Serum tests for gonorrhoea

Method	Authors	Country	Percentage positivity in gonorrhoea		Percentage positivity in controls
			Males	Females	
Complement-fixation Multiple heat-killed strains	Magnusson & Kjellander (1965)	Sweden	21	50	6
Fractionated known virulent organisms	Reising *et al.* (1969)	USA	28	88	0–4
Flocculation Protoplasm Bentonite	Lee & Schmale (1970) Wallace *et al.* (1970)	USA Canada	69 77	86 78	11–12 4
Microprecipitin	Chacko & Nair (1969)	India	60–100		8
Haemagglutination Composite antigen Lipopolysaccharide antigen	Logan *et al.* (1970) Ward & Glynn (1971)	USA UK	62 46	76 84	0–12 0–4
Immunofluorescence	Under investigation				

Against the gonococcus the prospects of an immunizing procedure are far more remote. Although antibodies can be demonstrated in infected patients of both sexes there is no indication that these are protective. It is hard to understand how injections of dead organisms could produce immunity when 20–30 natural attacks will not.

Nevertheless, the sobering thought remains that in the unlikely event of both syphilis and gonorrhoea being either eradicated from our society, or 'tamed' to the point of representing no problem, there would still remain a long list of other sexually transmitted diseases to create personal, marital and social problems and anxiety, unless—as the problem of trichomoniasis was solved in cattle—sexual intercourse is replaced by artificial insemination!

References

American Soc. Hlth. Ass. (1971). Todays VD control problem, New York.

Amies, C. R. (1969). Sensitivity of *Neisseria gonorrhoeae* to penicillin and other antibiotics—Studies carried out in Toronto during the period 1961–1968. *Brit. J. vener. Dis.*, **45**, 216–222.

Arya, O. P. and Phillips, I. (1970). Antibiotic sensitivity of gonococci and treatment of gonorrhoea in Uganda. *Brit. J. vener. Dis.*, **46**, 149–152.

Chacko, C. W. and Nair, G. M. (1969). Sero-diagnosis of gonorrhoea with a microprecipitin test using a lipapolysaccharide antigen from *N. gonorrhoeae*. *Brit. J. vener. Dis.*, **45**, 33–39.

Department of Health and Social Security. Report of the Chief Medical Officer for the year 1968. (See also *Brit. J. vener. Dis.* (1970), **46**, 76).

Fischnaller, J. E. *et al.* (1968). Kanamycin sulfate in the treatment of acute gonorrheal urethritis in Men. *J. Amer. med. Ass.*, **203**, 909–912.

Glynn, A. A. and Watt, P. J. (1971). Serological diagnosis in gonorrhoea. *Post Grad. Med. J.*, **48**, Suppl. 1, 23.

Gundersen, T., Odegaard, K. and Gjessing, H. C. (1969). Treatment of gonorrhoea by one oral dose of Ampicillin and Probenecid combined. *Brit. J. vener. Dis.*, **45**, 235–237.

Guthe, T. and Willcox, R. R. (1970). The international incidence of venereal disease. Excerpt from International Health Conference, Edinburgh, September.

Hatos, G. (1970). Treatment of gonorrhoea by penicillin and a renal blocking agent (Probenecid). *Med. J. Australia*, **1**, 22, 1096–1099.

Idsoe, O. and Guthe, T. (1967). Geneva, Tourism International Bureau, **43**, 227–243.

International Civil Aviation Organization Bulletin (1969).

Keys, T. F. *et al.* (1969). Single dose treatment of gonorrhoea with selected antibiotic agents. *J. Amer. med. Ass.*, **210**, 857–861.

Lee, L. and Schmale, J. D. (1970). Identification of the gonococcal antigen important in the human immune response. *Infect. and Immunity*, **1**, 207–208.

Leigh, D. A., Le Franc, J. and Turnbull, A. R. (1969). Sensitivity to penicillin of *Neisseria gonorrhoeae*—relationship of the results to treatment. *Brit. J. vener. Dis.*, **45**, 157–153.

Lloyd's Register of Shipping (1969). Statistical Tables.

Logan, L. C., Cox, P. M., and Norins, L. C. (1970). Reactivity of two gonococcal antigens in an automated micro-haemaglutination procedure. *Appl. Microbiol.* **20**, 907.

Lynn, R. *et al.* (1970). Further studies of penicillin resistant gonococci. *Brit. J. vener. Dis.*, **46**, 404–405.

McKenzie Pollock, J. S. (1970). Physician reporting of venereal disease in the USA. *Brit. J. vener. Dis.*, **46**, 114–116.

Magnusson, B. and Kjellander, J. (1965). Gonococcal complement-fixation test in complicated and uncomplicated gonorrhoea. *Brit. J. vener. Dis.*, **41**, 127–131.

Reising, G. *et al.* (1969). Reactivity of two selected antigens of *Neisseria gonorrhoeae. Appl. Microbiol.*, **18**, 337–339.

Wallace, R. *et al.* (1970). The bentonite flocculation test in the assay of *Neisseria* antibody. *Canad. J. Microbiol.*, **16**, 655–659.

Wilcox, R. R. (1964). *Textbook on Veneral Infectious and Treponematoses*. London, Heinemann Medical Books.

Willcox, R. R. (1971). Paper presented at American Social Health Association and Pfizer Laboratories International Venereal Disease Symposium, St Louis, Mo., USA, May. (In press.)

World Health Organization (1963). WHO Expert Committee on Gonococcal Infections. Geneva [*WHO Techn. Rep. Ser.*, No. 262].

World Health Organization (1970). Treponematoses Research. Report of a WHO Science Group. Geneva [*WHO Techn. Rep. Ser.*, No. 455].

Chapter 16

Alcoholism as a Reflection of Environment

E. B. Ritson

Alcohol addiction arises from the interaction of a psychologically, and possibly biologically predisposed individual with an environment which encourages or precipitates excessive drinking. The aim of this chapter is to discuss environmental influences on problem drinking.

The relationship between the individual personality, the drug alcohol, and the environment is complex. There is no evidence of any particular pre-alcoholic personality type, but it does seem clear that emotionally disturbed and damaged personalities are more likely to try to relieve anxiety by drinking, and to develop alcohol addiction. The choice of alcohol as a means of escape or relaxation is largely culturally determined. In some parts of the world other drugs such as cannabis would be chosen to fulfil the purpose for which alcohol is currently used in so many Western countries. Climate determines the availability of the raw material from which alcohol is produced, although almost every society has evolved some means of fermenting naturally occurring sugars to produce alcohol. Ready availability of alcohol is clearly an important variable influencing drinking behaviour. When excessive drinking is uncommon, then only the more distressed and deviant members of that society will drink heavily, whereas when it is socially acceptable, as it is in Britain, more people will drink to excess and enter the population at risk to becoming alcoholic.

By considering environmental influences alone, we must not imagine that we are concerned with a static model. Von Wartburg (1971) has analysed the dynamic interrelations of the individual, the drug and his environment in both biochemical and psychological terms. He views an individual as a self-regulating system, employing homeostatic mechanisms at every level of functioning to maintain an equilibrium. He states that 'total biological man, his physiology and biochemistry, his stored information, both phylogenetic and ontogenetic' is a 'system which responds to inputs from the total environment. This contains the elements of physical, spiritual, social and cultural forces, together with the direct behaviour of other men. The response of the individual to these inputs from the environment, in time, is a change of state, an action which we term behaviour. This behaviour then becomes part of the environment, interacting with the established situation to modify or create an input pattern for the individual.' Thus we are reminded that man is not a passive recipient of his environment but continuously influences and changes it.

The attitude a society adopts towards alcohol is the product of numerous influences:

(i) availability, this itself determined by geographic, economic and political factors;
(ii) the value attached to alcohol as food or relaxant;
(iii) competition from other drugs;
(iv) taboos and penalties attached to excessive drinking;
(v) religious beliefs.

Even the definition and meaning of alcoholism itself reflects the prevailing attitudes of that society: some regard alcoholism simply as a moral weakness, others as a disease; some ignore the alcoholic's predicament while other cultures provide institutions specifically for his care.

208

DEFINITION

The definition of alcoholism itself is strongly influenced by the social context within which alcohol dependence is viewed. The World Health Organization has defined alcoholics as:

'Those excessive drinkers whose dependence upon alcohol has attained such a degree that they show a noticeable mental disturbance or an interference with bodily or mental health, interpersonal relations, smooth economic functioning, or the prodromal signs of such development; they therefore require treatment.' (World Health Organization, 1952).

Most studies of the incidence of alcoholism rely on indicators of problem drinking which are themselves dependent on prevailing cultural attitudes. If habitual drunkenness is viewed purely as an offence then alcoholics will be found largely in criminal statistics; if it is seen as a symptom of underlying psychological disorder then psychiatric clinics will report alcoholism amongst their patients.

The more one looks for it the more alcoholism is a disease which appears with increasing frequency. The extent and the way in which a society looks for its alcoholics will affect the number that come to attention. Because of the shame attached to the diagnosis in the past, patients came only reluctantly, often when their illness had reached an advanced and chronic stage. As the disease concept of alcoholism becomes more acceptable, patients seek help earlier and more readily. In view of this it is unwise to make assertions about increases in the incidence of alcoholism which may tell us more about changing diagnostic indices and attitudes towards the disease, than about changing drinking practices. It is exceedingly difficult to apply traditional epidemiological techniques to the socio-clinical kind of definition which alcoholism attracts.

INCIDENCE

The most widely used formula for estimating the prevalence of alcoholism in a population was devised by Jellinek (1952). He has expressed doubts about its accuracy, (Jellinek, 1959). The formula has frequently been used to estimate the size of the alcohol problem in a country. The formula is derived from the relationship between the total number of chronic alcoholics in a given area at a given time, the number of hepatic cirrhosis deaths reported, the percentage of these attributed to alcoholism and the population of all alcoholics who die of cirrhosis of the liver. Usual estimates suggest that 25 per cent of alcoholics suffer from physical complications: post mortem findings indicate that only 9 per cent of these have liver cirrhosis, which was the cause of only 8 per cent of the deaths of cirrhotic alcoholics (representing only 0·2 per cent of alcoholics alive at the time.) (Wallgren and Barry, 1970).

The formula is dependent on the constancy of the relationship between these estimations; it relies on the accuracy of the hospital records and post mortem reports. It is likely that variables such as habitual drinking patterns of a society, beverage choice and the general nutritional state of the population will affect the number of alcoholics who become cirrhotic. Using the Jellinek formula it was estimated that there was in England and Wales an incidence of 11 alcoholics per 1,000 of the population aged 20 and over.

There have been several field surveys of the prevalence of alcoholism in this country during recent years. The most detailed was conducted by Moss and Davies (1968) in Cambridgeshire. They checked 13 different sources for clients known to have drinking problems. The sources were—hospital records, consultant psychiatrists' private patients, general practitioners, police, probation, Salvation Army, Church Army, Children's Department, NSPCC, Marriage Guidance, Alcoholics Anonymous, Samaritans, and Local Health Authority. The list is interesting not only in showing the thoroughness of the search, but also the widespread nature of the agencies who may know of the alcoholic.

Moss and Davies asked these agencies to report all clients who (using established criteria, such as hospital admission or arrest for drunkenness) were thought to suffer from alcoholism. Using this method they estimated a prevalence of 6·2 per 1,000 males and 1·4 per 1,000 females (aged 15 and over). All studies comment on the preponderance of male alcoholics presumably because heavy drinking is commoner and more acceptable in men. Social attitudes towards female drinking are changing rapidly in a way which may result in more alcoholic women in the future.

Amongst males the disease was most common in those who were widowed or divorced. It was particularly common in single men over 25 compared with those who were married. With women it was again more common amongst the widowed and divorced, but with the exception of the middle aged, married women outnumbered those who were single. For the most part, Moss and Davies found alcoholism to be more common in urban areas although they found the opposite trend in professional people. They observed a particularly low prevalence of alcoholism amongst unskilled farm workers in rural Cambridgeshire (4·6 per thousand) which contrasted with the excess amongst unskilled labourers in towns. (14·8 per thousand.)

While this recent study is the most detailed of its kind yet reported in Britain, it can only serve as a guide to the incidence of alcoholism in the rest of the country.

There have been no field studies of the prevalence of alcoholism in Scotland, but available evidence suggests that it is a greater problem there than in England, and that the further north one looks the more prevalent alcoholism becomes. The first admission rate of male alcoholics to Scottish psychiatric hospitals is approximately six times that for England (Morrison, 1964). Admissions to mental hospitals in Inverness (the most northerly psychiatric hospital in Britain) were three times the Scottish average (Whittet, 1967).

The reasons for this trend are not known. It has often been suggested that the cold and at times harsh climate encourages drinking to promote inner warmth, but there is no objective evidence to support this. Whisky is relatively more popular in Scotland than in England, and it is often suggested that this preference for the stronger drink is a significant factor. A preference for spirit drinking is also evident in Scandinavian countries. The more potent beverage provides more rapid intoxication which could account for more behavioural problems associated with alcohol but not necessarily greater dependence as France, a predominantly wine-drinking country, exceeds any spirit-drinking country in its alcoholism rates.

Most research shows that alcoholism is more common in urban than in rural areas. In this respect the north of Scotland appears as an exception. A national sample of 2,746 adults in the USA were questioned about their drinking practices (Cahalan and Cisin, 1968). They found that the inhabitants of the city centre contained the highest proportion of heavy drinkers. The suburbs had the largest number of regular drinkers (87 per cent) whereas rural areas had only 43 per cent reporting regular drinking.

Rural areas involved in the production of alcohol are exceptions to this trend. The north of Scotland has a relative excess of persons employed in the drink trade which again probably contributes to the excess of alcoholism in this particular rural area.

A rather different approach to the epidemiology of alcoholism has been adopted by de Lint and Schmidt (1971). They view alcoholism as one extreme on a continuum of drinking behaviour rather than a distinct and separate pathological entity. There certainly seems little doubt that the incidence of alcoholism is related to the drinking habits of a population.

Ledermann (1956) originally postulated that there was a fixed relationship between consumption averages and alcoholism prevalence. Individual consumption ranges from very small quantities to near lethal amounts. Within this range each consumption level occurs with a certain frequency. It has been demonstrated for a variety of populations that the distribution of these levels closely approximates to a smooth, skewed curve known as the logarithmic normal curve. (Fig. 58). It is reasonable to assume that alcoholics are to be found in the upper reaches of this curve. In support of this view it has been shown for several countries that there is a significant correlation between rates of liver cirrhosis deaths and per capita alcohol consumption.

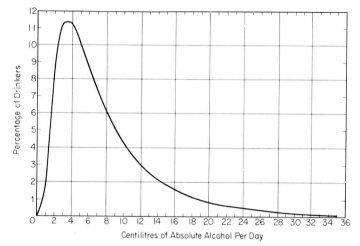

Fig. 58   The distribution of drinkers according to average daily volume of consumption in a population with an annual per drinker consumption of 30 litres of absolute alcohol. (de Lint and Schmidt, 1971.)

Using an arbitrary definition of the alcoholic as a person drinking more than an average of 15 cc's of absolute alcohol daily, the number of such drinkers in England and Wales is calculated as 19·5 per thousand aged 15 and over. It seems reasonable to assume that environmental factors which influence the national average consumption of alcohol will affect the incidence of alcoholism.

NORMAL DRINKING HABITS

From the above it seems logical to look for the origins of alcoholism amongst the normal drinking practices of a culture.

Beverage choice is influenced by local availability of different kinds of drink. Where vineyards are common, wine is usually the drink of choice. In Britain beer is the most popular drink: more than 5½ thousand million litres of beer are drunk each year in Britain compared with 112 million litres of spirits. It is commonly believed that one cannot become alcoholic on beer or wine alone and that spirits are the problem. Clinically there is evidently no truth in this assertion. Beverage preference amongst alcoholics does not depart much from that of the drinking population at large. There is no evidence that high rates of alcoholism are limited to a preference for spirits. For instance, in France, which has an exceedingly high incidence of alcoholism, spirits contribute only 13 per cent to total alcohol consumption. In Scandinavian countries where spirit drinking is common, the incidence of alcoholism is lower.

The average British household in 1970 spent 3·46 per cent of its weekly income on alcohol. It is more difficult to obtain an accurate picture of normal drinking practices themselves.

The British Market Research Bureau publishes reports on the drinking habits of a random sample of the population aged 15 or over. These show that beer is the most popular drink in younger age groups of both higher and lower social classes (Zacune and Hensman, 1972). Edwards (1972) has conducted a door-to-door survey of drinking habits in a sample of adults in six housing estates in one London borough. The overall frequency of drinking is shown in Table 43.

TABLE 43

Frequency of drinking last year in a London Borough
(from *Hensman and Zacune*, 1971).

Frequency of drinking	Males $n = 408$	Females $n = 520$
Every day	11%	4%
'Most days'	11	4
Weekends only	19	11
Once or twice a week	23	15
Once or twice a month	12	20
Once or twice in six months	7	14
Once or twice a year	10	22
Never	7	10
Total	100%	100%

The research reveals that younger people appeared to drink more heavily than older. As in other reports, men of Scottish or Irish background accounted for a disproportionately larger number of heavy drinkers.

Children incorporate the cultural values of their environment. In most countries generally accepted cultural attitudes towards alcohol have been established. It is open to each succeeding generation to accept or reject these values. In many Western societies adolescence is the time when these values are first challenged.

The reasons for starting to drink are strongly influenced by parental attitudes, peer group norms and more general societal influences such as advertising. Where drinking is strongly condemned by elders, teenagers may well be less likely to drink, but should they do so, then that drinking is more likely to be pathological. Presumably alcohol is much more emotionally charged when it is taken in the face of strong cultural taboos than when it is condoned or encouraged. In this situation there is an absence of models of moderate alcohol use with which the teenager can identify. Straus and Bacon (1953) studied drinking amongst American college students and noticed an unusually high incidence of drinking to intoxication in Mormons amongst whom alcohol is condemned.

In a study of 'how French children learn to drink', Anderson (1969) records that children often start drinking diluted wine at the age of two and that alcohol is associated with 'vigour, vitality and good health'. In the village studied, the adult men drank 1·58 quarts of wine daily and the women 0·5 quart. French see alcohol as a health-giving food and a recent Government campaign to shift the French away from this concept has largely failed.

Several authors, e.g. Glatt (1970), and Snyder (1958), have commented on the low incidence of alcoholism amongst Jews. This is particularly true of orthodox Jews. Snyder postulated that the low rates were associated with their religion because the intake of small amounts of alcohol is a part of certain religious rituals and because a sense of Jewish identity involves strong inhibitions against drunkenness.

The Irish culture is often taken as an example of one which encourages heavy drinking. There is a higher incidence of alcoholism amongst Irishmen in the USA than in any other ethnic group, and alcoholism is a very serious problem in Ireland (Walsh, 1969). Drinking often to drunkenness is accepted and common amongst Irishmen. In a study of drinking habits in Oakland, Knupfer and Room (1967) found that 36 per cent of Irish drank daily, as compared with 26 per cent of white Protestants, and 18 per cent of Jews. Drunkenness was most common amongst Irishmen.

Some of the sociological and cultural variations in the incidence of alcoholism seem to be most plausibly attributable to differences in the effectiveness of social controls. Many Western cultures are intensely ambivalent in their attitude towards alcoholism. Protestant and Irish cultures often combine official moral condemnation of all alcohol while using drunkenness as a means of dissolving conscience and permitting the expression of sexual and aggressive impulses. It seems that a society which permits a limited use of alcohol and frowns on drunkenness is the cultural environment best suited to moderate alcohol use.

Late adolescence is now one of the peak periods for heavy drinking. This is reflected both in the survey of drinking habits discussed earlier and in the dramatic increase in drunkenness offences in young people. Advertisers and publicans now make conscious appeal to this large group of youthful customers. Attitudes towards alcohol formed during the late teens have a profound effect on subsequent alcohol use. For many, the abnormal drinking patterns which develop later into established alcoholism, are evident by the age of 20. (Hassall, 1969). We should, therefore, be concerned about the way in which alcohol is presented in the teenage subculture.

McCord and McCord (1960) have pointed out that the implicit or explicit association between virility and being able to hold one's liquor 'like a man' may be an important incentive to drink heavily amongst those young people who have doubts about their own male image. The heavy drinker is widely represented in Western literature and culture as 'tough', and it is easy to see how an insecure boy would seek status by demonstrating his capacity for alcohol.

Adolescents are particularly conscious of peer group pressures and the habit of drinking round for round with a group of friends is often an important ritual binding the group together. In such a context, refusing a drink may be interpreted as weakness or disapproval. In older age groups customs such as the cocktail hour before dinner and the ritual of sealing business deals over several rounds of drinks, serve to establish drinking norms from which it is often difficult to deviate. Edwards' (1971) survey shows interesting class differences in drinking practices—middle class men tending to drink regularly in moderation, while working class men were more likely to have binges often at weekends rarely drinking earlier in the week.

OCCUPATION

It is clear that some occupations afford more opportunity for drinking than others. In view of the hazard of excessive drinking, it is surprising that more concern is not shown for those who are placed at particular risk. Workers in the drink trade itself—publicans, distillers, brewers, barmen, are particularly exposed to developing alcoholism. The prevailing industry of a district may in this respect influence the prevalence of alcoholism. For instance, the high incidence in the north of Scotland may in part be due to the preponderance of workers in the drink trade in that area. Expense-account businessmen who regard entertaining as a part of their job have every opportunity to drink excessively. Excessive drinking is common amongst casual labourers perhaps because drinking during the working day is condoned or encouraged. Furthermore, there is rarely any taking up of references for such casual employment so that regular drunkenness would not be known to a new employer.

AVAILABILITY

The degree to which alcohol is available has a very significant effect on use. As already discussed, alcoholism is common in those occupations in which alcohol is readily available. It was shown by de Lint and Schmidt (1971) that per capita consumption is correlated with the incidence of alcoholism. Control of availability is one of society's means of influencing per capita consumption. The State commonly controls availability either by legislation governing the sale of alcohol (licensing laws) or by taxation which affects the individual's financial capacity to purchase alcohol. Both methods have been adopted in this country from time to time in the past 500 years.

Glatt (1958) has outlined the history of problem drinking in this country in an article entitled 'The English Drink Problem: Its Rise and Decline through the Ages.' This interesting review illustrates the effect of control measures on drinking behaviour. During the 18th century foreign brandy was making such an impact on British markets that the production of cheap gin from English corn was encouraged: all controls on the sale of gin were removed. The outcome is well known and illustrated in Hogarth's print of 'Gin Lane' portraying the lethal effects of cheap gin. As the public became aware of the damage being caused by this product, Acts were introduced in the mid 18th century which increased the price of spirits and controlled their sale. These, coupled with other social changes, led to a reduction in the national use of gin. Legislation is an effective exercise in social engineering.

A similar crisis was evident at the onset of 1914–18 War when Lloyd George is quoted as saying that this country had three deadly foes: 'Germany, Austria, and drink.' People were appalled at the effect of drunkenness on productivity at a time of need. Taxation on alcohol was increased and the opening hours of licensed premises were severely curtailed. Largely as a result of these measures, drunkenness was diminished by half within eighteen months.

Similar evidence for the effect of licensing and taxation on the consumption of alcohol can be demonstrated in other countries. For instance, in Denmark (Nielsen, 1965) a steep rise in the tax on spirits led to a marked decline in the consumption of spirits and a rise in the use of beer. In Finland an increase in the number of liquor outlets was followed by a rise in per capita consumption. (Kuusi, 1957). Even prohibition in USA which had such disastrous social consequences was associated with a transient fall in the incidence of deaths from cirrhosis of the liver.

Those who would wish to control the availability of alcohol in any society have two useful tools in licensing laws and taxation increases. Both are means of influencing availability in the potential consumer's environment. On the other hand, those who produce and distribute alcohol have a vested interest in increasing availability and hence consumption of alcohol.

Between 1959 and 1969 production of spirits in the UK rose from 94·9 million gallons to 199·1 million gallons: equivalent figures for beer production were 914 million to 1,148 million gallons. Expenditure on beer has risen from £563 million to £1,058 million in 1969 (Zacune and Hensman, 1972). There has been a steady rise in the per capita consumption of beer and wine in this country during the past ten years. The same trend is not apparent for spirits. It may be that the high cost of spirits has had some effect in limiting consumption.

Those involved in the drink trade take every opportunity of stimulating consumption by:

(i) increasing the ease with which alcohol may be obtained—alcohol sales in supermarkets are a recent example;

(ii) attempting to attract more people into licensed premises—there has been a sustained attempt to make public houses more attractive and acceptable to women and younger people;

(iii) prolonging licensing hours—often on the grounds that this will encourage tourism or reduce drunkenness by promoting prolonged rather than intense episodic drinking;
(iv) Advertising—in 1967 the national advertising bill for alcoholic drinks was £25·5 million.

Advertising is clearly an important environmental factor in determining both beverage choice and attitude towards alcohol. In this country the Independent Television Companies Association give some guidance concerning the character of promotion of alcoholic drinks. For instance, 'Advertisements should not be transmitted in the commercial breaks immediately before, during or after children's programmes. Advertisements should not directly encourage young people to drink alcohol. Actors and actresses appearing in advertisements should clearly be at least in their twenties. In relation to the advertising of stronger types of drink, they may be required to be more mature still. Advertisements should not dare people to try a particular drink or imply that they will prove themselves in some way if they accept the challenge offered by the drink. Advertisements should not be based on any theme emphasizing the stimulant or sedative effect of any drink.' These and similar recommendations show some awareness of the potential hazards associated with promoting alcohol, but many advertisers set no such standards.

It is common to see advertisements which show alcohol being taken by teenagers with the clear suggestion that it enhances enjoyment. In an attempt to expand their market, drinks such as beer, which have hitherto been taken mostly by men, are shown being enjoyed by women. Some advertisements suggest a strong association between sexual pleasure and alcohol; others associate virility or sophistication with a particular drink. It is interesting that there has been little or no debate about the morality of advertising alcoholic drinks comparable with measures recommended for cigarette advertising.

STRESS

It is often said that alcohol is taken to relieve anxiety or as an escape from problems. This has led to the concept that the incidence of alcoholism will be related to the overall level of stress or anxiety in a culture. That is, given a society that both fosters drinking and has a high level of stress, then alcoholism will be common. Such a combination seems to exist in many Western countries at present.

The concept of stress is not easy to apply to populations or sub-groups within a culture because of the variation in an individuals perception of what constitutes a stressful situation. A stress to one man is a challenge to another.

Horton (1943), an anthropologist, examined the drinking habits of 118 primitive cultures in Asia, Africa and the Americas. He showed that drunkenness was more common amongst those groups where there was a high overall level of anxiety, for instance about food supplies, or who were undergoing a period of rapid social change. Horton's hypothesis that there is a correlation between difficult living conditions and an intensive use of alcohol was not supported by studies conducted in isolated parts of Northern Norway (Irgens-Jensen, 1970), and in Finland (Sariola, 1956).

As we move from somewhat atypical isolated and primitive communities towards a more familiar urban setting, it is even more difficult to talk of levels of stress or anxiety in any measurable way because of the complexity of these communities and the number of variables involved. Lynn and Hampson (1970) showed that ranked anxiety levels assessed by psychological tests amongst male students in a number of Western Universities were closely similar to the measured prevalence of alcoholism in these countries. France topped the list for both student anxieties and alcoholism.

The heavy drinking seen in Britain during the Industrial Revolution was seen not only as a response to increased availability of alcohol but more also particularly as an escape from the appalling conditions and exploitation of workers in the new factories. It was assumed that improved working conditions introduced during this century would reduce the need to seek oblivion in alcohol. This, amongst other factors, probably contributed to the decline in the overall level of drunkenness in Britain during the first half of this century. In 1949 a Lancet editorial felt able to talk of 'a general recognition that alcohol was no longer the pressing social problem that it had been in former years.'

This optimism was short lived as the consumption of alcohol and incidence of alcoholism has again risen steadily since this time. Although the physical conditions of the environment are perhaps less stressful in material terms, it may be that new problems are causing people to drink. Relative affluence coupled with increased leisure leads to more time devoted to drinking. Excessive drinking may be another by-product of the leisure problem in a culture which is not yet equipped to utilize extensive and diverse leisure. The psychological depression associated with unemployment often leaves men with little recourse but to spend their time in the pubs. In a similar way, retirement leaves a man with 'time on his hands' which is often filled by drinking. The stress and loneliness of old age leads to a considerable, often hidden, problem of alcoholism in the elderly. Thus we see a tendency for some to relieve the stress engendered by common crises by excessive drinking. When the psychological consequences of some of the stresses can be foreseen, a caring society should offer concern and counsel for those exposed.

Migration within countries and between social classes, exposes certain individuals to stress and removes them from the support of familiar surroundings and institutions. Several authors have commented on the high incidence of problem drinking amongst Irish and Scots who migrate to the industrial areas of England. (Edwards *et al.*, 1968). Similar observations have been made about Irish and Italian emigrants in the USA.

Of the Irish and Scots in London, Edwards (1970) states that research 'indicates that the social origins of much chronic public drunkenness is the young Irishman or young Scot who, emigrating to London as an unskilled labourer, has his social controls removed, money in his pocket, no healthy or attractive leisure alternative to the pub, and no invitation to spend his money on other than beer.'

It is clear that this group is exposed to stress and that society has provided nothing to help them cope with it.

SKID ROW

In the centre of most large cities one finds a bleak, often semi-derelict area where vagrants and other down-and-outs gather. Such a district is often known as skid row. The original skid row is situated in a part of Seattle where logs were skidded into the river. The grim environment of the skid-row districts of cities tends to attract social misfits. However, urban renewal programmes, designed to improve the environment, usually cause this group to move elsewhere in search of some other blighted area in which to settle.

This is the 'end of the line' for the most socially deteriorated alcoholic who has been forced to sleep rough or at best to live in lodging houses or reception centres; many are reduced to drinking cheap wine or 'meths'. It is important to recognize that they constitute less than 7 per cent of the alcoholic population, but they are an important minority.

It would also be misleading to suggest that all down-and-outs were alcoholics. One quarter were thought to be alcohol addicts in a recent study, but 45 per cent of this group had been arrested at least once for public drunkenness. (Edwards *et al.*, 1968).

Almost all of the people living in such conditions in London were found to have moved into the area from far afield, often from Scotland or Ireland.

Edwards *et al.* (1966) interviewed 51 alcoholics who were regular attenders at soup kitchens in Stepney; 37 per cent of the sample were Irish and 27 per cent Scottish: 74 per cent were Roman Catholic. Most came from loosely knit, unstable, working class families and had severed these tenuous ties with home early in life. Their lives were characterized by instability in work and human relations. They provoke anxiety in the minds of residents who feel that they contribute to the further decay of the area. For instance, the Health Committee of Southwark Borough Council reported that 'these people urinate, defaecate and vomit wherever they may be and their clothing is filthy. Public places, subways and highways are worse for their presence.'

It is only recently that concern has been shown towards this group in the report on Habitual Drunken Offenders (1971). This report recognizes that the men themselves require a concerted therapeutic approach and that merely patching up the surroundings in which they are found does not solve the problem.

Edwards (1966) states: 'A man's arrival in Skid Row is not due only to his path thither being unobstructed; an important part of the explanation is that he took the path because the whole Skid Row milieu answered so many of his needs. He has found a pseudo-adjustment which makes his anxieties tolerable, he has opted out of the society which demands close emotional engagement and which expects him to make decisions and bear responsibilities.'

CONCLUSION

The environment within which an individual develops moulds his attitude towards alcohol, and also determines its availability. The individual who seeks solace in alcohol will often have received earlier insults from an unkind fate, or harsh environment, that make him psychologically at risk to develop alcoholism. Such a line of thought brings our present subject within the larger issue of the influence of the environment on the mental health of a community—measures designed to improve the quality of life should result in improved mental health within that community with resultant decline in disorders such as alcoholism.

Such holistic, global and somewhat utopian concepts are less useful at this stage than attempts to focus on practical problems and groups at risk, for instance, the recent measures recommended for the homeless drunk offender described above. We should make provision for those occupations which are exposed to special risk and provide them with early detection services. Those individuals who are in our society exposed to stresses such as loneliness, retirement or bereavement, merit more caring services than they currently receive so that relief may be obtained from a relationship rather than from alcohol.

Attitudes change slowly within a population. Some of the ways in which a culture's religious and social evaluation of alcohol influence the way in which an individual perceives drinking have been discussed. Advertising is a relatively new environmental force seeking to influence drinking behaviour and it could be controlled. The most effective form of social control that is readily applicable is legislation to control availability. It is important that the social consequences of tampering with such legislation are recognized.

References

Anderson, B. G. (1969). How French children learn to drink. *Transaction*, **7**, 20.
Cahalan, D. and Cissin, I. H. (1968). American drinking practice. *Quart. J. Studies Alcohol*, **29**, 130.
Chandler, J., Hensman, C. and Edwards, G. (1971). Determinants of what happens to alcoholics. *Quart. J. Studies Alcohol*, **32**, 349.
Department of Employment and productivity (1970). *Family Expenditure Survey*. London, HMSO.

Edwards, G. (1970). Place of treatment professions in society's response to chemical abuse. *Brit. Med. J.*, **2**, 195.

Edwards, G., Hawker, A., Williamson, V. and Hemsman, C. (1966). London's Skid Row. *Lancet*, **1**, 249.

Edwards, G., Williamson, V., Hawker, A., Hensman, C. and Postoyan, S. (1968). Census of a reception centre. *Brit. J. Psychiat.*, **114**, 1031.

Glatt, M. M. (1958). The English drink problem: its rise and decline through the ages. *Brit. J. Addict.*, **55**, 51.

Glatt, M. M. (1970). Alcoholism and drug dependence amongst Jews. *Brit. J. Addict.*, **64**, 297.

Hassall, C. (1969). Development of alcohol addiction in young men. *Brit. J. Prev. Soc. Med.*, **23**, 40.

Home Office (1971). *Habitual Drunken Offenders*. London, HMSO.

Horton, D. (1943). The functions of alcohol in primitive societies: a cross-cultural study. *Quart J. Studies Alcohol*, **4**, 199.

Irgens-Jensen, O. (1970). The use of alcohol in an isolated area of northern Norway. *Brit. J. Addict.*, **63**, 181.

Jellinek, E. M. (1952). Phases of Alcohol Addiction. *Quart. J. Studies Alcohol*, **13**, 673.

Jellinek, E. M. (1959). Estimating the prevalence of alcoholism. *Quart. J. Studies Alcohol*, **20**, 261.

Knupfer, G. and Room, R. (1967). Irish, Jewish and Protestant drinking. *Quart. J. Studies Alcohol*, **28**, 676.

Kuusi, P. (1957). *Sales Experiment in Rural Finland*. The Finnish Foundation for Alcohol Studies.

Lancet (1949). Editorial, **1**, 310.

Ledermann, S. (1956). *Alcool—Alcoolisme—Alcoolisation: Données Scientifiques de caratère physiologique, économique et social*. Inst. National détudes demographique, Cahier 29, Presses Universitaires de France.

de Lint, J. and Schmidt, W. (1971). Consumption averages and alcoholism prevalence. *Brit. J. Addict.*, **66**, 97.

Lynn, R. and Hampson, S (1970). National anxiety levels and prevalence of alcholism. *Brit. J. Addict.*, **64**, 305.

McCord, W. and McCord, J. (1960). *Origins of Alcoholism*. Stanford, Stanford University Press.

Morrison, S. L. (1964). Alcoholism in Scotland. *Health Bulletin*, **22**, 1.

Moss, M. and Davies, B. (1968). A survey of alcoholism in an English county. *Geigy. (U.K.)*.

Nielsen, J. (1965). Delirium tremens in Copenhagen. *Acta Psychiat. Scand.*, **41**, 1.

Sariola, S. (1956). *Drinking Patterns in Finnish Lapland*. Helsinki, The Finnish Foundation for Alcohol Studies.

Snyder, C. R. (1958). *Alcohol and the Jews*. New Brunswick, Rutzers.

Straus, R. and Bacon, S. D. (1953). *Drinking in College*. New Haven, Yale University Press.

Von Wartburg, J. P. (1971). International Conference of Alcohol and Drug Addiction. Dublin.

Wallgren, H. and Barry H. (1970). *Actions of Alcohol*. Amsterdam, Elsevier.

Whittet, M. (1967). Highland and Island psychiatric reflections. *Brit. J. Med. Psychol.*, **40**, 1.

Walsh, D. (1969). Alcoholism in the Republic of Ireland. *Brit. J. Psychiat.*, **115**, 1021.

World Health Organization (1952). Technical Report Series, No. 48. Geneva, WHO.

Zacune, J. and Hensman, C. (1972). *Drugs, Alcohol and Tobacco in Britain*. London, Heinemann.

Chapter 17
Drug Dependence in Contemporary Societies

J. L. Reed

It cannot be known for certain how long man has used stimulants or sedatives to relieve physical stress in his everyday life, to combat the anxieties which are a natural feature of man's existence, or as part of magic or religious rituals. However, evidence from archeological investigations suggests that the use of these substances can be dated back into pre-history. All the substances used mediate their effects principally through the central nervous system and alter in some way the users' relationship with the reality of their environment. Over the years it has been noted that some individuals use such substances in a way that is deleterious either to themselves or to those who are living in the same group. It is the knowledge that such substances may cause damage either to society or to the individual that has led to the concept of 'drug addiction', and the publicity that has been accorded to it recently merely represents a new emphasis on an old problem.

It is important in considering drug abuse to attempt to define some terms. The first is the word 'drug' which has several different meanings. Apart from the pharmacological or therapeutic meanings, in a more limited way 'drugs' can mean those substances which produce dependence in the person who takes them, and in an even more limited sense the word 'drug' is restricted to a dependency-producing substance of which the society in which it is being used disapproves. This last reservation is of some importance as, while there is no doubt that tobacco may produce a dependency syndrome, within Western cultures it is not disapproved of and is not, therefore, generally considered a drug and certainly not a drug of addiction. Similarly, although alcohol fulfils all the criteria for producing a severe dependency syndrome, yet because of its social acceptability within Western civilizations, the problem of alcohol abuse is generally considered as one separate from the non-socially approved drugs such as heroin, cocaine and the hallucinogenic drugs.

It has proved impossible to make any clear distinction between 'habituation' and 'addiction', and in 1964 the World Health Organization recommended that the words addiction and habituation be replaced by the term 'dependence'. They held that drugs capable of producing dependence had the single common property of producing psychological dependence so that an individual experienced a desire for the continual or periodic administration of the drug for pleasure or to relieve discomfort. Beyond this the various dependency producing drugs should be regarded as having different dependency syndromes which might or might not involve physical dependence and hence physical withdrawal symptoms. It was recommended that any mention of drug dependence should specify the type of drug involved, e.g. one should speak of 'drug dependence of the morphine type' or 'drug dependence of the alcohol type'. Abuse of the drug was held to occur when the dependence produced a detrimental effect on the individual or on society. The replacement of the term 'addiction' by 'dependence' has in many ways been helpful not least in displacing a word which has considerable emotive connotation. It has not been generally accepted, however, and it is difficult, without being pedantic, to discuss problems of drug abuse without using the term addiction.

Surveying the extent of drug abuse in different societies, there is in most of them considerable abuse of one or more drugs which are socially acceptable. Although the abuse of these drugs may present a considerable strain on the society, because of the high degree of social acceptability of the drug involved, relatively little attention is paid to the problem despite the number of people affected. Against this background of widespread but culturally accepted drug abuse one may see smaller numbers of cases of abuse of other drugs which are to a greater or lesser extent socially proscribed. To this relatively small number of drug abusers an apparently disproportionate amount of attention may be paid whether medical, sociological or legal. This attention presumably reflects the society's disapprobation of the practice of taking a particular drug and also its fear of socially deviant behaviour in itself rather than an indication of the number of people actually being harmed by the drug.

Socially disapproved drug taking often occurs in outbursts in a manner very similar to an epidemic of an infectious disease, and Rathod and his co-workers (1969) have shown how 'infection' may spread from case to case.

DRUGS LIABLE TO ABUSE

In considering the different drugs involved in abuse it is a difficult and certainly a pointless task to draw up any kind of 'league table' of those which are most widely used or those which appear to be most dangerous in their abuse. No adequate statistics of drug usage are available for many countries and, as has already been discussed, what may be considered abuse in one country may be accepted as normal in another.

Tobacco

Abuse of tobacco is certainly one of the most widespread problems and has shown an extraordinarily rapid dissemination in a relatively short period of time. Since its introduction into Europe in the 16th century smoking has virtually become a worldwide habit and shows no signs of waning despite increasing medical disapprobation. The great increase in tobacco consumption since the introduction of the cigarette at the end of the 19th century illustrates how a changed, easily consumed presentation of a drug may influence its use. Although it had been clear to many people for a long time that tobacco was a drug which produced dependence, its use was so widespread and its effects apparently so relatively harmless that there was little concern until the recent evidence of the association between smoking and carcinoma of the lung and coronary artery disease. As these are very late complications of its use it is extremely unlikely that tobacco will ever be looked upon as a 'drug' in the same way that the narcotics, which rapidly produce socially undesirable effects, are regarded. The principal cause for surprise about tobacco is not that its use is so common but, as Blum points out, that some societies such as the Yami in Taiwan, or the Senussi in Libya have avoided tobacco use when nearly all other cultures have become involved.

Alcohol

In general Western societies approve the use of alcohol, at times even its immoderate use. This approbation is not general and Bales (1959) has suggested that one of four attitudes towards drinking may be found in a society. Firstly, abstinence may be the socially approved pattern, usually on religious grounds such as is found in Muslim countries. Secondly, alcohol may be approved but within strictly limited settings, usually in ritual and religious ceremonies, as in orthodox Judaism. The third attitude towards drinking is that labelled by Bales as 'convivial' in which alcohol is used on social occasions because it makes for greater social ease and facilitates social group interaction. The fourth attitude is labelled 'utilitarian' when drinking is aimed at making the individual feel better in himself and within his group. These two latter groups correspond to the attitudes of most Western societies and also include some

Eastern cultures such as the Japanese. Bales recognizes that the four attitudes are not mutually exclusive, and in particular convivial drinking may progress to utilitarian drinking among individuals or groups pre-disposed to greater drug abuse. A study by Sargent (1971) of drinking patterns in universities in Sydney and in Japan has shown some relationship between these cultural attitudes to drinking and the degree of alcoholism within the different groups.

It is difficult to form accurate measures of incidence of alcohol abuse in different countries partly on account of these cultural attitudes, which may lead to the concealing or ignoring of alcohol problems. Two methods are however available; the direct method of population survey or the more widely used method which is an indirect estimation based on a formula devised by Jellinek (WHO, 1951). This formula $A = PD/K$ where A is the total number of alcoholics in a given year, D the number of deaths due to cirrhosis of liver, P the proportion of these deaths attributable to alcohol and K the percentage of all alcoholics with complications who die of cirrhosis of the liver, is based on the assumption that the relationship between alcoholism and cirrhosis of the liver is a constant. Although this formula has been criticized on several occasions and indeed revised by Jellinek it is still widely used and Table 44, giving

TABLE 44

Prevalence of alcoholism, rates per 100,000 population, aged 20 years or more, selected countries

(*from WHO Technical Report Series, 1967*)

Place	Year	Jellinek Method Estimate	Independent Method Estimate
France	1951	5200	7300
USA	1953	4390	—
Chile	1950, 1953	3610	4150
Ontario, Canada	1961	2460	2375
Switzerland	1953, 1947	2100	2700
Denmark	1948	1950	1750
Finland	1951–57	1120	1330
England and Wales	1948, 1960–63	1100	865

the prevalence of alcoholism per 100,000 population in various countries, shows that for the most part there is a fairly close relationship between the numbers found by independent survey and those derived from the Jellinek formula. The countries shown are not necessarily those with the highest rates of alcoholism but are those from which data are available. Similarly some indication of the degree of alcohol abuse in different countries can be estimated from the relationship between the *per capita* intake of alcohol and the death rate from cirrhosis of the liver. On this index France shows a high death rate (over 30 per 100,000) and a high alcohol intake of 25 litres *per capita* (15 years and over) per year. Italy also shows high rates although approximately half those of France. The difference between these two countries, both major wine producing areas in which alcohol is widely used, suggests some factors which may relate to alcohol abuse. In France there is no stigma attached to drinking, drunkenness or alcoholism. Regular drinking throughout the day is the norm towards which there is considerable social pressure. In Italy, although drinking is acceptable, drunkenness is proscribed and drinking is almost exclusively at meal times and in moderation. Jellinek (1960) suggests that three economic factors as well as social influences are significant—the economic state of the individual, the general prosperity of the country and the belief in alcohol production as a national asset. He suggests particularly that the latter may apply in France where a third of the electorate is dependent, wholly or in part, on the production and distribution of alcoholic drinks.

Although cultural and economic factors may influence national rates of alcoholism or its incidence in members of certain sub-cultures such as emigrant Irish labourers or publicans in whom a high alcohol consumption is an accepted part of life, they are not sufficient to explain the behaviour of individuals, and in their case independent psychological factors must also be invoked.

Narcotic Drugs

Opium, its derivatives and synthetic analogues have created a problem of drug abuse in many countries throughout the world for centuries despite vigorous international attempts at control through legislation. Eating and smoking opium has been a chronic problem in the Middle and Far East whence it spread to most Western countries. Just as the development of the cigarette led to a rise in the consumption of tobacco so did the extraction of morphine and the production of heroin lead to a greater problem of narcotic abuse. The causes of this increase in abuse are not simple and relate to social and economic factors as well as to the availability of the drug. The fact that narcotic dependence remains a minor problem among the less developed countries of Africa and South America may partly lie in the unsuitability of some areas for cultivation of the opium poppy together with their distance from the main centres of illegal distribution. Furthermore, their greater degree of poverty makes it a commercially unviable proposition to develop a black market for the drugs.

Information about patterns of narcotic abuse in the Middle and Far East is not always very reliable, but certain trends can be detected which may have relevance to problems in Western societies. Reports from Assam suggest that initially far more people use opium because of association with and the example of relatives and friends than for the pleasant effect of the drug. In Hong Kong, 'one of the most heavily addicted countries of the world', opium could be legally obtained up to 1946 after which its sale was banned. Since then heroin has come into common use; moreover the smoking of heroin which was to some extent socially acceptable has now changed to administration by injection, a technique proscribed by law.

In the USA the problem of heroin abuse has gradually developed throughout this century from a background of abuse of opium, laudanum and paragoric. Increasingly severe penal sanctions without effective associated facilities for treatment have had little effect in slowing the increase in the number of people dependent on heroin. Current estimates suggest that there are over 300,000 narcotic addicts in the USA and that the great majority of these are dependent on heroin. Addicts come predominantly from deprived minority groups. The situation there has been compared on many occasions with that in the UK where there is a very much smaller number of addicts and a different approach to control of the spread of heroin abuse. Whereas in the USA the possession of heroin is illegal, it has, until recent amendments to prescribing regulations in 1968, been legal for any doctor in the UK to supply heroin to an addict. For many years this free approach appeared to result in very satisfactory control. The number of addicts in Britain was small and appeared to be on the wane; indeed in 1945, there were fewer than 400 narcotic addicts, nearly all of whom were middle aged people dependent on morphine and who had become addicted during the course of treatment for a painful illness or who, because of their profession (doctor, dentist or nurse), had freer than usual access to drugs. However, this policy of free prescribing proved to be a double edged sword with the arrival in Britain of addicts from North America who came here to obtain supplies legally and to escape the harsh legal sanctions in their native countries. Lack of experience and some lack of care led to considerable over-prescribing of heroin which became available for distribution to non-addicts. This has been at least a part cause of the sudden rise in the number of addicts from just over 400 in 1958 to nearly 1,400 in 1966. The rise in numbers was accompanied by a change in the type of person being reported as dependent since almost all the new cases were 'non therapeutic or medical professional addicts' who had started to take drugs obtained illicitly. Also the age incidence dropped. In 1959 there were no known addicts under 20; by 1966 there were 317. Unlike the USA all social groups became affected and not merely deprived minority groups. However, instead of banning the prescription of heroin, which would have undoubtedly resulted in the establishment of a true black market, legislation was introduced limiting the right to prescribe heroin and cocaine to addicts to a small number of specially licensed doctors working within the National Health Service. Considerable precautions were taken to ensure that addicts received adequate but not over-generous prescriptions and at the same time that they had as few opportunities as possible for altering or forging prescriptions. Although four years is clearly a very short time in which to judge the effectiveness of a system of treatment for a condition so unpredictable as heroin dependence, this method does seem to have met with some success in that the number of addicts attending centres is now remaining stationary and that the amount of narcotics being prescribed is falling. Moreover, no very extensive black market supply has yet developed perhaps because as long as drugs are legally available on prescription, it would not be a satisfactory commercial proposition to establish the necessary machinery for illegal distribution.

The problem of narcotic addiction in the UK can by no means be regarded as solved. This is shown, in one way, by the increasing number of people who present as being primarily dependent on methadone. Unfortunately, this narcotic is in some circles apparently regarded as being considerably less addictive and less harmful than heroin, a view not supported by many of those who have to treat primary methadone addicts. Similarly more cases of misuse of other synthetic drugs both stimulant and sedative are being reported, suggesting substitution of drugs within a culture if not within an individual.

Programmes of maintenance and blockade on methadone such as have been pioneered by Dole and Nyswander have proved to be an extremely useful form of therapy in cultures in which there is an extensive black market constituting the principal source of narcotic supply and in which supply of heroin by legal means is impossible. However, the appropriateness of such a programme where legal supplies of narcotics—both heroin and other drugs—are available, remains to be proved. In the author's opinion it is not necessarily always the case that methadone even in injectable form, is a suitable substitute for heroin, and one must face the fact that some people are addicted to heroin itself and will take this whenever the opportunity arises even if methadone is more freely available. Both on clinical and experimental evidence the two drugs cannot be viewed as being interchangeable.

Stimulant Drugs

Natural Stimulants

This group comprises vegetable products consumed without any elaborate chemical preparation and includes tea, coffee, betel, kava and khat as well as coca and cocaine. The consumption of both tea and coffee has a very long history and their use now is generally socially acceptable, problems arising only from excessive ingestion with associated dyspepsia and sleeplessness.

Cocaine. For thousands of years the leaves of the coca plant have been used as a stimulant and to allay pain, hunger and fatigue among the indigenous population of Peru. A recent survey of a Peruvian village showed that 12 per cent of the inhabitants were regular chewers of the coca leaf and that the principal reason given for this was to relieve fatigue and hunger. The coca chewers have poorer physical health and personal hygiene and a worse work record than those who do not chew coca. An individual may chew coca for the relief of hunger and stress; this leads to loss of appetite and a vicious circle of malnutrition and apathy about hygiene leading to hookworm infestation and further deterioration in health.

The active element of the coca plant, cocaine, was isolated in the 1850's. As with many other addictive substances it was suggested as a cure for morphine addiction, and consequently by the end of the 19th century Germany and many other European countries were faced with a sizeable problem of cocaine addiction. After this initial epidemic the use of cocaine became less frequent in advanced societies and it now appears to be limited almost exclusively to some groups of narcotic addicts. Because of the 'addict folk lore' belief that heroin addicts must always increase their dose of heroin, cocaine was given to addicts. In theory this was to offset the drowsiness induced by increasingly large doses of heroin, but in fact it sustained any remaining positive pleasure derived from the use of heroin. Now that it is recognized that it is possible to maintain a heroin addict on a suitable dose of narcotic the use of cocaine in this group has virtually died out.

Other natural stimulants. Of the remaining naturally occurring stimulants kava is drunk widely in different islands in Oceania. Having originally fulfilled largely a religious and ceremonial function, its use is now becoming increasingly related purely to personal pleasure though it is still believed to have value as a folk medicine. Some authors ascribe its continued use to a resistance against changes enforced by Christianity and other Western influences and point to a deterioration in social behaviour that has coincided with the substitution of alcohol for kava.

Betel nut or leaf is widely chewed from Western Africa to Oceania. It is employed for many purposes from aiding digestion to its use among Burmese monks as an aid to greater self-awareness, and there is little proscription on its use. Consumption of khat, another mild stimulant, is principally confined to Eastern Africa. It is used socially and for personal pleasure; however, opposition to it has grown over the years both because its continued use leads to self neglect and also on religious grounds because of its intoxicating effect.

Synthetic Stimulants

The principal synthetic stimulants are related (either chemically or in action) to amphetamine which was first produced in 1887. Since their first introduction a small number of chronic abusers have been reported. Latterly, however, a succession of countries have experienced explosive epidemics of synthetic stimulant abuse. The first reported was in Japan where the use of amphetamine had become routine practice among ammunition workers and some of the armed forces during the Second World War. At the end of the war manufacturers were left with large stocks, and these were advertised for sale at terms that would naturally be attractive to a nation which had suffered defeat and was undergoing dramatic social changes. By 1954 over half a million abusers of amphetamines were estimated to exist and the final figure may have been three times as high as this. Drastic legal sanctions against production. supply and possession, resulted in a very rapid decline in the number of people affected, and is an indication of how legal measures in association with a treatment programme may be effective in a community which is geographically isolated from other sources of drug supply. Shortly afterwards Sweden experienced a very similar epidemic. There had been a low level of abuse from 1940, but between 1960 and 1968 there was a sudden explosion of abuse, principally of phenmetrazine although amphetamine was also used. In 1965 every fifth male taken into custody by the police was found to have been taking stimulants intravenously and by 1968 the figure had risen to every other male. Recently there are signs that the epidemic has been controlled again largely by increased legal sanctions controlling import and drastically increased sentences for offenders. The number of addicts has now apparently become stable though there is some evidence of substitution of other drugs such as heroin and the hallucinogens for the amphetamines.

Britain also has experienced a wave of abuse though this has not been limited to any one particular member of the group of synthetic stimulant drugs. The extent of abuse has been extremely difficult to ascertain though a study as long ago as 1962 suggested that 500 people in a town of a population of 270,000 were dependent on amphetamines. There seems little reason to doubt that the figure has increased since then, and the author's personal experience suggests that in some of the more deprived areas of London virtually all children between the ages of 12 to 16 will have experimented with drugs at one time or another although by no means all will be dependent. Such a high extent of abuse should perhaps not be taken as typical, as another survey conducted in 1968 in London suggested that less than six per cent of school children had taken amphetamines within the last six months. In England in the author's experience it has proved difficult if not impossible effectively to treat an epidemic of this proportion without control of supply of the drugs being abused. The long term outcome in terms of future drug abuse for a group of adolescents growing up in a culture where illicit drug taking is at least statistically a norm, must be a matter for great concern.

'Hallucinogenic' Drugs

There is a wide group of substances both naturally and synthetically produced which when consumed produce great alterations in mood, consciousness and appreciation of the environment. Although usually called 'hallucinogenic' this term is not entirely satisfactory as all may not produce true hallucinations. All the naturally occurring substances such as peyote, *datura stramonium* and *amanita muscaria* have a very long history of use in many different cultures mainly in a mystical and religious setting or in association with healing ceremonies. Although a very wide range of substances are still employed, remarks in this section will be confined to these drugs most widely used at present.

Peyote. Peyote, produced from a cactus and other plants indigenous in Mexico, has been used in Central and North America certainly since the time of the Aztecs and is still widely employed among certain groups of Central and North American Indians. Whereas previously its use was mainly in religious ceremonies it has become increasingly used for personal pleasure and also as a medicine particularly as an analgesic. Among some groups of Plains Indians one report suggests that it is taken as freely as white Americans take aspirin. Numerous studies of the recent spread of the peyote cult among American Indians shows that since the middle of the last century the number of tribes involved in the use of the drug has steadily increased, and several authors have professed to discern a link between the increased influence of white Americans on the Indians and the increased use of the drug. Aberle, in a study of the Navajos, has pointed out how the drug may be used to allow a race to accept an inferior status and a falling economic position by providing satisfactions which would previously have been achieved through more traditional Indian activities. Not all groups of the tribe have, however, come to use peyote and those groups which are close to a source of supply are most likely to become affected. Amongst such groups, those who have had enforced upon them the greatest amount of change from their indigenous way of life with the consequent increased degree of anxiety and social disturbance are the most vulnerable. Many authors view the use of peyote as a means of resistance against increasing domination by the whites. Peyote has also come to be employed by some white Americans—primarily artists, intellectuals and students closely associated with the indigenous users—and it is now used by young people particularly by 'hippy' colonies.

The use of other naturally occurring hallucinogenic drugs has become less common in recent years. However, the eating of Fly Agaric has been reported until quite recently amongst isolated tribes in many areas as far apart as Siberia, Mexico, New Guinea and Borneo.

LSD (Lysergic acid diethylamide). This synthetic drug, first produced in 1938, was initially used therapeutically in psychiatric practice. Since then, its non-medical use has developed, initially being confined to groups of intellectuals seeking to broaden their experiences and to achieve some greater degree of self-awareness. Although it is still in vogue within such circles, its use in 'advanced' societies has now spread to include a much wider group of people; these tend to be of a younger age and include not only the 'hippies' but also many other socially deviant groups. Indeed it is amongst these 'non intellectual' users that the use of LSD appears to be most frequent now in Britain. Under these circumstances the drug is taken not for any supposed intellectual benefit, but purely because of its ability to alter the person's relationship with reality and the resultant pleasure derived from this alteration. Other synthetic 'hallucinogens' such as DMT (dimethyltryptamine), DOM (STP, dimethoxymethamphetamine) *etc.* have a similar action though their intensity and duration of effect are different from that of LSD.

Cannabis

The use of cannabis appears to have spread throughout the world almost as rapidly and as widely as that of tobacco, and in the UK and the USA its use can no longer be regarded as being confined to any special sub-group. With such a widespread degree of dissemination it is of interest to speculate as to why in some cultures it should be accepted and in others so strongly disapproved of. Several authors such as Carstairs and Murphy have suggested that acceptance or rejection depends on the society's attitude towards aggressive activity. In societies where forcefulness and aggression are the accepted way of life then cannabis, which has a pronounced tranquillizing effect, is socially unacceptable. In other cultures where calmness and passivity are considered virtues, then the use of cannabis is acceptable and alcohol may be rejected as provoking aggression. The current rapidly extending wave of cannabis use within most Western cultures and the urbanized areas of Africa and Asia tends to support this argument. The users see themselves as withdrawing temporarily from an increasingly aggressive and competitive society albeit in a socially non acceptable way.

Facts about the degree of damage if any produced by the use of cannabis are very difficult to ascertain. Certainly some users do enter a state of mind in which their lives become centred round the drug; social and personal deterioration follow, the greater part of the individual's life being devoted to the securing of supplies and to consuming the drug. More carefully controlled studies in areas where it is widely used, e.g. the North African littoral, and more exacting psychopharmacological studies of the active principles of the drug may help to answer the problem of what dangers, if any, cannabis poses for society.

NORMAL DRUG USE

Any discussion of drug abuse must take note of the vast quantity of psychoactive drugs potentially capable of being abused and which are consumed every year within the limits tolerated by Western societies. In 1965 over one million prescriptions for barbiturates were issued in England and Wales. The vast majority of these would have been for middle aged people who had become habituated to the use of drugs as hypnotics; only a small fraction of the prescriptions would be used by young addicts. The quantity of acetylsalicylic acid sold in a year far outweighs the amount that could possibly be necessary on the grounds of relief of organic pain only. An extremely interesting study by Blum has shown that a normal American urban population could be divided into four groups on the basis of their drug experience (see Fig. 59). Only one per cent had had no experience of any psychoactive drug.

	0	10	20	30	40	50	60	70	80	90	100	Per cent of Pop. Use (rounded)
CLASS I												
Aspirin-type compounds										X		95
Beer, wine										X		95
Alcoholic spirits										X		93
Tobacco										X		88
Painkillers								X				66
Laxatives								X				64
CLASS II												
Anxiety-control agents							X					46
Sleeping aids							X					43
Health and appearance aids						X						38
Wake-ups and stay-awakes					X							34
Allergy					X							26
Birth control*					X							22
Weight control				X								19
Amphetamines				X								17
Antidepressants			X									11
CLASS III												
Marijuana		X										9
Proprietary remedies for 'kicks'		X										7
Heroin, cocaine, etc.	X											4
Hallucinogens	X											4
Volatile intoxicants	X											3
Sex-increasing	X											3
Sex-decreasing	X											2

* Regarding proportion of female population only.

Fig. 59   Frequency of reported one-time or greater use of psychoactive drugs (from Blum, 1969).

Five per cent of the sample had confined their drug taking to conventional and proprietary drugs such as alcohol, tobacco, laxatives and simple analgesics. Two thirds of the population had used both the conventional drugs mentioned above and also had taken other psychoactive drugs usually prescribed medically for the control of anxiety, depression or insomnia. The fourth group, which comprised a sixth of the sample, had, in addition, taken other non-medically prescribed drugs including compounds generally disapproved of on social grounds e.g. cannabis, heroin, cocaine and 'hallucinogens'. The six per cent constituting the so-called 'drug conservative' group were mainly older people. The third group of 'high normal users' were spread throughout the entire population, whilst the fourth group were mostly younger and contained a high proportion of students.

CONCLUSION

Certain features common to different drug using cultures are apparent. Man seems incapable of resisting the pleasure of using substances that alter his relationship with reality. Drugs originally used in religious ceremonies are now widely employed for personal pleasure. Whilst some societies are more prone to use drugs than others it would be a mistake to relate the stresses of living in a highly sophisticated Western society causally to an increased use of drugs. 'Stress' is very difficult to quantify and the lives of many more primitive people would seem to be highly stressful. We do not have adequate information to be sure that some sociological changes which are claimed to reduce or abolish drug abuse really do so. Firmer evidence for the low level of drug abuse in the USSR, for the reduction of opium abuse in the People's Republic of China and for the reduction of cannabis use in Communist Cuba would be welcome. The high level of 'normal' drug use found by Blum may reflect financial prosperity and easy drug availability rather than a response to 'stress' or 'alienation' from society. Most societies in the modern world do not have aspirin or other drugs freely available to them. If they did it is possible that remote hill tribes in South Eastern Asia might have a drug intake comparable with of Americans living in the Bay area of San Francisco.

References

Advisory Committee on Drug Dependence (1969). *Cannabis*. London, HMSO.

Advisory Committee on Drug Dependence (1970). *The Amphetamines and Lysergic Acid Diethylamide (L.S.D.)* London, HMSO.

de Alarcon, R., Rathod, N. H. and Thomson, I. G. (1969). Observations on heroin abuse by young people in Crawley New Town. In *Scientific Basis of Drug Dependence*, editor Steinberg, H., London.

Bales, R. F. (1959). Cultural differences in rates of alcoholism. *Drinking and Intoxication*, editor MacCarthy, R. G., New Haven.

Bewley, T. (1969). Drug dependence in the USA. *Bull. Narcot.*, **21** (2), 13–30.

Blum, R. H. (1969). *Drugs*. Vol. I, Drugs and Society. San Francisco.

Canadian Government Commission of Enquiry Interim Report. *The Non-Medical Use of Drugs*. Ottawa 1970. (Harmondsworth 1971).

Esbjornson, E. (1971). The drug problem in Sweden from the police point of view. *Bull. Narcot.*, **22** (1), 15–21.

Grinspoon, L. (1969). Marihuana. *Scientific American*, **221**, 17–23.

Jellinek, E. M. (1960). *The Disease Concept of Alcoholism*. New Haven.

Lewis, A. J. (1970). *Amphetamine, Barbiturates, LSD and Cannabis—Their Use and Misuse*. Public Health and Medical Subjects No. 124, London, HMSO.

Lindesmith, A R. (1966). *Narcotic Addiction*. New York.

May, A. R. (1972). *Patterns of Drug Dependence in Council of Europe Member States*. Strasbourg.

Nagahama, M. (1968). A review of drug abuse and counter measures in Japan since World War II. *Bull. Narcot.*, **20** (3), 19–24.

Office of Health Economics (1967). Drug Addition. Studies in current health problems, No. 25, London.

Office of Health Economics (1970). Alcohol Abuse. Studies in current health problems, No. 34, London.

Sargant, Margaret B. (1971). A cross cultural study of attitudes towards alcohol and drugs. *Brit. J. Sociol.*, **22**, 83–96.

Spear, H. B. (1969). The growth of heroin addiction in the United Kingdom. *Brit. J. Addict.*, **64**, 245–255.

Vaillant, G. E. (1969). The natural history of urban narcotic addiction—some determinants. *Scientific Basis of Drug Dependence*, editor Steinberg, H., London.

World Health Organization (1951). WHO Technical Report Series, No. 42.

World Health Organization (1967). WHO Technical Report Series, No. 363.

Chapter 18

Urbanization, Stress and Mental Health: the Arguments and the Research Evidence

Alex Robertson

The expansion of cities and the general trend towards urban living has been one of the major facts of life in most Western nations during the present century. This growth has been accompanied by undeniable changes in the pace and conditions of life of the inhabitants of cities. It is not infrequently asserted that these changes have produced an increase in the overall amount of stress which city-dwellers must endure—an increase which is in turn held to reflect itself in the growing numbers of people who seek psychiatric treatment. Plunkett and Gordon (1960) give voice to a common sentiment:

> 'The heritage of modern man carries along a substantial substratum of mental illness. The human central nervous system evolved under the simple conditions of a community life close to nature. As the environment grew more complex, positive adaptations became more difficult. To whatever degree it has been successful, this process has left residual strains, especially evident in certain personality types.'

Two kinds of evidence may initially be cited to support the notion that urbanization has a bearing on mental health. First, by contrast with the inhabitants of rural areas, urbanites seem always to have shown a greater propensity for indulging in such practices as criminality, alcoholism, suicide, and unorthodox forms of sexual behaviour. Even here, however, the evidence on urban-rural differences in rates of mental disorder is rather more equivocal. Second, and of equal relevance, is the fact that substantial discrepancies exist between different areas within any one city, in the incidence of a variety of types of deviant or pathological behaviour (see Giggs, 1970) including mental disorder. Much of the ensuing discussion will therefore be addressed to these two facts.

Basic approaches to the problem

In his entertaining book on the 'Arcadian Myth', P. J. Schmitt challenges the 'anti-urban' bias of much of the writing about city life, and the romantic and idealized picture which is often painted (by urbanites) of rural existence (see Schmitt, 1969). It is certainly true that the bulk of the writing and research on the relationship between urbanization and psychological stress has tended to assume that it is more 'natural' for human beings to live in rural surroundings, an assumption that is made explicit, for example, in the above quotation from Plunkett and Gordon.

While explanations vary as to why removal to the 'alien' environment of the city should threaten the psychiatric well-being of the individual, they can usefully be reduced to two essential types. On the one hand are those accounts which argue that the *physical* environment of the city—with its population density, its patterns of housing, its noise-levels, etc.—creates demands and pressures that the human organism is not equipped to meet. On the other, are those which claim that the urban *social* environment is more impersonal and disintegrated than that existing in rural areas. As such, it is held to provide less definite standards or guides

229

to behaviour, and less satisfying personal relationships, both of which are claimed to give rise to psychological distress.

Within social psychiatry, these latter types of explanation have perhaps proved the more influential. It is to the consideration of these that we shall therefore first turn. But before doing so, we must consider the other major variable in our equation—that of mental disorder.

Research definitions of mental disorder

The concepts of 'mental health' and 'mental illness' are notoriously difficult to define or quantify. Leaving aside the question of whether psychiatric disorders may validly be regarded as 'diseases', the major problem in formulating a research definition of psychiatric impairment is an operational one. How does one identify a psychiatric 'case'? Should one use *incidence* or *prevalence* as the criterion for estimating the amount of mental disorder in a given population?

Incidence measures. These define disorder in terms of the number of persons in a given population who *become* ill within a given period of time. In practice, this normally means those persons who are referred for psychiatric treatment within, say, a year. This introduces two possible kinds of bias. First, it is by no means certain that everyone who becomes disturbed seeks formal psychiatric treatment. Thus, it is possible that variations in the incidence of mental disorder in different populations are due rather to the greater ability of certain groups to tolerate disturbed behaviour or to handle it in other (non-medical) ways, than to any 'true' differences in the occurrence of psychiatric impairment. A second source of bias lies at the level of psychiatric diagnosis. Some evidence suggests, for example, that psychiatrists are more likely to diagnose middle-class patients as 'neurotic', while working-class referrals attract the label 'psychotic'.

Prevalence measures. Prevalence estimates are largely free of these problems, but present special difficulties of their own. Essentially, these try to gauge the number of people who may actually be judged to *be* ill at one particular time. This involves interviewing samples from the general population, and the major difficulty here is obviously that of developing a set of reliable criteria on which to base a diagnosis. For example, Srole *et al.* (1962) estimated twenty-four per cent of the population of central Manhattan to be suffering from 'marked', 'severe', or 'incapacitating' symptoms of mental disorder. High prevalences of disorder were also discovered in the 'Stirling County' study, to be described below. Assuming such estimates to be accurate, one is faced with the important question of why it is that only certain people seek treatment, while others with equally strong reasons do not. The issues in this whole area have been examined in some detail by Dohrenwend and Dohrenwend (1969).

These differences obviously make it rather difficult to draw confident conclusions from studies using different definitions of mental disorder. But bearing these strictures in mind, what can be said about the relationship between urbanization and psychopathology?

MENTAL HEALTH AND THE URBAN SOCIAL ENVIRONMENT

In a famous essay, the American sociologist Louis Wirth (1938) suggested that three features of the urban community—namely its size, its density and its heterogeneity—serve to distinguish it from that in rural areas. Wirth further argued that from these features there emerges a peculiarly urban social structure and way of life.

The variable of size is important in the obvious sense that an increase in the numbers resident in any one place will render it impossible for anyone to have an intimate knowledge of, or close personal interest in, the affairs of all his fellow citizens. As a result, individuals become more independent of each other. City-dwellers are accordingly freed to some extent from the emotional ties and social pressures that are a feature of life in smaller communities.

The variables of density and heterogeneity tend to reinforce these effects. Increased density gives rise to greater diversification within the population. Work, for example, becomes more specialized; and work as an activity is separated from the other parts and activities of a man's life. The work place becomes separated from the home. Work itself involves behaviours and relationships with others which are unlikely to be repeated outside the work situation. As the community becomes more heterogeneous, beliefs and standards of behaviour grow more secular and diverse. As a result, moral codes become less definite and less universally accepted. The control of behaviour comes increasingly to rely on formal rules and specially appointed agents (such as a police force), in contrast to the personal ties and informal pressures on which a substantial part of the mechanism of social control in smaller and more homogeneous communities depends.

For Wirth, the whole trend of city life was therefore towards more tenuous, more impersonal social relationships, with fragmented social roles, and poorly-defined behavioural norms. While Wirth did not himself explicitly speculate on the psychiatric consequences of these trends, this kind of formulation has inspired much rhetoric, and a certain amount of research, on the potential effects on mental health, of urbanism and the 'urban way of life'.

The Research Evidence

Urban-rural Differences

These assumptions received something of a setback in Eaton and Weil's (1955) study of the Hutterite communities in midwestern Canada and the United States. Based on their religious principles, the Hutterites pursue a self-sufficient, agrarian way of life, in which particular importance is attached to family loyalties. Independent reports have testified to the stability and cohesiveness of these communities. Admissions to mental hospital had been noted to be very uncommon among this sect.

Using a measure of what they termed 'lifetime prevalence' Eaton and Weil judged that one in every 43 Hutterites had at some time in his life been psychiatrically unwell. Of these, 53—or six in every thousand of the total population—were diagnosed as psychotic. This rate was substantially the same as that obtained in nine other surveys conducted in various parts of the world. When these findings were compared with those for cities within the USA, the prevalence of psychosis among the Hutterites proved slightly higher than that for a predominantly slum area of Baltimore, and slightly lower than the incidence of all mental disorders in New York State.

While a mere five of the Hutterites had received formal psychiatric treatment, these findings obviously do not support the idea that small, cohesive communities have lower rates of mental disorder than towns and cities. The interpretation normally placed on these results is that well-integrated social groups such as the Hutterites are more tolerant and supportive of disturbed members, so that they can continue to function in the community. Of equal interest in this connection is the finding that the rate of *recovery* from mental illness seems much higher in Hutterite communities than in the USA as a whole.

In a recently published study, Fortes and Mayer (1969) compare their observations of village life in Northern Ghana at the present time, with Fortes' recollections of an extended period of field work in the same area some 30 years before. They conclude that the recent industrial developments in Ghana are having profound repercussions on this African community. These changes, they further argue, are creating stresses which have given rise to a higher prevalence of mental disorder in this district at the present time (see also, World Federation for Mental Health, 1965).

During his first sojourn in the area, Fortes (a social anthropologist) encountered only three persons who could be judged mentally abnormal. By contrast, in 1963 Mayer (a psychiatrist) interviewed and treated 13 persons whom she diagnosed as psychotic. Given the available estimates of the population of the area at these two different periods, the three cases noted by Fortes would represent a prevalence of roughly one in 11,000 while Mayer's 13 psychotics would constitute an incidence of about one in 4,000. While it should be noted that Fortes was not, during his first period of observation, making a systematic attempt to assess the rate of mental illness in this tribe, the fact that the 13 psychotics seen by Mayer in 1963 'were overt cases (who) had all been identified as such by their families' may render the two sets of findings more comparable with each other.

In explanation of these findings, Fortes draws attention to the fact that a dichotomy now exists between the values and expectations of the traditional family unit, which remains the dominant social unit in this area, and the demands of life outside the family, in which individuals are now frequently required to perform roles which lack the guidance of—and may at times be in direct conflict with—traditional norms. It is also of interest that, in common with the Hutterite investigators, Mayer found the rate of recovery to be appreciably higher among her subjects than is normally the case among psychotics living in more urban and industrialized societies.

The most ambitious attempt to test an explanation of mental breakdown derived from such assumptions is probably that of the Leightons and their co-workers, of the relationship between social integration and mental health in a number of communities in Nova Scotia—the so-called 'Stirling County' study. Within the geographical area of 'Stirling County', five rural communities were the subjects of an intensive sociological and psychiatric investigation. Two of these communities were described as 'integrated' and three as 'disintegrated'. In addition, interviews were conducted on random samples of the adult population of the whole County, and of 'Bristol' its largest town. From this information, independent psychiatrists made a number of assessments concerning the respondent's psychiatric history, including an estimate of the probability that at some time in his life (including the time of the interview), the individual could be judged psychiatrically 'ill'.

Their results lend support to the Leightons' hypothesis that the mental health of an individual is directly related to the degree of social integration of the community in which he lives. The inhabitants of 'disintegrated' communities had a greater prevalence of psychiatric disorder than those in 'integrated' communities, with different prevalences in these communities by sex and age. The prevalence of psychiatric disorder in the town of Bristol was, however, no higher than in more rural areas of the County.

The results of these studies seem to indicate that life in towns and cities is not in itself any more conducive to mental breakdown than life in rural communities. The balance of evidence tends to support those social-anthropological studies of South American peasant communities which conclude that living in inescapable proximity to others may be as productive of anxiety and personal difficulties as the impersonal nature of urban existence and relationships.

Three findings of these investigations do, however, seem worthy of particular note. First, there does appear to be a relationship between social integration and psychiatric impairment, although the nature of this relationship remains obscure (see Kunitz, 1970). The evidence does not suggest, however, that urban communities are necessarily more disintegrated than rural communities. Second, the study by Fortes and Mayer (1969) suggests that rapid social change causes an increase in psychiatric disorder. Finally—and perhaps related to this—

small rural settlements seem better able to support and tolerate 'eccentric' individuals. As noted above, this might explain the fact that the Hutterite and American urban populations had similar prevalences of psychiatric impairment, while the Hutterites had a much lower rate of hospitalization for mental disorder. It may also account for the better rate of recovery among both Hutterite and African psychotics.

Dunham (1964; 1965) argues that the notion of 'social selection' seems best to fit existing epidemiological evidence on schizophrenia. The social selection hypothesis assumes that, in any social system, there is a more or less regular proportion of individuals who, for a variety of reasons, are predisposed to psychiatric breakdown. (The similarity between the Hutterite and Baltimore prevalence rates, for example, might tend to support this assumption). Such persons enjoy a relative degree of immunity from psychological crisis when a social system is in a state of equilibrium, because they are not required to adapt to numerous changes in the roles and routines which they are called upon to perform at work, within the family, etc. When, however, changes (such as may follow technological innovations) take place parts of the existing role structure begin to break down. Behavioural norms begin to shift, and the boundaries of behaviour become less well defined. In such a situation, predisposed individuals lose their immunity and tend to break down, or to behave in ways which lead to their being defined as 'odd' and requiring special handling.

This explanation would accordingly imply that it is social change, rather than urbanization *per se*, which produces the conditions that give rise to pathological behaviour in predisposed individuals. Psychopathology, as indicated by the investigation by Fortes and Mayer (1969), will equally increase in *rural* communities when widespread social changes are taking place.* It should also be noted that this explanation says nothing of those factors which may make for psychiatric predispositions within the individual.

Stress and Differential Environments within the City

A number of investigations have replicated the principal finding of Faris and Dunham's pioneering Chicago study, that the incidence of schizophrenia is highest in the poorer areas of a city, although some contradictory evidence also exists. Ever since their research was first published in 1939 a major debate has centred on the interpretation Faris and Dunham chose to place upon their findings. Using independent criteria, these authors defined the areas with high rates of schizophrenia as 'disorganized', and argued that social disorganization gave rise to impoverished patterns of communication between the persons living in such areas, which they in turn held to be a major causal factor in schizophrenia. The principal alternative to this explanation has come to be known as the 'drift' hypothesis. In essence, this attributes the concentration of schizophrenia in certain areas to the 'unconscious' drift of persons with developing psychotic conditions to those parts of the city where life is held to be more anonymous, and social relationships therefore less demanding.

The results of a recent painstaking attempt to solve these issues by a comparative study of the distribution of schizophrenia in two areas of Detroit have led Dunham to reject the notion that schizophrenia is caused by the kind of social environment that exists in 'high-risk' groups or areas. His results indicate rather that the large number of schizophrenics he discovered in the lowest social class was due to their moving down the social hierarchy. Their

---

* This interpretation might also gain support from the finding by Srole *et al.* (1962, pp. 261–265) that, in Midtown Manhattan, immigrants born in low-class families in European *rural* communities had a much higher prevalence of psychiatric impairment than immigrants coming from middle-class families in European *towns and cities*. Murphy (1961), however, argues that the existing research evidence indicates that the relationship between social change and mental breakdown is consistent only in African countries undergoing 'Westernization'.

social class position might therefore be presumed to be a function of their schizophrenia, rather than *vice-versa*. Similarly, his evidence on the differences in rates of psychosis between the two areas indicated that the preponderance of schizophrenia in the poorer area was due to the fact that those families which produced schizophrenics had tended to move into that area. Dunham explains these trends by a rather complex variant of the social selection hypothesis, based on the notions of personality predispostion and what he terms 'life chances.' The details of this explanation need not concern us here. The important point to note is that Dunham's findings indicate that 'disorganized' urban areas do not *cause* psychosis; but rather, influence its *distribution* within the city.

MENTAL HEALTH AND THE PHYSICAL ENVIRONMENT OF THE CITY

Possibly the most potent source of evidence concerning the adverse consequences for the organism of certain physical circumstances of urban life, has stemmed from a sizeable corpus of research in the fields of ethology and animal psychology. And within this research, undoubtedly the greatest focus of interest has been on the effects of (over) crowded living conditions (see for example, Calhoun, 1962; Loring, 1967). From such work, three consistent findings have emerged. First, groups of animals living in overcrowded conditions seem more susceptible to epidemics of disease. The major question here is, of course, whether crowded conditions lower individual resistance to disease; or whether (as perhaps seems more likely) the very proximity of animals to one another makes it easier for diseases to spread by contagion. But more relevant to the present discussion are the two remaining sets of findings. Overcrowding has been noted to produce marked changes in the physiological level of animals: in particular, animals living in crowded conditions secrete increased quantities of adrenalin, which reflects itself in increases in the average body weight of these animals. Finally, crowding has been found to cause serious disruption of such apparently instinctual behaviour as mating and the demarcation of 'territories', which in turn is often associated with an increase in aggression between animals.

Compelling analogies have been drawn between these findings and the situation in modern cities. Extrapolations from these animal studies to man are, however, fraught with difficulties, and ignore potentially important differences in the attributes of animals and humans. The whole concept of 'instinct' for example has rather dubious relevance for human behaviour, and it is by no means certain that human beings are motivated by the same kinds of territorial imperative as seems to underlie the behaviour of some animals. Moreover, ethological work has itself demonstrated that there are important divergences between *different animal species* in the way in which they respond to similar situations.

The Research Evidence

It should first be noted that those city areas with the highest incidence of social problems tend also to have a higher incidence of overcrowding than do areas with lower rates of pathology. One cannot, however, argue from this kind of evidence that social pathology and overcrowding are causally connected. Analysis of the extent to which overcrowding *causes* pathology requires investigation of the relationship between these two in individuals within these areas.

Existing evidence cannot be said to support the argument that overcrowded living conditions cause mental disturbance in human beings. For example, Dunham's measure of crowdedness as a possible stress factor in schizophrenia yielded no significant differences between schizophrenics and other diagnostic groups, though it should be noted that this comparison did not include a 'normal' (non-psychiatric) group. Other recent investigations covering general population samples again give negative results. In the course of a study of some 300 teenage boys, for example, the present author found no relationship in any social class or area of residence within the city between the size of the family a boy came from, and his level of anxiety, neuroticism, or introversion (see Robertson, 1971).

But by far the most impressive evidence is contained in a recent study by Mitchell (1971). In his preliminary review of the literature, Mitchell underlines the almost totally negative nature of existing evidence on the relationship between residential density, housing design, and psychopathology. He also draws a useful distinction between 'density' (or the amount of *physical space* available to each person in a dwelling unit), and 'congestion' (which refers to the extent to which *activities* are restricted, through competition between those within the same dwelling for the use of limited resources).

Mitchell's survey covered a very large random sample of families from the high-density housing developments in Hong Kong and attempted to assess the prevalence of different types of 'emotional strain' among people living in different housing conditions. Two of his measures have particular relevance for the present discussion. First were two indicators of what he defined as 'somewhat superficial' levels of strain—namely, 'unhappiness' and 'worry'. Second, was a set of standard psychosomatic items, measuring two 'separate and more severe' levels of strain—'hostility' and 'emotional illness'. In addition, information was obtained on the extent to which respondents complained of lack of space and privacy.

Mitchell's findings indicated that people living in crowded conditions (as measured by the number of square feet of living space per person in the household) were (*a*) conscious of their lack of space and (*b*) more likely to register high scores on his measure of 'superficial' strain. However, he also discovered a significant relationship between income level and emotional strain, with respondents from low-income families reporting significantly higher levels of worry and unhappiness than those from families with a higher income. When level of income was statistically controlled, it emerged that only in families with a low income did density have a bearing on the level of strain reported by the informant. Mitchell accordingly concludes that overcrowding does not *of itself* generate worry and unhappiness. Only when it exists alongside other potential stress factors, such as poverty, does it increase these kinds of strain.

But 'worry' and 'unhappiness' cannot validly be equated with psychiatric impairment. When he considered the 'more severe' types of strain reported by his informants, Mitchell discovered no relationship between crowded living conditions and levels of 'hostility' and 'emotional illness'. Nor did such other physical characteristics of the dwelling unit as the number of rooms and the number of amenities available within the unit, appear to have any bearing on these deeper levels of strain. Hostility and emotional illness were sensitive, however, to the interactions between two other variables—namely, the number of *families* sharing the same dwelling unit, and the floor level on which that unit was situated. When he examined the relationship between these, Mitchell discovered that among persons living in units shared by more than one family, there was a correlation between the severity of the symptoms of 'emotional illness' they displayed, and the floor level on which their dwelling was situated. Thus, persons living on the ground floor showed identical (low) levels of emotional illness, whether or not their dwelling was shared with another family. But persons living in shared dwellings on the sixth floor and above showed significantly more emotional illness than persons living on the same floor level in dwellings occupied by only one family. Much the same pattern emerged in the case of hostility. Although the hostility levels of all respondents tended to increase in proportion to the number of floors they lived above ground level, the trend was significant only for those respondents who lived in a dwelling shared by more than one family.*

* The calculation of this was done by the present author on the data presented by Mitchell.

The burden of these results would therefore seem to be that there is no direct cause-and-effect relationship between physical aspects of the housing environment and the kinds of stress which may be associated with mental breakdown. The association seems to depend on a more complex interplay between these physical features and certain characteristics of the dwelling's occupants. Mitchell explains his findings by suggesting that sustained social interaction with persons who are not members of one's own immediate family, is more likely to generate internal stress and conflict than is interaction with relatives. When a dwelling is shared by more than one family, and situated on the upper floors of a multi-storey building, it is less easy for the person to escape or avoid a stress situation by moving out of the home when he wishes. Mitchell concludes that:

> 'Housing affects patterns of social relationships, and individuals respond to the system of social relationships that housing conditions have helped to create.'

The findings of a number of other surveys tend to support the notion that the physical environment does not directly influence the psychological condition of the individual. For example, Wilner *et al.* (1962) studied poor negro families who had been re-settled in a new housing development in Baltimore. When these were compared with a control group of families who had not been moved, the authors found no significant changes over a period of three years after the move, in levels of nervousness, anxiety and 'self-esteem', but did find that over the same period, those who had moved to the new development became more optimistic, more satisfied with the *status quo*, and less 'aggressive toward authority figures' than their controls. Bearing in mind that these were low-income families, and that in Kennedy's study it was only in the lowest income groups that overcrowding was related to levels of 'worry' and 'unhappiness', the Baltimore findings lend support to the idea that housing conditions must co-exist with other stress factors, of which poverty seems particularly important, before the psychological state of the individual is adversely affected. Even then, however, it is only at a relatively superficial level that the individual's psychological functioning is impaired. Exceptional circumstances, such as those described by Mitchell, must exist before pathological changes take place.

Evidence to the effect that the physical environment has no direct influence on psychiatric health comes also from the comparisons of old and new housing areas in this country, by Hare and Shaw (1965) and Taylor and Chave (1964). Using both incidence and prevalence measures, the former authors elicited no important differences in the mental health of the populations of a new and an old area of Croydon, but found that in each population there was a group of persons who were particularly prone to both physical and mental ill-health. Similarly, Taylor and Chave discovered no difference between an old London borough and a planned new town in the *prevalence* of 'sub-clinical neurosis' among their residents, although the *incidence* of psychosis in the new town population was significantly below the national average for England and Wales. The same authors were also led to the conclusion that those new town residents who complained of feeling bored and lonely did so *because* they had poor mental health, rather than *vice versa*. The finding concerning psychosis rates obviously complicates the picture somewhat, but requires to be tested in a much more detailed fashion before any definite conclusions can be drawn from it.

As in the case of the social environment, effective exploration of the relationship between mental health and the physical aspects of urbanization would therefore seem to demand revision of the rather simplistic assumptions which have underlain the bulk of the research in this field to date. To summarize, existing evidence suggests that the physical environment of the city does not *per se* affect the psychiatric condition of the individual. People seem, for example, to be adversely affected by the stress of inadequate housing only when this co-exists with other potential stressors, such as poverty. Even then, however, the effect is to make people less happy or optimistic about life, rather than psychiatrically unwell. Only in very exceptional circumstances—such as those described by Mitchell—do living conditions seem actually to impair the mental health of the individual.

Further Reading

Calhoun, J. B. (1962). Population density and social pathology. *Scientific American*, **206**, 139–149.

Dohrenwend, B. P. and Dohrenwend, B.S. (1969). Social Status and Psychological Disorder: a Causal Inquiry. New York, Wiley Interscience.

Dunham, H. W. (1964). Anomie and mental disorder. In Clinard, M. B. (ed.), *Anomie and Deviant Behaviour*. New York, Free Press.

Dunham, H. W. (1965). Community and Schizophrenia. Detroit, Wayne State University Press.

Eaton, J. and Weil, R. (1955). Culture and Mental Disorders. New York, Free Press.

Ewald, W. R. (ed.) (1967). Environment for Man: the next Fifty Years. Bloomington, Indiana, Indiana University Press.

Fortes, M. and Mayer, Doris, Y. (1969). Psychosis and social change among the Tallensi of northern Ghana. In S. H. Foulkes and G. S. Prince (eds.). *Psychiatry in a Changing Society*. London, Tavistock.

Giggs, J. A. (1970). Socially disorganised areas in Barry: a multivariate analysis. In *Urban Essays: Studies in the Geography of Wales*. H. Carter and W. K. D. Davies, Eds. London, Longmans.

Gist, Noel P. and Fava, Sylvia F. (1964). Urban Society (5th Edn.). New York, Crowell & Co.

Hare, E. H. and Shaw, G. K. (1965). Mental Health on a New Housing Estate. (*Maudsley Monograph No.* 12). London, Oxford University Press.

Kunitz, S. J. (1970). Equilibrium theory in social psychiatry: the work of the Leightons. *Psychiatry*, **33**, 312–328.

Leighton, D. C. and Harding, J. S., *et al.* (1963). The Character of Danger: Psychiatric Symptoms in Selected Communities (Vol. 3 of the *Stirling County Study*). New York, Basic Books.

Loring, W. C. (1967). Comment on Dyckman. In W. R. Ewald (Ed.), *op. cit.*

Mitchell, R. E. (1971). Some social implications of high-density housing, *Amer. Sociol. Rev.*, **36**, 18–29.

Murphy, H. B. M. (1961). Social change and mental health. *Milbank Memorial Fund Quarterly*, **39**, 385–434.

Plunkett, R. J. and Gordon, J. E. (1960). Epidemiology and Mental Health. New York, Basic Books.

Robertson, A. (1971). Social class differences in the relationship between birth order and personality development. *Soc. Psych.*, **6**, 172–178.

Schmitt, P. J. (1969). Back to Nature: the Arcadian Myth in Urban America. London & New York, Oxford Univ. Press.

Srole, L. and Langer, T. S. *et al.* (1962). Mental Health in the Metropolis (Vol. 1 of the *Midtown Manhattan Study*). New York, McGraw-Hill.

Taylor Lord and Chave, S. (1964). Mental Health and Environment. London, Longmans.

Wilner, D. M. and Walkley, R. P. *et al.* (1962). The Housing Environment and Family Life. Baltimore, Johns Hopkins University Press.

Wirth, L. (1938). Urbanism as a way of life. *Amer. J. Sociol.*, **44**, 1–24.

World Federation for Mental Health (1965). Technical Assistance, Urbanisation and Social Change.

Chapter 19
The Geography of Genes

D. F. Roberts

After the existence of the ABO blood groups was discovered by Landsteiner at the beginning of the present century, early studies indicated that all four types (A, B, AB and O) were to be found in all the samples tested, and there seemed to be little difference among the various nationalities examined. During the First World War, however, Dr and Mrs Hirszfeld, working in Salonika, examined large numbers of soldiers, prisoners of war and others from different countries and showed conclusively that the frequencies of the ABO groups varied from one population to another. Many millions of tests were subsequently done, covering almost every country, and the results up to 1938 were published by Boyd in a massive compilation. In the meanwhile the A_1 and A_2 subdivisions of the ABO groups had been established in 1911, the MN and P groups had been discovered in 1927, the differing abilities to taste phenylthiocarbamide in 1931, and the differing abilities of people to secrete the ABH blood group substances in their saliva in 1932. The Rhesus blood groups and their subsequent subdivisions were discovered during and soon after the Second World War, which heralded a period of great expansion of knowledge of human genetics. The Lutheran blood groups were discovered in 1945, the Lewis and Kell in 1946, the Ss subdivisions of the MN groups in 1947, the Duffy groups in 1950, Kidd in 1951, the Lewis saliva groups in 1954, the Diego in 1955. Attention to human serum demonstrated the genetic variation in haptoglobin type (1955), the Gm groups (1956), the transferrins (1957) and the group-specific components (Gc) in 1959. The great group of discoveries concerning the enzymes commenced with the variants of serum cholinesterase in 1957, of glucose-6-phosphate dehydrogenase in 1958, red cell acid phosphatase 1963, phosphoglucomutase and 6-phosphogluconate dehydrogenase 1964, adenylate kinase in 1966, and this field is vigorously increasing today. Thus it is that a wealth of normal variants under simple genetic control is known in man, in whom the geography of genes can be explored.

The intrinsic difficulties of the simple procedure of mapping of genes, as with other biological variables, are often overlooked, and a somewhat different approach is needed from more usual cartographical procedures. The gene frequencies in various populations are calculated and plotted, and then isogenic lines drawn in the same way as contour lines. It is obviously impossible to examine all individuals in a population, so that the population frequencies are estimated from samples drawn at random. There is thus some random sampling error to disturb any underlying pattern that may exist. It is as if, in producing an orographic map, the various triangulation points were independently and at different rates moving up and down, although always within fixed limits round their averages, the movement being slowest in the neighbourhood of the averages and fastest at the extremes. To ensure the randomness of the samples themselves is difficult. It is impossible to see literally that one population possesses more individuals with say blood group A gene than another, and technical discrepancy in examination provides another possible source of incomparability of sample. All samples then, and the methods used, must initially be subject to vigorous scrutiny. As regards the method of representation, an isopleth method, based on samples, implies that all areas are populated, and all populations are sampled, and moreover tends to identify sample mean with population mean. But in view of the sampling variability, who is to say whether the isopleth passes through the point of location of the sample or on which side of it? Care therefore is necessary in interpreting isogenic maps. However some of these problems can be solved topologically. For instance, the ABO blood groups are controlled by three allelic genes; if the distributions of frequencies for all three are drawn on the same map, where the isogenes for the two alleles cross, that for the third must cross them at the same point.

WORLD DISTRIBUTION

Blood Groups

Mourant's (1954) pioneer compilation first clearly demonstrated the importance of geographical factors in gene frequency distribution. He provided maps of the ABO, MN, and Rhesus gene frequencies in indigenous populations, and the ABO maps were revised in 1958 in a comprehensive publication devoted specifically to them (Mourant *et al.*, 1958).

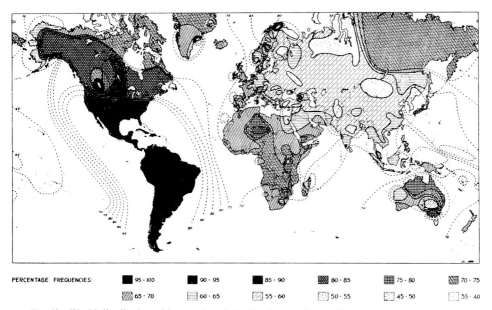

PERCENTAGE FREQUENCIES: 95 - 100 90 - 95 85 - 90 80 - 85 75 - 80 70 - 75 65 - 70 60 - 65 55 - 60 50 - 55 45 - 50 35 - 40

Fig. 60   World distribution of frequencies of the blood group O gene (from Mourant *et al.*, 1958).

For the ABO blood groups populations with a high frequency of the group 0 gene tend to have a peripheral distribution about the great land mass of the Old World (Fig. 60). They are found on the northwest fringes of Europe (Wales, Scotland, Ireland, Iceland), in southwest Africa, in parts of Australia, and most strikingly among the Indians of south and central America who today are almost entirely of blood group O, and may have been exclusively so before the coming of the Europeans. High O frequencies tend also to occur in populations in isolation, for example in mountain areas (e.g. the Caucasus, the Yemen) and in islands. The distribution of the B blood group gene is fairly regular, with maximum frequencies in central Asia and northern India, from which there are gradients of diminishing frequency outwards in all directions (Fig. 61). The B gene is almost completely absent from the American Indians and Australian aborigines, and it is probable that there was no B in either population before the coming of outsiders. A lower maximum of B occurs in West Africa. The gradient of B is particularly clear across Europe, where it diminishes steadily from the borders of Asia to a minimum among the Basques, from whom again it may well have been absent until relatively recently. However there is a slight but significant rise along the so called Celtic fringe of the extreme north-west of the continent.

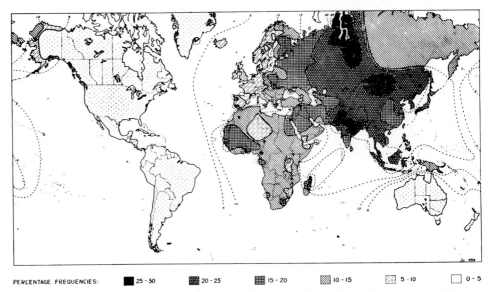

PERCENTAGE FREQUENCIES: ■ 25 - 30   ▨ 20 - 25   ▦ 15 - 20   ▨ 10 - 15   ▨ 5 - 10   ☐ 0 - 5

Fig. 61   World distribution of frequencies of the blood group B gene (from Mourant *et al.*, 1958).

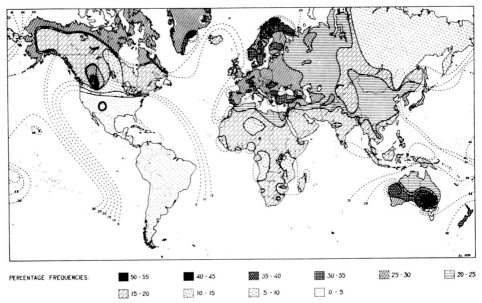

PERCENTAGE FREQUENCIES: ■ 50 - 55   ■ 40 - 45   ▨ 35 - 40   ▦ 30 - 35   ▨ 25 - 30   ▤ 20 - 25
▨ 15 - 20   ▨ 10 - 15   ▨ 5 - 10   ☐ 0 - 5

Fig. 62   World distribution of frequencies of the blood group A gene (from Mourant *et al.*, 1958).

The distribution of the A gene is much more patchy and irregular than either that of O or B (Fig. 62). Frequencies of A are high in Europe, especially Scandinavia and the mountain systems of southern Europe, and southwest Asia, but otherwise high frequencies occur only in Australian aboriginal populations and in a few other areas, e.g. the plains Indians, Greenland Eskimos. The A_2 gene reaches its greatest frequency in Europe; it is present in Africa and to a lesser extent in Western Asia but is rare or absent among indigenous populations elsewhere.

In the MN blood groups the frequencies tend to be rather less variable. The highest frequencies of the M gene occur in American Indians, and particularly in the lower latitudes of the continent, and in Eskimos (Fig. 63). They diminish in eastern Asia, and continue to diminish as one passes through Indonesia into New Guinea and Australia; the lowest values of

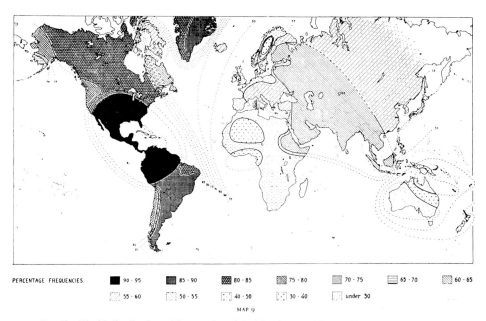

PERCENTAGE FREQUENCIES.   ■ 90 - 95   ▦ 85 - 90   ▨ 80 - 85   ▨ 75 - 80   ▥ 70 - 75   ▤ 65 - 70   ▧ 60 - 65

▨ 55 - 60   ⬚ 50 - 55   ⬚ 40 - 50   ⬚ 30 - 40   ⬚ under 30

MAP 9

Fig. 63   World distribution of frequencies of the blood group M gene (from Mourant, 1954).

the M gene occur in New Guinea. Over western and southwest Asia, southern Asia and eastern Europe the M frequency lies between 60–65 per cent, and then again it diminishes westwards into Europe and southwards into Africa, where a second minimum occurs in west Africa. Closely linked to the MN genes are those of the Ss blood groups. In Europe about half of the M genes carry S and half s, while of N genes about 5/6ths are associated with s. In Africa where the MN frequencies are similar to those of Europe, the S frequencies diminish going southwards, reaching a minimum in the south of the continent, and S is almost equally divided between M and N. The S frequency also declines going eastwards across Asia, being particularly low in the aborigines of Malaysia, and in some islands of the western Pacific; it is completely absent from Australian aborigines. It rises in frequency again in the interior of north America, but in tribal south America tends to be particularly variable.

Of the rhesus blood group systems, the gene D which gives rhesus positive status is at its lowest in Europe, and particularly among the Basques in whom it diminishes to approximately 50 per cent. It increases in frequency eastwards and southwards to approximately 80 per cent over almost all of Africa south of the Sahara. In eastern Asia and in Australia and Indonesia it often attains 100 per cent. The same holds for American indigenous populations, in many of whom the D frequency is 100 per cent. The other genes of the rhesus system also show numerical patterns of interest. The gene C has its maximum frequencies of over 90 per cent in New Guinea and Indonesia, from which focus it diminishes in all directions— south into Australia, north into Asia and across into the Americas, and steadily west across southern Asia into Europe, and particularly into Africa south of the Sahara where the lowest frequencies of under 10 per cent occur (Fig. 64). The gene E of the rhesus blood group system reaches maximum frequencies in the Americas and diminishes westwards into eastern Asia with a further diminution southwest into Africa. There is a secondary increase in frequency in central Africa, and a conspicuous increase in Australia and New Zealand. The range of frequencies of the E gene, less than 10 to over 50 per cent, is very much less than that of the C gene (less than 10 to over 90 per cent) (Fig. 65).

Data for other blood group systems are still insufficient for maps to be attempted, although in general a similar distinction between continents tends to hold. Synthesis of P group frequencies is bedevilled by technical testing difficulties, particularly in the earlier samples, but it seems that over Europe the P_1 frequency is about 49–54 per cent, is considerably higher in Africa south of the Sahara, moderate to high in American Indians and lower in southeast Asia. The Fy^a gene of the Duffy system, present at about 40 per cent in Britain and western Europe except for the high frequencies among the Lapps, shows low frequencies in Africa and rises to high frequencies in eastern Asia and in American Indians. The Kidd gene Jk^a is at moderate frequencies in Europe (50 per cent), but rises to high frequencies in Africa and eastern Asia. Of the low-frequency blood group variants, the Lutheran Lu^a is at 4 per cent frequency in northwest Europe and is absent in southeast Asia and Australian aborigines. The K gene of the Kell system is similarly low (5 per cent) in Europeans, may be somewhat higher in some American Indian peoples, but is absent from southeast Asia.

Serum Protein Variants

Data on the geographical distribution of the haptoglobin genes were recently drawn together by Giblett (1969) and Walter (1969). The Hp^1 gene frequency is remarkably similar throughout the European continent, between 35 and 43 per cent, except for the Lapps in the north and the southernmost Italians in the south. In Africa the Hp^1 frequency is much higher than it is in Europe, except in Ethiopia (40 per cent) and in the Bushmen (30 per cent). There is a general decline across southern Asia, to the lowest values observed in some peoples of southern India; it increases again in eastern Asia, is at a general frequency of about 30 per cent among the Eskimos, and then generally increases southwards through the Americas to frequencies approaching 80 per cent in Chile, Andean and occasional isolated tribal populations. The frequencies in Australian aborigines are low (17–24 per cent), but generally in the Pacific area the Hp^1 frequencies tend to be rather high. There are few data on the geographical variation in the subtypes of haptoglobin, but it appears that the Hp^{1F} gene is virtually nonexistent in mongoloid populations including American Indians, while in Africa its frequency may be higher than that of Hp^{1S}, the reverse of the situation in Europeans.

For the Gm antigens, the geographical distribution was reviewed by Steinberg (1967). Most studies have been limited to Gm(1), and this antigen occurs at its lowest frequency in southern European populations (less than 50 per cent) and increases northwards from Italy to Sweden and Finland. There is a similar gradient for Gm(2). Outside Europe Gm(1) occurs at nearly 100 per cent frequency, while Gm(2) is virtually absent in Africa but of variable frequency in Asia. Very unusual alleles have been detected in some populations. Striking features of the Gm gene distribution are not only that four of the continental groups tested have alleles not present in others, but also that no two racial groups have the same

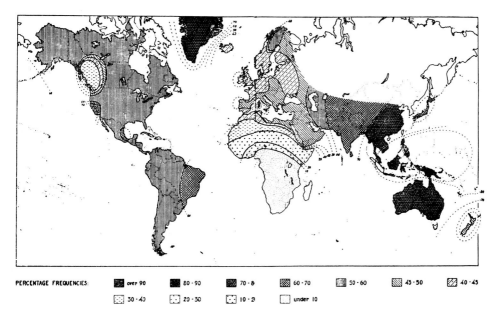

PERCENTAGE FREQUENCIES: ⬛ over 90 ⬛ 80 - 90 ⬛ 70 - 8 ▨ 60 - 70 ▦ 50 - 60 ▨ 45 - 50 ▨ 40 - 45 ⬚ 30 - 40 ⬚ 20 - 30 ⬚ 10 - 2 ☐ under 10

Fig. 64   World distribution of frequencies of the blood group C gene (from Mourant, 1954).

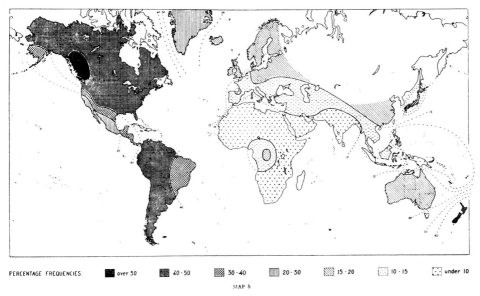

PERCENTAGE FREQUENCIES ⬛ over 50 ⬛ 40 - 50 ▨ 30 - 40 ▥ 20 - 30 ▨ 15 - 20 ⬚ 10 - 15 ☐ under 10

MAP 8

Fig. 65   World distribution of frequencies of the blood group E gene (from Mourant, 1954).

array of alleles. Few other known human genetic systems have so distinctive a distribution of alleles.

In the transferrins, geographical differentiation occurs in the types of variant that are present, rather than in the frequencies; only four of the aberrant types—CD_1, CD_{chi}, B_2C and $B_{0-1}C$—have a frequency of 1 per cent or more in the populations tested. The most common molecular variants are D_1 and D_{chi}, the former having two different population reservoirs in the aborigines of Australia and New Guinea on the one hand and in Africa south of the Sahara on the other, and it is of interest that the D_1 variant is chemically similar in both groups. The variant D_{chi} is found in southeast Asia and south American Indian populations. The fast moving B variants are virtually absent from the populations of Africa and the Pacific islands, the highest frequency reported occurring in the Indian tribes in Mexico and central America, but even here the gene frequency is less than 4 per cent, by comparison with about 1 per cent in most western European groups.

The geographical distribution of the genes governing the group specific component, (Gc) were summarized by Schultze and Heremans (1966). Among Europeans the Gc^2 gene frequency is fairly constant at about 26 per cent, although in some north European isolated populations there is considerable variation. The Gc^2 frequencies diminish southwards into Africa, where they rarely exceed 10 per cent, and into Asia, although again there are local variations. The highest frequencies occur in some Indian tribes of south America (69 per cent in Xavante and 56 per cent in Caingang). In Australian aborigines Gc^2 has a relatively low frequency in most tribes, though there is an apparent increase from the central desert area towards the north and east coast peoples.

Enzyme Variants

For almost all these gene systems of recent identification, population studies are too few to allow the distribution to be mapped. For the pseudocholinesterase phenotypes, the E_1^a gene appears to reach its maximum frequency (5 per cent) in Jews in the Middle East. Otherwise the highest frequencies of up to 3 per cent occur in Europe, and diminish southwards into Africa and eastwards into Asia, and the gene is virtually absent from Japanese, Chinese, Philippinos, Eskimos, South American indigenous and Pacific populations.

For red cell acid phosphatase, recently summarized by Giblett (1969), in western Europe P^a gene frequencies are fairly constant in the range of 26 to 37 per cent while P^c frequencies do not exceed 10 per cent. P^a diminishes in frequency in Africa, but remains much the same over Asia except for the northeast where there is some elevation among Japanese and Aleuts approaching the higher frequencies amongst Eskimos and Athabascan Indians. Amongst other American Indians the frequency again is lower, and there may well be a southward gradient. The gene is almost as infrequent in the Australian region.

Information on 6-phosphogluconate dehydrogenase gene distribution has recently been drawn together by Tills *et al.* (1971*a*), In Europe the PGD^c gene is least frequent in Ireland (1 per cent) but elsewhere lies between 2 and 4 per cent. It is generally at higher frequency in Africa, and particularly in the northeast among the Amhara and Beja, and elsewhere in the continent is approximately double the European frequency. The high zone from northeast Africa extends into southwest Asia, affecting particularly south Arabian Arabs and Yemenite Jews, and remains at about double the European frequency elsewhere. In eastern Asia it is at low to moderate frequencies, but these rise in the sub-Himalayan regions of Nepal and Bhutan. It is between 4 and 6 per cent in Australian aborigines but is virtually absent from the aboriginal American populations so far examined.

Red cell phosphoglucomutase, an important enzyme which reversibly catalyses the transfer of phosphate from the first to the sixth position of glucose, shows several variant forms, under the control of genes at three loci. Only for the gene frequencies at locus 1 is sufficient known of the geographical variation for present discussion. Frequencies of the PGM_1^2 gene appear to increase from south to north in Scandinavia, and are particularly high in the Lapps. They are fairly constant over most of Europe, though a north/south gradient may exist, and over Africa, they rise slowly into India and diminish again eastwards to under 10 per cent amongst the Ainu. Frequencies are low in Australian Aborigines and among all American populations tested except the Greenland Eskimo.

The distribution of the genes governing variants of the enzyme adenylate kinase were recently summarized by Tills *et al.* (1971*b*). The AK^2 gene is mainly found in European populations (up to 5 per cent) with the exception of the Lapps in whom it is absent. The gene is virtually absent from African populations south of the Sahara. Southwest Asia resembles or is slightly lower than Europe in its frequencies, but there is a rise to a high frequency of up to 15 per cent in India. This is quite different from the lower or zero frequencies in eastern Asia, and the gene appears to be absent from Australian Aborigines and American Indians.

For adenosine deaminase, the lowest frequencies of the ADA^2 gene occur in Africa, moderate frequencies (4–9 per cent) in Europe with a suggestion of an increasing north/south gradient, higher frequencies in southern Asia, and the highest so far recorded from New Guinea (12–17 per cent). There are many other enzyme variants for which patterns of geographical interest may be expected to emerge.

Discussion

This summary of the distribution on a world scale of several normal genetic variants shows:

1. The variations in gene frequency are far from random. In the spatial distribution of the frequencies, general geographical patterns may be discerned. Similarly, the frequencies recorded are far from a random array of all possible frequencies. For instance, the population frequencies of the genes of the ABO groups cluster towards one corner of the triangle represented by all possible frequencies.

2. Of the geographical patterns discernible, many of the characters show strong clonal arrangements, though with some variation in frequency about the gradient itself; this is perhaps sometimes due to sampling, but sometimes it is due to isolation of the populations or their subdivision.

3. These gradients indicate that geographical distance between populations is an important determinant of their gene frequencies.

4. Where major steps in a gradient exist, they usually separate one major continental group of man from another. Sometimes these coincide with geographical barriers, for example the marked discontinuity of the Sahara separating Mediterranean frequencies in the rhesus system from those of sub-Saharan Africa.

5. The terminations of the gradients, i.e. the maximum and minimum frequencies reported, differ from character to character, so that, for example, the maximum frequency of blood group gene B occurs in central Asia, that of gene M in central America, and that of gene C in the New Guinea region. Hence no one human population can be picked out as being conspicuously different from all others; this implies that all have advanced similar distances along the evolutionary pathway but in slightly different directions.

There is probably no simple uniform explanation of the gradients. The most likely is as follows. A variant gene arises by mutation in a particular population. Most such new mutants are lost, either by chance or because they confer some disadvantage on their possessor. Sometimes, however, they confer advantage, and under the action of natural selection they increase in frequency. They also diffuse through neighbouring populations by migration and intermixture, giving rise to gradients of diffusion. Sometimes they prove to be still more advantageous in a new population in which they arrive, either because of the latter's environment or because of its existing pool of genes in interaction with which the new gene operates; in each circumstances selection for the new genes is greater and they increase in frequency to a still higher level. If such advantage varies in intensity over a distance then a gradient of selective advantage may occur. Thus the analysis of the geography of gene frequencies may provide a valuable pointer to the nature of the selective advantage, the identification of the environmental factor in relation to which the advantage operates. Here examination of regional and local distribution patterns may be particularly relevant.

REGIONAL AND LOCAL DISTRIBUTION

The contrasts that such distribution patterns present are well illustrated by the *abnormal haemoglobins*. A number of variant forms of the haemoglobin molecule under simple genetic control are known. The most widespread is haemoglobin S, the sickle cell haemoglobin, which is essentially an African characteristic. Its distribution in Africa may be analysed in terms of three components:

(a) *Regional*. It occurs mainly across a broad band in the middle of the continent, bounded on the north by the Sudan grasslands and on the south by the Zambesi.

(b) *Local*. Within this belt there is considerable local variation, higher frequencies occurring in the moister, lower lying areas than in the drier higher areas.

(c) *Ethnic*. Within the same area two peoples living side by side of different ethnic affinity often have different frequencies.

In other continents, wherever African peoples have entered to any marked degree into the composition of the present population, appreciable haemoglobin S frequencies are found, for example in the American negro, and the black Caribs of Honduras; where there has been some slight African influence, the gene is generally present at very low frequency *e.g.* around the shores of the Mediterranean, although here there can occur pockets of high frequency, *e.g.* around Lake Kopais in Greece where it is not possible to attribute this phenomenon to African immigration. Other areas of high frequency occur in the south of India and the southern tip of Arabia, of which the former at least is independent of any massive African immigration.

Examining in more detail the local component of variation as manifest in East Africa, peoples with higher frequencies of the gene are found in the moister, low lying areas of the coastal plain, the lacustrine plateau, and the river valleys, while the gene is rare or absent amongst those living in the dryer higher areas. Fig. 66 shows the distribution of heterozygote frequencies in relation to the pattern of endemicity of malignant tertian malaria, suggesting the implication of malaria in the gene frequency levels. The mechanism is as follows. Homozygous HbS individuals (i.e. those possessing two haemoglobin S genes, one from the father and one from the mother) tend to die young from haemolytic anaemia, resulting in a steady elimination of the haemoglobin S gene from the population. The gene is retained in the population by a counterbalance, and this is provided by a similar disadvantage of the normal homozygotes (i.e. those with two normal haemoglobin A genes) relative to the heterozygotes (i.e. those with one S and one A gene). The greater the disadvantage of the normal homozygote, the higher the equilibrium gene frequency in the population. Although the details of the balance are complex, in essence the disadvantage of the normal homozygote lies in its greater susceptibility when assailed by malignant tertian malaria, this having been well established by a series of experimental and empirical studies. Thus were a new haemoglobin S gene to appear in a population previously lacking it but living in a malarious habitat, either by mutation or by introduction from outside, the heterozygote advantage would tend to retain that gene in the population and increase its frequency by natural selection until the equilibrium level for that habitat was reached.

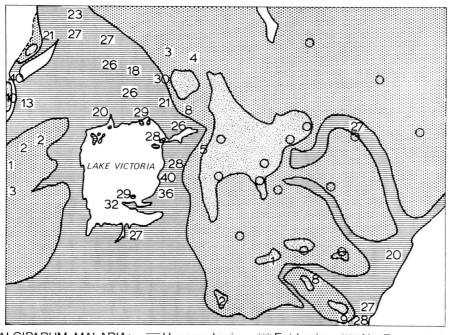

FALCIPARUM MALARIA: ≡Hyperendemic   ⠿ Epidemic   ⠿ No Transmission

Fig. 66   Frequencies of haemoglobin S heterozygotes and the transmission of malignant tertian malaria in East Africa (from Allison, 1955).

In this case the details of the local geographical distribution provided important early evidence of the biological significance of haemoglobin S. A contrasting, much more restricted, distribution pattern is shown by the gene for haemoglobin C. This attains maximum frequency in the populations of the northernmost part of Ghana and adjacent territories, from which there are gradients diminishing outwards in all directions (Fig. 67a). Outside west Africa this gene is present only at trace frequencies, again where immigration from west Africa is known to have occurred. It appears that the original mutation of haemoglobin C occurred in this locality; its advantage under local conditions encouraged its increase, and it spread outwards so that its isogenes appear like ripples on a pond. Since its advantage is as yet unidentified it is not clear whether it is continuing to spread or whether the distribution is fixed. From fitness studies the former appears more likely. Haemoglobin E has a similar restricted focus in south east Asia (Fig. 67b).

Similar distribution patterns of interest emerge for the sex-linked gene responsible for red cell *glucose-6-phosphate dehydrogenase deficiency*. Males with this deficiency tend to suffer acute haemolytic reactions to drugs such as primaquine and sulphanilamide, and to contact with broad beans (favism). The gene is of wide distribution, and there are amazingly high

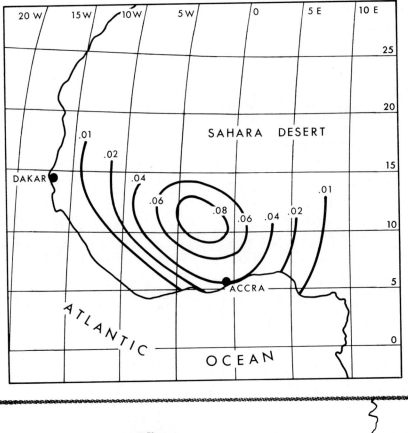

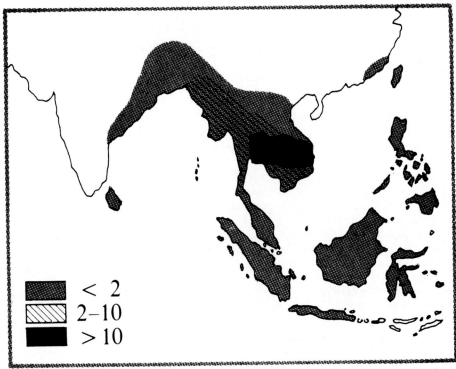

Fig. 67(*a*)   Distribution of haemoglobin C gene frequencies in West Africa (top). (*b*) Frequencies of the haemoglobin E gene in South East Asia (bottom) (*from Cavalli-Sforza and Bodner, 1971*).

frequencies in some small groups such as Kurdish Jews. There are many different variants of the gene known—the type found in Africans is biochemically different from that found in the Mediterranean and southeast Asia; this suggests multiple origins and considerable selective importance. Again the geographical distribution has led to the suggestion that this gene may also confer some protection against malaria, and this is supported in local studies as in Sardinia (Siniscalco *et al.*, 1961).

The suggestion that *thalassaemia*, a name given to various types of anaemia of different degrees of severity due to other disorders of haemoglobin structure and synthesis, was also related to malaria, similarly came from the study of its geographical distribution. The most detailed work has been done in Italy, where the incidence of thalassaemia minor varies from 0·5 per cent in some large cities of the north to about 10 per cent in the region of Ferrara and to almost 20 per cent in a few localities, while the incidence of thalassaemia major may rise as high as 1 per cent. In Sardinia a strong association with altitude, where low lying villages show the highest incidence and mountain villages the lowest, correlate fairly closely with the distribution of malaria before it was eliminated by public health measures.

These illustrations have been restricted to genes controlling physicochemical erythrocyte variations in relation to malaria. There is ample scope for further detailed investigation of distribution of other gene frequencies as a clue to their biological significance.

GEOGRAPHICAL FACTORS IN GENE FLOW

Apart from being of use in identifying possible ecological factors of significance in gene frequency determination, study of the regional and local geography of genes demonstrates very clearly the dynamic nature of gene frequencies and the fluidity of the genetic structure of populations. Not only are gene frequencies modified by natural selection; they are also modified by random events which may be influenced by geographical factors as, for example, in the appreciable changes in genetic constitution of Tristan da Cunha which resulted from the reductions in population size at the middle and end of last century (Roberts, 1968). Of particular importance also is that they are modified by gene flow. Wherever human beings move then gene flow occurs. If they subsequently reproduce then the genes they have brought with them remain in the new area, and if hybridization with a pre-existing population occurs then they appear in the resulting population. Hence all geographical factors that affect the movement of individuals also affect genetic evolution.

Routeway. The importance of a line of communication and of geographical distance along it is well illustrated by the distribution of the haemoglobin S gene in the upper Nile valley (Fig. 68). There is a considerable incidence of falciparum malaria throughout this region so that on the malarial hypothesis the gene, once introduced, would be expected to increase in frequency until equilibrium were reached. From a heterozygote frequency of 40 per cent to the south of Lake Albert, there is a steady diminution northwards until the gene is absent from the main body of the northern Nilotics, amongst whom it is just beginning to appear in the southern-most groups. The frequency falls in samples living inland away from the Nile, suggesting either that the Nile is its main channel of penetration, or that there is a lower selective pressure away from the Nile with consequent slower elevation of frequency. Occasional intermixture of individuals from the river peoples seems the most likely means by which this penetration is brought about, a suggestion supported by the apparent breaks of gradient at the points of major ethnic distinction.

Configuration of a coastline. Whether a coastline is concave or convex influences the manner in which genes introduced on the coast flow into the interior. In north Australia, the

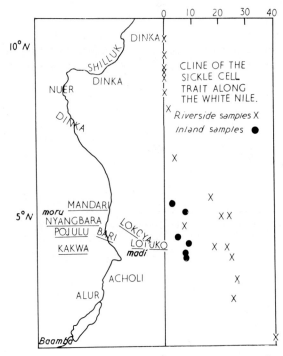

Fig. 68   Penetration of the haemoglobin S gene in the upper Nile valley.

coast lying between the Fitzmaurice river on the west and the Roper river on the east is convex, while between the Roper river round the south of the Gulf of Carpentaria to the Holyrod river on the east it is concave. By way of illustration, if alien genes were introduced into the coastal tribes by massive intermixture to a frequency of 0·50, the bounding rivers acting as total barriers, immigration then ceased and panmixia of coastal and inland tribes developed in each section, then in the absence of other disturbing factors, the final gene frequency in the convex section would be 0·26, and in the concave section 0·16; the frequency of the introduced gene is some 65 per cent higher in the first section, solely as a result of this particular coastal configuration. This model was applied by Birdsell (1950) to interpret the biological features of the northern aborigines (Fig. 69).

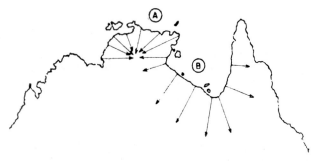

Fig. 69   The effect of coastal configuration on gene penetration.

Settlement pattern. In a fairly densely and evenly settled area with no barriers to gene flow, not only will geographical distance between sections of the population affect the spread of genes from one to another, but the rate of spread will be influenced by the settlement pattern itself. The Shilluk are linearly distributed along the Nile, and among them the effective size of the population is about 9,000 breeding adults. Inland the settlements of the Dinka are areally distributed, and the Dinka size of isolate is consequently less (about 1,400, Roberts, 1956). Hence a gene will take longer to penetrate a section of say 100,000 Dinka than the same size group of Shilluk, i.e. its rate of spread will be slower among the Dinka.

Geographical barriers. These may prevent the spread of genes between populations on either side of them, they may affect the intensity and nature of selection pressure, or by restricting population numbers they may promote random genetic drift. Barriers are of course traversed on occasion. If the barrier is a mountain range, passage will probably be via a pass. Mating of incomers will probably occur more frequently with those members of the indigenous population living nearest to the point of arrival than with those farther away. So that gene frequency gradients may be set up which will indicate the point of entry of the alien gene. If the barrier is a desert, the point of entrance on the other side may be the end of caravan trails across it; north African tribes living round the terminations of the trans-Sahara caravan routes have higher frequencies of typically sub-Saharan genes than those in the more remote areas. If the barrier is a body of water, the point of entrance on the other side may be related to sheltered locations, harbours *etc*; the slightly higher blood group O frequencies in parts of south Lancashire are probably due to the Irish immigrations through Liverpool of last century. The high A gene frequency on the peninsula where Pembroke stands probably indicates the Viking colonization attracted there by the harbour.

Of the relatively few studies sufficiently detailed to demonstrate this process in operation, that from north Australia is of interest. The gene for blood group B is practically absent from Australian aborigines. Birdsell (1950) plotted the distribution of the few aboriginal individuals who were found to possess the B gene. They all occurred in a relatively narrow band inland from the Gulf of Carpentaria and on the Cape York peninsula (Fig. 70). This suggests to him the incipient stages of gene flow, resulting from some coastal hybridization with Papuan and Malayan fishermen. The B gene is just beginning to appear in the continent.

It is important to recognize that what is apparently an effect of a geographical barrier may in fact be attributable to some other factor. An intensive study of blood groups on the Greek island of Tinos was made on a 7 per cent sample of the total population (Roberts *et al.*, 1965). The terrain is difficult, and the gene frequencies examined on a local basis showed distinct and highly significant differences between different parts of the island. Study of the movements of individuals for marriage showed very limited migration between regions. It is tempting to attribute the genetic differences to the inhibiting effect on gene flow of geographical barriers, via the marriage pattern. But were the geographical influence direct, then within a restricted locality within a sample area where there is no difficulty of communication, a more regular pattern of movement between settlements would be expected. This is in fact not so. Marriages between villages within a locality showed conclusively that the tendency to remain within the village of birth for marriage was almost as strong as the tendency to remain within the locality of birth. Thus the retention of genetic differences between local populations, which at first sight could well be interpreted as being due to the effect of geographical barriers, appears on analysis to be due to an underlying cultural factor. This may of course in its turn derive ultimately from a localization of interest owing much to geographical conditions, but there is no evidence that it is to be attributed primarily to these

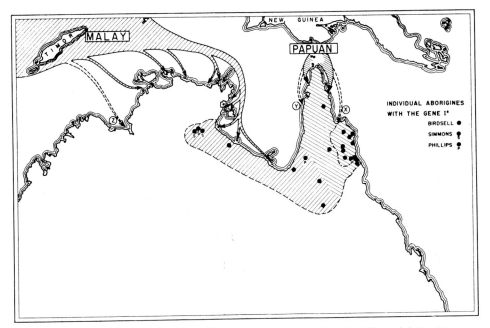

Fig. 70   Penetration of the blood group B gene among Australian Aborigines (from Birdsell, 1950).

rather than to, say, historical factors. There are many cultural variables in gene flow to interact with or negate the effects of geographical factors.

Population structure and density. In interior Australia, it is the tribal population that approximates the genetic isolate within which mating occurs, so that tribal boundaries represent the main barriers to gene flow. On account of geographical variations in the generosity of the habitat, tribes of differing density are spread out over tribal areas of differing sizes, so that as a regulator of gene flow the number of tribal boundaries to be crossed becomes more important than geographical distance in miles. Suppose alien genes are introduced on the south coast among the Nauo of the Eyre peninsula, and by normal gene flow processes spread outwards until eight tribal boundaries have been crossed. In one direction the tribe outside the eighth boundary is the Targari, in another the Warkawarka. The actual geographical distances from the point of gene origin are respectively 1,400 and 450 miles as the crow flies. Thus, other things being equal, it would take genes just as long to reach these two peoples despite their different geographical distances.

As a final illustration the relative importance of *language* and geographical barriers in gene flow was examined by Hulse (1957) among three Indian groups in the Pacific northwest. He examined gene frequencies in the Okanagan and Yakima in the interior, 200 miles apart, within fairly easy access of each other, and the Swinomish on Puget sound separated from the Yakima by the difficult terrain of the Cascades, but across which aboriginal trails led. Geographical factors therefore suggest that there should have been more intercommunication, gene flow and hence genetic similarity between the two interior groups than between either and the Swinomish. However the Swinomish and Okanagan both speak Salishan dialects, while the Yakima belong to the Sahaptin linguistic stock. The gene frequencies of the

Okanagan are much more similar to the linguistically similar coastal Swinomish, than to the other inland group the Yakima. In this case then the linguistic barrier so far appears to have been stronger than the geographical barrier in maintaining genetic differentiation.

DISCUSSION

Contrasting distribution patterns, and how they may be of relevance in detecting the functional significance of genetic variation, are illustrated by genetic variants of red cell structure, the abnormal haemoglobins, thalassaemia, and glucose-6-phosphate dehydrogenase deficiency, and their relationship to malarial susceptibility. Distributions appear of particular help when they are intensively studied on a local scale. However, they should be regarded essentially as pointers for more detailed experimental and other specific investigations.

Distributions may also be of value in establishing the effects of geographical factors on gene frequency variation and rate of gene penetration, e.g. of lines of communication, of different types of settlement pattern and of coastal configuration. The effects of geographical barriers, and where they are penetrated, can be shown by intensive surveys, but other non-geographical variables, especially cultural, may interact with or indeed override geographical influences. Nevertheless a geographical approach can convey the dynamic nature of gene frequency variation. There is ample scope for further detailed surveys of frequencies of genes governing both normal and pathological variants in relation to geographical factors.

References

Allison, A. C. (1955). Aspects of polymorphism in man. *Cold Spring Harbor Symposium on Quantitative Biology*, **20**, 239.

Birdsell, J. B. (1950). Some implications of the genetical concept of race in terms of spatial analysis. *Cold Spring Harbor Symposium on Quantitative Biology*, **20**, 259.

Boyd, W. C. (1959). Blood groups. *Tab. biol.*, **17**, 113.

Cavilli-Sforza, L. L. and Bodner, W. F. (1971). The Genetics of Human Populations. Reading and San Francisco, W. H. Freeman & Co.

Giblett, E. R. (1969). *Genetic Markers in Human Blood*. Oxford, Blackwell.

Hirszfeld, L. and Hirszfeld, H. (1919). Serological differences between the blood of different races. *Lancet*, **ii**, 675.

Hulse, F. S. (1957). Linguistic barriers to gene flow. *Amer. J. Phys. Anth.*, **15**, 235.

Mourant, A. E. (1954). *The Distribution of the Human Blood Groups*. Oxford, Blackwell.

Mourant, A. E., Kopec, A. C. and Domaniewska-Sobczak, K. (1958). *The ABO Blood Groups*. Oxford, Blackwell.

Roberts, D. F. (1956). Some genetic implications of Nilotic demography. *Acta Genetica et Statistica Medica*, 1956–1957, **6**, 446–452.

Roberts, D. F. (1968). Genetic effects of population size reduction. *Nature*, **220**, 1084–1088.

Roberts, D. F., Luttrell, V. and Pasternak Slater, C. (1965). Genetics and geography in Tinos. *Eugen. Rev.*, **46**, 185–193.

Schultz, H. E. and Heremans, J. (1966). *Molecular Biology of Human Proteins*, **1**, 424. New York, Elsevier.

Siniscalco, M., Bernini, L., Latte, B. and Motulsky, A. G. (1961). Favism and thalassaemia in Sardinia and their relationship to malaria. *Nature*, **190**, 1179.

Steinberg, A. G. (1967). Genetic variations in human immunoglobulins. The Gm and Inv types. In Greenwalt, T. J. (ed.), *Advances in Immunogenetics*, p. 75. Philadelphia, Lippincott.

Tills, D., van den Branden, J. L., Clements, V. R. and Mourant, A. E. (1971a). The distribution in man of genetic variants of 6 phosphogluconate dehydrogenase. *Human Heredity*, **21**, 302.

Tills, D., van den Branden, J. L., Clements, V. R. and Mourant, A. E. (1971b). The world distribution of electrophoretic variants of the red cell enzyme adenylate kinase. *Human Heredity*, **21**.

Walter, H. and Steegmuller, H. (1969). Studies on the geographical and social distribution of the Hp and Gc polymorphisms. *Human Heredity*, **19**, 209.

Chapter 20

The Documentation of Environmental Medicine: Conventional and Computerized Information Retrieval Methods

Helen E. C. Cargill Thompson and *Edith Frame*

The first scientific journals appeared in 1660, and have continued to make their appearance in a steady stream ever since. At first scientists kept in touch with each other by correspondence and by scanning the journals. However, from the beginning the need for some form of abstracting service was realized, and the first, *Journal des scavans*, was published in 1665. Throughout the 18th century a number of short-lived abstracting journals were published. Initially, abstracting and indexing journals were introduced for well defined scientific fields, and journals such as *Chemisches Zentralblatt* (1830–1970), *Chemical Abstracts* (1904 to date), *Physics Abstracts* originally published under the title *Science Abstracts* (1897 onwards) and *Index Medicus* formerly *Index catalogue of the library of the Surgeon-General's Office, United States Army* (1880 onwards) have continued to prove their usefulness to scientists.

Current trends in research are now moving towards interdisciplinary topics. Accordingly the purpose of this chapter is to outline some of the major abstracting and indexing services available in the English language, and to indicate how these may be used to maximum advantage in the interdisciplinary field of environmental medicine. Due to limitations of space only a few services can be discussed here, but fuller lists can be found in the National Lending Library's **KWIC index to the English language abstracting and indexing publications .* . . . (1969), the International Federation for Documentation (FID) *Abstracting services* (1969) and R. L. Collison's *Abstracts and abstracting services* (1971).

In general, the services to be described herein have been prepared manually, and the computerized service is a by-product created by using the computer as a means of producing the printed volume. A useful list of computerized services can be found in R. Finer's *A guide to selected computer-based information services* (1972).

GENERAL SCIENCE

Two journals, *Pandex*, published by Crowell Collier and Macmillan, and *Science Citation Index*, published by the Institute for Scientific Information (ISI) lead in the field of general science. The principle behind their publication is that since some 80 per cent of the major articles appear in about 10 per cent of the journals, indexing a limited number of key periodicals is considered sufficient to locate the majority of significant papers written.

Pandex

Pandex, first published in 1969, is primarily an alerting journal with a computer produced KWOC (Key Word Out of Context) subject index. The main section of the index consists

* KWIC = Key Word In Context.

simply of a list of articles arranged under the journal title. The journals are grouped in one of eighteen subject areas such as Clinical Medicine or Multidisciplinary. This allows for easy scanning within a broad subject area. The subject index is limited to words contained in the title of the article together with a few supplementary terms. (Words with the same meaning are grouped under a single subject heading, e.g. BIRTH which covers such terms as birth, natal, premature, stillborn, etc.). As with all computerized services, certain words are treated as stop (i.e. not used) words. These include not only words such as 'an', 'of', 'and', 'the', etc., but also non-specific nouns, for example 'animal' or 'method'. The value of the index is limited by the precision of wording of the titles of the articles listed. Several articles are irretrievable by *Pandex* either because the title consists solely of stop words or because it is insufficiently expressive. A good example is "An An and Chi Chi" which was the title of an article, published in *Nature*, on the mating of giant pandas. The *Pandex* index is cumulated every 3 months and appears in microfiche form. As from December 1972 publication of the hard copy form will cease and the journal will appear only on microfiche.

A computerized SDI (Selective Dissemination of Information) service is available from *Pandex*, but its high cost would not recommend it to workers in environmental medicine.

The major advantage of *Pandex* is its broad subject base which enables the searcher to scan papers on such broad topics as 'pollution' from journals in fields as far apart as chemistry, engineering and the life sciences.

Science Citation Index

Science Citation Index (SCI) depends on a principle long accepted in the field of law but new to the sciences. Basically, the principle is that if there is a paper of accepted validity, knowledge of the researchers subsequently referring to it is of considerable importance. This introduces a completely new concept to literature searching in the basic sciences in that, given a point in time, it is possible to search forwards to the present rather than to use the more traditional back searching techniques employed hitherto. Practical application of this concept to the wide field of science and technology became possible only with the development of computers.

Science Citation Index now covers some 2,500 journals and appears in three parts–*Citation Index*, *Source Index* and *Permuterm Index*. The *Citation* and companion *Source Indexes* have been published since 1964, with a 5-year cumulation for the period 1965–1969. Further 5-year cumulations are projected. The *Permuterm* (subject) *Index* was introduced in 1966; cumulations of this index are not envisaged at present. A *Social Sciences Citation Index* is to commence publication in 1973, including psychology and geography in its coverage.

It is possible to approach *Science Citation Index* either by subject or by a known reference. Either one requires two parts. For a subject approach the *Permuterm Index* together with the *Source Index* are needed; for a citation approach the *Citation Index* and the *Source Index* are required. The diagram (Fig. 71) shows an example from *Science Citation Index* for January–March 1971.

The *Source Index* may be used independently to check for recent publications by known authors.

Since *Science Citation Index* is entirely computer-produced the creation of a computer-based retrieval service is a commercially viable development. The service known as ASCA (Automatic Subject Citation Alert) is a weekly current awareness service. References, either by subject, by citation (i.e. known reference), by journal or by author may be retrieved in this way. The service is available either programmed to individual research requirements or more cheaply as a group profile known as ASCATOPICS, for example 'Heavy metals in the environment' or 'Population control'.

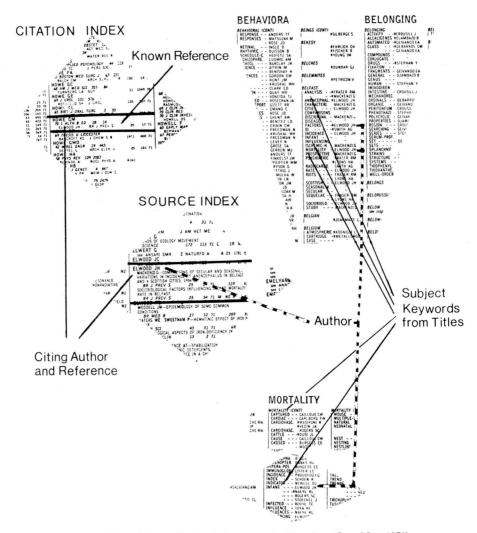

Fig. 71 Use of Science Citation Index (example taken from Jan.–Mar. 1971).

Another alerting service published by the Institute for Scientific Information is the series of weekly journals entitled *Current Contents*. Each copy of *Current Contents* consists simply of photo-reduced facsimiles of the contents pages of current issues of journals, together with an author address list. Some 4,500 journals are covered by the 5 existing *Current Contents*:

Agriculture, biology and environmental sciences;
Behavioral, social and educational sciences;
Engineering and technology;
Life sciences;
Physical and chemical sciences.

It is proposed to commence publication of a *Current Contents: Clinical Practice*, initially covering 700 journals, in January 1973. From January 1973, the issues of *Life Sciences, Physical and Chemical Sciences* and *Clinical Practice* will contain a single key word subject index, the key words being derived from the titles of the articles.

MEDICAL SCIENCES

There are two major indexing journals within the field of medicine. These are *Index Medicus* and *Excerpta Medica* and they compliment each other.

Index Medicus

Index Medicus is prepared on a co-operative basis under the overall control of the National Library of Medicine (NLM), Bethesda, Maryland. It is an indexing journal covering over 600,000 articles per year in some 5,000 journals. In the early 1960's the production of *Index Medicus* nearly came to a halt because of the sheer bulk of material to be processed. Accordingly the editors considered the possiblity of computer production, and in 1964 the input to *Index Medicus* was transferred to a computer.

Each article scanned for *Index Medicus* is indexed in depth by a subject specialist according to a strict code of rules and using a rigorously controlled vocabulary published as Medical Subject Headings (MeSH). The number of terms applied varies from one to over thirty, although the average is around ten, of which about three are 'print' terms, i.e. terms under which a reference appears in each monthly issue. Of the several categories of terms applied, specific subject headings may be 'print' or 'non-print', but terms which indicate geographical area, species studied, age group of subjects, type of experiment and article type are always designated as 'non-print'.

The medically-biased research worker can locate rapidly the major articles in his field by searching the printed *Index Medicus* under the appropriate search terms. In addition there is available a MEDLARS (MEDical Literature Analysis and Retrieval Service) search: the retrieval service created as a by-product of computerization. MEDLARS searches are particularly useful when the search topic is of a complex nature or involves a large number of search concepts or 'non-print' terms. In this way the worker in environmental medicine can locate papers dealing not only with a specific subject but also with a specific area. Because of the rigorously controlled and hierarchical nature of the vocabulary, searching of MEDLARS tapes is relatively fast and so retrospective searches are not only feasible but also reasonably inexpensive. Experiments are in progress to develop an on-line search facility to be known as MEDLINE.

Excerpta Medica

The Excerpta Medica Foundation was established in 1946 to make information on all significant basic research and clinical findings available to the medical and related professions. Over 3,000 journals are scrutinized and informative abstracts are made of relevant articles. Currently abstracts are published in forty-eight separate sections, as for example *Anthropology, Gerontology and geriatrics*, and *Public health, social medicine and hygiene*. Within each section the abstracts are presented in a classified arrangement, and each monthly issue contains an author and a subject index. The indexes are cumulated annually on completion of a volume.

The great advantage of *Excerpta Medica* over *Index Medicus* is that the abstracts are so carefully prepared that in many cases one need not read the original paper. This is a particularly useful facility when the paper is in a foreign language. The disadvantage of *Excerpta Medica* is its slowness of publication.

ENVIRONMENTAL MEDICINE

The working environment exposes individuals within the community to the potential health hazards of dust, fumes, heat, etc. for about one third of each day. This area of environmental medicine is served by the International Occupational Safety and Health Information Centre (CIS), set up in 1959 at the International Labour Office in Geneva.

Occupational Safety and Health Information Centre

The CIS card index service provides a back-file from 1959, continually updated by the issue of more than 2,000 abstracts each year. The abstracts are compiled primarily from periodical articles, although monographs are also included. The Centre issues a list of the periodicals searched, the latest edition (Jan 1972) listing 1484 titles predominantly of periodicals produced in the English speaking world, western and eastern Europe and the Soviet Union, and a few from Japan. As articles written on occupational safety and health are mostly compound subjects, the CIS has developed its own faceted classification which is explained in *CIS Classification—guide to the card service* (3rd edn. July 1966) (Fig. 72).

Before making a search from the CIS cards the reader is advised to consult the *Alphabetical Chain Index*, now in its 12th cumulative edition (1972), in order to evolve the most effective search strategy.

CIS also produces a current awareness service twelve times per year in the form of the *Occupational Safety and Health Bulletin*. This *Bulletin*, available shortly before the batches of cards, reproduces the card-index abstracts, three per page and in booklet format. It is thus more suitable than the cards for inter-office circulation. CIS bibliographies on selected subjects are occasionally published in pamphlet form.

The CIS photocopy and microfilm service ensures that all published work, abstracted by CIS, is available to the reader direct from the Centre. This is especially helpful should there be difficulty in obtaining the particular periodicals and monographs locally.

Environmental Health

In 1971 the Excerpta Medica Foundation brought out a new abstracting journal entitled *Environmental Health*. The source material for this journal was some 3,400 biomedical periodicals already regularly scanned for the *Excerpta Medica* series together with over 15,000 other journals covering technical, meteorological, agricultural, legal, scientific and other aspects relating to environmental health. About 6,000 abstracts were listed in 1971. *Environmental Health* is published in the usual *Excerpta Medica* style with a detailed classified arrangement together with author and subject indexes in each monthly issue. The indexes are to be cumulated for each volume.

GEOGRAPHICAL SCIENCES

The International Geographical Union's Commission on Medical Geography was established in 1952 and has since met every four years at International Geographical Congresses; both the American Geographical Society and Britain's Royal Geographical Society have sponsored research programmes in medical geography.

The tasks of the medical geographer are to portray data which relate to spatial variations of disease incidence, to assess the statistical significance of the distribution shown, and to trace the apparent association of disease distribution with specific environmental factors. Although areal variations in disease incidence may provide a pointer to a previously unsuspected environmental relationship, care is taken to distinguish between causation and simple

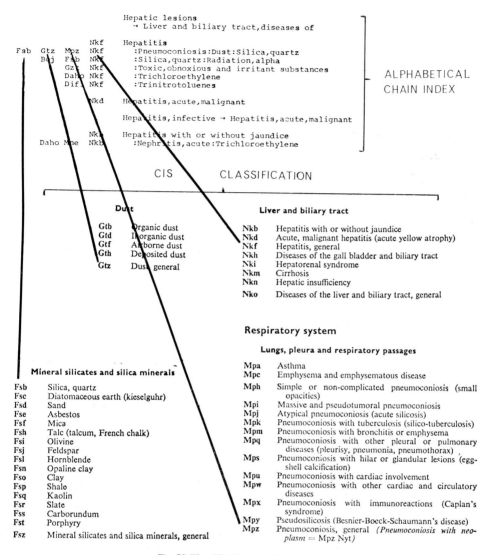

```
                                    Hepatic lesions
                                     → Liver and biliary tract,diseases of

                           Nkf      Hepatitis
Fsb  Gtz  Mpz  Nkf                    :Pneumoconiosis:Dust:Silica,quartz          ┐
         Boj  Fsb  Nkf                :Silica,quartz:Radiation,alpha              │   ALPHABETICAL
              Gzk  Nkf                :Toxic,obnoxious and irritant substances    │
              Daho Nkf                :Trichloroethylene                          │   CHAIN INDEX
              Dif  Nkf                :Trinitrotoluenes                           │

                   Nkd      Hepatitis,acute,malignant                            ┤

                           Hepatitis,infective → Hepatitis,acute,malignant        │

                   Nkb      Hepatitis with or without jaundice                    │
Daho Mne  Nkb                :Nephritis,acute:Trichloroethylene                  ┘


                        CIS        CLASSIFICATION
```

```
            Dust                              Liver and biliary tract

      Gtb   Organic dust                Nkb   Hepatitis with or without jaundice
      Gtd   Inorganic dust              Nkd   Acute, malignant hepatitis (acute yellow atrophy)
      Gtf   Airborne dust               Nkf   Hepatitis, general
      Gth   Deposited dust              Nkh   Diseases of the gall bladder and biliary tract
      Gtz   Dust, general               Nki   Hepatorenal syndrome
                                        Nkm   Cirrhosis
                                        Nkn   Hepatic insufficiency

                                        Nko   Diseases of the liver and biliary tract, general
```

Respiratory system

Lungs, pleura and respiratory passages

```
                                        Mpa   Asthma
      Mineral silicates and silica minerals    Mpc   Emphysema and emphysematous disease

Fsb   Silica, quartz                    Mph   Simple or non-complicated pneumoconiosis (small
Fsc   Diatomaceous earth (kieselguhr)             opacities)
Fsd   Sand                              Mpi   Massive and pseudotumoral pneumoconiosis
Fse   Asbestos                          Mpj   Atypical pneumoconiosis (acute silicosis)
Fsf   Mica                              Mpk   Pneumoconiosis with tuberculosis (silico-tuberculosis)
Fsh   Talc (talcum, French chalk)       Mpm   Pneumoconiosis with bronchitis or emphysema
Fsi   Olivine                           Mpq   Pneumoconiosis with other pleural or pulmonary
Fsj   Feldspar                                  diseases (pleurisy, pneumonia, pneumothorax)
Fsl   Hornblende                        Mps   Pneumoconiosis with hilar or glandular lesions (egg-
Fsn   Opaline clay                              shell calcification)
Fso   Clay                              Mpu   Pneumoconiosis with cardiac involvement
Fsp   Shale                             Mpw   Pneumoconiosis with other cardiac and circulatory
Fsq   Kaolin                                    diseases
Fsr   Slate                             Mpx   Pneumoconiosis with immunoreactions (Caplan's
Fss   Carborundum                               syndrome)
Fst   Porphyry                          Mpy   Pseudosilicosis (Besnier-Boeck-Schaumann's disease)
                                        Mpz   Pneumoconiosis, general (Pneumoconiosis with neo-
Fsz   Mineral silicates and silica minerals, general    plasm = Mpz Nyt)
```

Fig. 72 The CIS Faceted Classification.

Each CIS card dealing with a compound subject bears a code, the various components or symbols of which correspond to the different aspects or facets of the subject under consideration. In the example, a paper concerned with hepatitis, silica dust and pneumoconiosis is represented by the symbols Fsb Gtz Mpz Nkf.

spatial association; this is necessary if the formation of inappropriate hypotheses is to be avoided.

A special contribution of medical geography is the mapping technique, a notable example being the *National Atlas of Disease Mortality in the United Kingdom* prepared by G. Melvyn Howe on behalf of the Royal Geographical Society. This form of mapping focuses attention on areas where detailed studies may be beneficial. The extension of medical mapping to disease incidence is hindered by the difficulty of obtaining detailed areal statistics on disease morbidity as opposed to disease mortality, and by the fact that in some countries public health statistics are based on large administrative units which cut across geographical regions. Within the field of geography the two most significant current awareness services are *Current Geographical Publications* and *Geo Abstracts. Current Geographical Publications* is issued ten times a year by the American Geographical Society, and is based on periodicals and books acquired by the Society's library in New York. *Current Geographical Publications* is classified by subject area, and the sections on *medical geography, man and the geographical environment* and *geography of population* are relevant sections to scan.

Geo Abstracts

The publication in January 1966 of *Geographical Abstracts* brought computer indexing into the documentation of geography. Issued six times a year, *Geographical Abstracts* was produced for the first six years in classified series A to D. In January 1972 the title changed to *Geo Abstracts* to coincide with an expansion to six series A to F. In both *Geographical Abstracts* and *Geo Abstracts* series *D Social geography* has been the series most likely to include items in the field of environmental medical interest since it contains sections on population distribution, population movement and change, and man and environment. About 800 journals are scanned in the preparation of *Geo Abstracts.* These include journals on soil science, ecology, hydrology, sociology and demography together with others which are more narrowly 'geographical'. Scrutiny of *Geo Abstracts* reveals articles in periodicals which are not normally seen by the majority of medical practitioners. The annual computer-produced index contains a journal list, author index and a subject index which takes the form of a KWIC index. This index, compiled from the significant words in the title of journal articles scanned, is supplemented by additional indexing terms, added from information given in the abstracts, when the title of the article is not sufficiently descriptive. Similar subject cumulative indexes are now available for each of the separate series A to D, for the years 1966–1970.

Now that a back file of tapes is available, computer searching is offered experimentally by *Geo Abstracts,* and a thesaurus of search terms is being compiled to assist a future computer service.

THESES AND ON-GOING RESEARCH

As theses accepted for higher degrees of universities are not 'published' in the normal sense, they are not available to organizations which compile abstracting and indexing services on a subject basis.

The two major English-language publications listing university theses are the Aslib *Index to theses accepted for higher degrees in the Universities of Great Britain and Ireland* and *Dissertation Abstracts International.* Neither is completely comprehensive, as not all universities co-operate.

The Aslib *Index to theses . . .* has been published annually since 1950–51. Normally British theses are deposited in the university library of each respective university and loans are made to other libraries, not to individuals. Requests to borrow theses should be made through inter-library loans services.

Dissertation Abstracts, listing US dissertations which are available as microfilm or xerographic copies, was published from 1938 to 1969, with a nine-volume cumulative index covering these years. Starting with volume 30, 1969, the word *International* was added to the title when several Canadian, Australian and European universities began to participate. Each monthly issue of *Dissertation Abstracts International* (A. Humanities and social sciences; B. Sciences and engineering) contains a list of co-operating institutions. All works listed in *Dissertation Abstracts International* since January 1971 are held in microfilm format at the National Lending Library, Boston Spa, Yorkshire and may be obtained speedily through inter-library loans services. Individual DATRIX computer searches can be obtained on works listed in *Dissertation Abstracts/International* from 1938 to date.

It is more difficult to find out about on-going research than completed theses. *Scientific research in British universities and colleges*, published annually in three volumes by HMSO, gives a comprehensive listing of research in progress in higher education establishments and, for the social sciences only, in government-sponsored research institutions.

In the USA there is no comparable government-sponsored research index. Some individual US Government departments publish a list of research which they sponsor. The worker in environmental medicine is fortunate in that the Department of Health, Education and Welfare publishes an annual two-volume *Research Grants Index*, which lists a major proportion of government funded environmental research.

In addition, the Smithsonian Institution, Washington, maintains a computer file which is said to be a 'national collection of current research information'. Searching of this data base is on individual profiles only, costing about £30 per search.

RECORDS AND RECORD KEEPING

TABLE 45: Elements required in citing references

Article in a Periodical	Article in a Book	Book	Conference Proceeding Paper	Conference Proceeding Complete proc.	Thesis
Author(s)	Author(s)	Author(s) or Editor(s)	Author(s)	Editor(s)	Author
Title of article	Title of article	Title of book	Title of paper	Title of proceedings	Title
		Edition	Date conference held Place conference held		Degree
Title of journal	Title of book		Title of proceedings		
Volume & part number	Editor(s) of book		Editor(s) of proceedings		
	Place of publication	Place of publication	Place of publication		Location of University
	Publisher	Publisher	Publisher		Name of University
Month & Year First and last pages	Year First & last pages of article	Year	Year of publication First & last pages of paper		Date

In this increasingly computerized age, it is essential that the research worker should keep full and accurate references together with a note of the source of the information. Unnecessary duplication of effort can be avoided if full notes are made initially and if the search terms used and the numbers of issues scanned are noted. This obviates effort when runs of abstracting/indexing journals have to be searched again as 'new' terms occur to one.

The three main types of reference are (1) to a book, (2) to an article in a book, and (3) to an article in a periodical. The essential elements of these references are shown in Table 45. Fuller details will be found in British Standard 1629:1950. It is common practice to abbreviate the titles of periodicals, but since so many titles are very similar it is important that the rules concerning the abbreviation of journal titles be obeyed in order to avoid ambiguity. Until recently there were certain minor differences between American and British practice but these are gradually being eliminated. Currently, users in the UK should follow the rules laid down in British Standard 4148: part 1: 1970. A list of accepted abbreviations is to be found in the UNISIST/ICSU-AB *International list of periodical title word abbreviations.*

SUMMARY

Existing research programmes are continually expanding and new ones are being started. Therefore, it is no longer realistic to rely solely on information from colleagues for news of on-going research. Similarly the proliferation of published material makes it impossible to keep up to date merely by scanning the limited selection of journals which come into a specialized library, or to which one subscribes personally.

Environmental medicine is an inter-disciplinary subject and relevant material appears in a wide range of 'non-medical' periodicals. It is only by making a systematic search of the abstracting/indexing services, followed by a rational use of inter-library loan facilities that one can benefit fully from relevant research initiated in related disciplines.

Ignorance of what has been done or is being done elsewhere can lead to duplication of research effort which is both wasteful in manpower and costly in research funds.

REFERENCES*

British Standard: 1629:1950 Bibliographical references. London, British Standards Institution, 1950.
British Standard: 4148: Part 1: 1970 Specification for the abbreviation of titles of periodicals: part 1: principles. London, British Standards Institution, 1970.
Collison, R. L. Abstracts and abstracting services. Santa Barbara, ABC Clio Press, 1971.
Finer, R. A guide to selected computer-based information services. London, Aslib, 1972.
Howe, G. M. National atlas of disease mortality in the United Kingdom: prepared on behalf of the Royal Geographical Society. London, Nelson. 2nd enl. rev. edn. 1970, [covers 1954–1963].
International Federation for Documentation (FID). Abstracting services, 2nd edn. The Hague, FID, 1969.
National Lending Library for Science and Technology (NLL). KWIC index to the English language abstracting and indexing publications currently being received by the National Lending Library, 3rd edn. Boston Spa, NLL, 1969.
UNISIST/ICSU-AB. International list of periodical title word abbreviations. Paris, UNISIST/ICSU-AB Secretariat, 1970.

Further Reading

Gates, J. K. Guide to the use of books and libraries, 2nd edn. New York, McGraw-Hill, 1969.
Jahoda, G. Information storage and retrieval systems for individual researchers. New York, Wiley-Interscience, 1970.
Mullins, L. S. Sources of information on medical geography. Bulletin of the Medical Library Association, v. 54, no. 3, July 1966, pp. 230–242.
The mapping of disease: a special issue. Special Libraries Association, Geography and Map Division Bulletin, no. 78, December 1969.

* The layout of the references for this Chapter conforms to British Standard: 1629, 1950.

Index